P9-CQZ-705

Falling Upwards

By the same author

One for Sorrow (poems)

Shelley: The Pursuit

Shelley on Love (editor)

Gautier: My Fantoms (translator)

Nerval: The Chimeras (with Peter Jay)

Mary Wollstonecraft and William Godwin:
A Short Residence in Sweden and Memoirs (editor)

De Feministe en de Filosoof

Dr Johnson & Mr Savage

Coleridge: Early Visions

Coleridge: Darker Reflections

Coleridge: Selected Poems (editor)

Footsteps: Adventures of a Romantic Biographer

Sidetracks: Explorations of a Romantic Biographer

Insights: The Romantic Poets and their Circle

Classic Biographies (series editor)

The Age of Wonder: How the Romantic Generation
Discovered the Beauty and Terror of Science

RICHARD HOLMES

Falling Upwards

How We Took to the Air

—————

Pantheon Books

New York

Copyright © 2013 by Richard Holmes

All rights reserved. Published in the United States by Pantheon Books,
a division of Random House, Inc., New York. Originally published
in Great Britain by William Collins, an imprint of HarperCollins Publishers, London.

Pantheon Books and colophon are registered trademarks
of Random House, Inc.

Library of Congress Cataloging-in-Publication Data
Holmes, Richard, [date]
Falling upwards : how we took to the air / Richard Holmes.
p. cm.
Includes bibliographical references and index.
ISBN 978-0-307-37966-5
1. Balloonists—History. 2. Ballooning—History. I. Title.
TL616.H65 2013   387.7'3209—dc23   2013011128

www.pantheonbooks.com

Jacket images: (hot-air balloon) Corbis; (border) © RMN-Grand Palais / Art
Resource, NY; (Paris background) Rue des Archives / The Granger Collection
Jacket design by Pablo Delcan and Peter Mendelsund

Printed in the United States of America
First United States Edition
2 4 6 8 9 7 5 3 1

To Eleanor Tremain and John Lightbody
with love and balloons

Contents

Voices Overhead

'A Cloud in a paper bag'
JOSEPH MONTGOLFIER, 1782

'Someone asked me – what's the use of a balloon?
I replied – *what's the use of a new-born baby*'
BENJAMIN FRANKLIN, 1783

'Practical flying we may leave to our rivals the French.
Theoretical flying we may claim for ourselves'
SIR JOSEPH BANKS, 1784

'I would make it death for a man to be convicted of flying,
the moment he could be caught'
WILLIAM COWPER, 1794

'O Thou who plumed with strong desire
Would float above the Earth – beware!
A shadow tracks thy flight of fire –
Night is coming!'
P.B. SHELLEY, 1818

'There's something in a flying horse,
There's something in a huge balloon'
WILLIAM WORDSWORTH, 1819

'No man can have a just estimation of the insignificance of his species,
unless he has been up in an air-balloon'

BENJAMIN ROBERT HAYDON, 1825

'Your balloon voyage so occupied my mind that I dreamt of it!'

J.M.W. TURNER, 1836

'Beautiful invention, mounting heavenward – so beautifully,
so unguidably! Emblem of our Age, of Hope itself'

THOMAS CARLYLE, 1837

'How should I manage all my business if I were obliged to marry –
I never should know French, or go to America, or go up in a Balloon'

CHARLES DARWIN, 1838

'To look down upon the whole of London as the birds of the air look
down upon it, and see it dwindled into a mere rubbish heap'

HENRY MAYHEW, 1852

'Chance people on the bridges peering over the parapets,
into a nether sky of fog, with fog all round them as if they
were up in a balloon and hanging in misty clouds'

CHARLES DICKENS, 1852

'The basket was about two feet high, four feet long ... to me it
seemed fragile indeed ... the gaps in the wicker work in the sides
and the bottom seemed immense and the further we receded
from the earth, the larger they seemed to become'

GEORGE ARMSTRONG CUSTER, 1862

'Poetry has described some famous descents into the subterranean
world ... But we have just had an ascent such as the world
has never heard of or dreamed of'

THE TIMES, 1863

'Next to the climbers of the Alpine Club, in order of utter uselessness
are the people who go up in balloons, and who come down to tell us
of the temperature, the air-currents, the shape of the clouds,
and amount of atmospheric pressure in a region where nobody
wants to go, nor has the slightest interest to hear about'
BLACKWOOD'S MAGAZINE, OCTOBER 1864

'Dear Nadar, I must beg you to renounce these terrible balloon-*antics*!'
GEORGE SAND, 1865

'I am an Ancient Mariner of the Upper Atmosphere'
CHARLES GREEN, 1868

'Paris is surrounded, blockaded, blotted out from the rest of the world!
– and yet by means of a simple balloon, a mere bubble of air, Paris is
back in communication with the rest of the world!'
VICTOR HUGO, 1870

'It has already done for us that which no other power ever
accomplished: it has gratified the desire natural to us all
to view the earth in a new aspect'
JAMES GLAISHER, 1872

'The spectacle was over by the time we gained the top of the hill. All the
gold had withered out of the sky, and the balloon had disappeared'
ROBERT LOUIS STEVENSON, 1878

'Is it not a little strange to be floating here above the Polar Sea?
To be the first that have floated here in a balloon! How soon,
I wonder, shall we have successors? Shall we be thought
mad or will our example be followed?'
SALOMON ANDRÉE, 1897

'Between my boots and the now distant earth there was nothing.
I wriggled my feet and laughed. I was walking on air ...
I really was walking on air!'
DOLLY SHEPHERD, 1904

'To be alone in a balloon at a height of fourteen or fifteen thousand feet is like nothing else in human experience. It is one of the supreme things possible to man. No flying machine can ever better it.
It is to pass extraordinarily out of human things'

H.G. WELLS, 1908

'The miracle is not to fly in the air, or to walk on the water, but to walk on the earth'

CHINESE PROVERB

1

The Falling Dream

———

1

My own flying dream began at a village fête in Norfolk. I was four years old. My uncle, a tall and usually silent RAF pilot, had bought a red party balloon from a charity stall, and tied it to the top button of my aertex shirt. This was my first balloon, and it seemed to have a mind of its own. It was inflated with helium, which is a gas four times lighter than air, though I did not understand this at the time. It pulled mysteriously and insistently at my button. 'Maybe you will fly,' my uncle remarked. He led me up a grassy bank so we could look over the whole fête. Below me stretched the little tents, the stalls, the show ring with its bales of straw and small dancing horses. Above me bobbed the big red balloon, gleaming and beautiful, blotting out the sun. It bounced off the top of my head, making a strange springy sound, full of distance. It tugged me impatiently towards the sky, and I began to feel unsteady on my feet. I felt that I was falling – upwards. Then my uncle let go of my hand, and my dream began.

2

Throughout history, dreamlike stories and romantic adventures have always attached themselves to balloons. Some are factual, some are pure fantasy, many (the most interesting) are a provoking mixture of the two. But some kind of *narrative basket* always seems to come tantalisingly suspended beneath them. Show me a balloon and I'll show you a story; quite often a tall one. And very frequently it is a story of courage in the face of imminent catastrophe.

What's more, all balloon flights are naturally three-act dramas. The First Act is the launch: the human drama of plans, hopes, expectations. The Second Act is the flight itself: the realities, the visions, the possible discoveries. The Final Act is the landing, the least predictable, most perilous part of any ascent, which may bring triumph or disaster or (quite often) farce. The ultimate nature of any particular balloon ascent – a pastoral, a tragedy, a comedy, a melodrama, even a sitcom – is never clear until the balloon is safely back on earth. Sometimes it is not clear even then.

Even the well-known fable of the Cretan engineer Daedalus and his young son Icarus, so often retold as the Genesis myth of flying, is curiously ambiguous in its outcome. It appears originally in Book VIII of Ovid's long poem *Metamorphoses*, 'The Transformations', completed two thousand years ago, around 8 AD. Having constructed wings for both of them, Daedalus and son launch into the empyrean together, but famously the impetuous Icarus flies too high; the wax joints of his feathered wings melt 'in the scorching heat of the sun', and he tumbles down into the sea. Yet this primal legend of flight is more complex than it might appear.

It is often forgotten that in the same Book VIII of Ovid's poem, Daedalus also has a twelve-year-old nephew (the son of his sister) called Perdix. Perdix is a brilliant and precocious child inventor, loved by all in Crete. But Daedalus, in a crazed fit of grief and jealousy after the death of Icarus, hurls Perdix 'headlong down from the sacred hill of Minerva'. Yet unlike Icarus, Perdix does not crash to earth and die. Instead, he takes to the air and flies with divine aid: 'Pallas Athene, the goddess who fosters all talent in art and craft, caught him and turned him, still in mid-air, to a fluttering bird and covered his body with feathers, so the strength of his quick intelligence sprang into his wings and feet.' He becomes Perdix, the partridge (*perdrix* in French), a child who has indeed learned to fly successfully – although unlike Icarus he always remains close to the ground, 'and does not build his nest in mountain crags'.[1]

What may happen while actually aloft is equally mysterious. Balloons have always given a remarkable bird's-eye or angel's-eye view of the world. They are unusual instruments of contemplation, and even speculation. They provide unexpected visions of the earth beneath. To the earliest aeronauts they displayed great natural features like rivers, mountains, forests, lakes, waterfalls, and even polar regions, in an utterly new light. But they also showed human features: the growth of the new industrial cities, the speed and violence of modern warfare, or the expansion of imperial exploration.

Long before the arrival of the aeroplane in the twentieth century, balloons gave the first physical glimpse of a planetary overview. Balloons contributed to the sciences and the arts that first suggested that we are all guests aboard a unified, living world. The nature of the upper air, the forecasting of weather, the evolutions of geology, the development of international communications, the power of propaganda, the creations of science fiction, even the development of extra-terrestrial travel itself, are an integral part of balloon history.

But there are also stranger, existential elements, far less easy to define. The mental release, the physical heart-lift, the calm perilous delight of ballooning – an early aeronaut described it as 'hilarity' – is an absolute revelation, but one not easily or convincingly described. I have tried to capture its spirit indirectly, by tying together this cluster of true balloon stories and colourful tales, from the vast 'history and lore of aero-station', in the hope that they will bear us aloft for a little while.

While airborne, they may also provide a new perspective. The vulnerable globe of balloon fabric is itself symbolically related to the vulnerable globe of the whole earth. There is some haunting analogy between the silken skin of a balloon, the thin 'onion skin' of safety, and the thin atmospheric skin of our whole, beautiful planet as it floats in space. This thin breathable layer of air is not much more than seven miles thick – as balloonists were the first to discover. In every way, balloons make you catch your breath.

3

Falling upwards by helium party balloon may sound unlikely. But on Sunday, 20 April 2008 a forty-one-year-old Catholic priest, Father Adelir Antonio de Carli, made a heroic ascent using a very similar method. Father de Carli was known locally as Padre Baloneiro – 'The Balloon Priest' – and he flew for charity. He took off from the port city of Paranaguá, in Brazil, strapped into a buoyancy chair suspended beneath a thousand small multicoloured helium balloons. They were grouped into five vivid clusters of pink, green, red, white and yellow. His aim was to raise money for a truckers' rest stop and spiritual centre. He was known for his human rights campaigns, and that January had made a successful four-hour charity ascent suspended beneath six hundred helium party balloons.

On this second flight, armed with a thermal flying suit, GPS system and satellite phone, he rose successfully to some nineteen thousand feet, near to the edge of the sustainable atmosphere, where the sky becomes dark blue, and human breath forms glittering ice crystals in the ever-thinning air. Here, close to heaven, he cheerfully reported back by phone to his flight control. He also gave a phone interview to the Brazilian TV channel Globo, in which he said he was 'fine, but very cold', and was having trouble operating his GPS device. But he was also being carried out to sea, and was now thirty miles off the coast. At 8.45 that Sunday evening, he lost contact with the coastguards. An air-sea rescue search was mounted early the following Monday morning, but without success. A surviving cluster of fifty balloons was found floating in the sea late on Tuesday, but without Padre Baloneiro attached. The Brazilian naval search was called off on 29 April. But his parishioners continued to believe in his miraculous survival, and prayed daily for him.

Three months later, on 4 July 2008, an oil-rig support vessel found
the remains of his body (lower torso and legs only) floating about sixty
miles off the Brazilian coast, still attached to his buoyancy chair. It
seems that part of the helium balloon rig must have separated or failed
in some way during the first twenty-four hours of his flight. Possibly
some of the balloons began bursting at high altitude, but this of course
would have automatically reduced his lift, much as planned, and
brought him back comparatively gently to earth. Except that now there
was no earth beneath him. It seems that Padre Baloneiro must have
spent some time meditating in the sea. Finally, he was probably eaten
by sharks. But he was a brave man, a daring balloonist, and possibly
even a saint. ⎶

4

The dash and eccentricity of so many of those who have flown balloons
since the first Montgolfiers of 1783 is strangely mesmerising. I find it difficult not to admire such figures as Sophie Blanchard, Charles Green, Félix
Nadar, James Glaisher, Thaddeus Lowe, Gaston Tissandier or Salomon
Andrée. Indeed, I find it difficult not to fall for them. The word 'intrepid'
is automatically used of balloonists; but almost always thoughtlessly. In
my experience, balloonists come in every shape and personality type:
meticulous, cautious, reckless, obsessive, sportive, saturnine, or

---

⎶ A transcendent cluster of philanthropic balloons also appears in the classic French
film The Red Balloon, but with happier consequences. Shot in the backstreets of Paris,
significantly in the time of recovery after the Second World War, it was directed by
Albert Lamorisse and won both an Oscar and the Palme d'Or for short films at Cannes
in 1956. A lonely little orphan boy discovers a beautiful red helium party balloon
mysteriously tethered to a lamp post in the poor district of Ménilmontant. The
balloon has its own magical life and personality, quickly befriends the child and
follows him down the street, first to school and then home, where it waits faithfully
all night outside his bedroom window. Later the pair are chased by a jealous gang of
street urchins, the red balloon is caught and, in an act of primitive savagery, stoned
to death on a piece of high waste ground overlooking the city. The film concludes with
a memorable visionary sequence, in which hundreds of other party balloons rise up
from all over Paris, flock down to console the little boy, and together lift him high into
the air and far away across the city, perhaps even to Heaven.

devil-may-care. Equally they seem to have every kind of motivation: professional, commercial, scientific, philanthropic, escapist, aesthetic, or just plain publicity-seeking.

But the one thing they never quite seem to be is down-to-earth. All of them seem to have one enigmatic thing in common, besides physical courage and a head for heights. This is a romantic dream of flying, a strange – an almost unnatural – longing to be airborne. There is something both exotic and magnetic about such people. A biographer is drawn to their enigma.

The balloons themselves are mysterious, paradoxical objects. They are both beautiful and ephemeral. They are a mixture of power and fragility in constant flux. They offer a provoking combination of tranquillity and peril; of control and helplessness; of technology and terror. They make demands.

Consider an earlier balloon flight for charity, which took place on the afternoon of 22 July 1785, when a full-size hydrogen balloon was seen flying at three thousand feet over the Norfolk fishing village of Lowestoft. (Indeed, very close to my village fête.) The balloon was heading rapidly eastwards, directly out over the North Sea, and its pilot was clearly unable to bring it back to earth. There was nothing between the balloon and the distant shores of the Baltic.

The man in the basket was thirty-three-year-old John Money, a half-pay officer from the 15th Light Dragoons. Major Money had taken off earlier that afternoon from Ranelagh Gardens in Norwich, to raise cash for the new Norfolk and Norwich Hospital, founded in 1772. It was a cause supported by the Bishop of Norwich and the local Norwich MP, William Wyndham, a friend of Dr Johnson's and also a balloon enthusiast. The Major knew a lot about horses, harness, and driving a coach and pair, but he had little practical experience of balloons. He was however a man of courage and resource, who enjoyed a gamble as well as supporting a good cause.

Money had originally joined the Norfolk Militia, then the 15th Light Dragoons, and subsequently went out to serve as a captain under General John Burgoyne for the British Crown in the American War of Independence. He was noted for his unfashionable objection to military flogging for desertion (often a lethal punishment), mildly suggesting that a neat tattooed 'D' on the upper right arm might prove more effective. He was captured in Canada after the Battle of Saratoga, but eventually bargained his way out of prison. It seems he was a cool customer in a tight situation.

He was now back home, riding his horses and kicking his heels on his small country estate at Crown Point, in the village of Trowse Newton, just south of Norwich. Balloons fascinated him, partly for their military possibilities, but also for their sheer if uncontrollable beauty. He regarded them as if they were a species of wild horse. Admittedly, he had only made one previous ascent, in London that spring, in what was known as the 'British Balloon'. This had been constructed as a patriotic rival to the already celebrated Italian balloons of Vincenzo Lunardi and the wealthy eccentric Count Zambeccari. Characteristically, Money had somehow convinced the owners of the British balloon to let him transport it to Norwich, and to fly it solo for this philanthropic ascent.[2]

The launch went fine, according to the local *Gazette*, attended by 'a large and brilliant assembly of the first and most distinguished personages in the city and county'.[3] The balloon rose easily above the stately copper beeches on the northern boundary of the gardens (their leaves barely stirring), and was then carried on a gentle summer breeze across the river Yare, in a north-westwards direction towards distant Lincolnshire. But as it gained height, an 'improper current' arose, and a brisk wind blew it back across the city – to more enthusiastic cheers – and then south-eastwards, still gaining height, towards the Norfolk coast, a mere fifteen miles away. By 6 p.m. the balloon was spotted sailing high over Lowestoft, and heading out over the North Sea. It was supposed that some problem had arisen with the valve of his balloon, and that Money was unable to vent sufficient hydrogen gas to bring himself down. He disappeared rapidly out over the sea and into the softening eastern haze of the summer evening.

Among the 'distinguished assembly' who witnessed the launch at Ranelagh Gardens was the Earl of Orford. He wrote anxiously to William Wyndham the following morning. 'I am sorry to inform you that a Major Money ascended alone under the British Balloon at 4 o'clock yesterday afternoon. The balloon rose to a great height and took a direction towards the sea. It was seen entering over the ocean about a league south of Lowestoft at a very great height at six o'clock. By which circumstance I am greatly apprehensive for his thus continuing in the air, but that by some accident perhaps the String which connects to the valve was broken ...'[4]

Orford's notion of a balloon controlled by a 'string' was a little simplistic; but he noted accurately that although the balloon 'was not half full', and that its lower part appeared to have suffered what he called 'a collapsion', it continued unchecked towards the horizon. Indeed, Money was struggling to rein in his balloon as if it were a runaway horse, but without success. It was only an hour later, when he was well out of sight of land, that the cooling night air finally deposited his balloon twenty miles off the Norfolk coast, in an area known on mariners' charts as Long Sand, notorious for its shoals and shipwrecks.〜 The balloon still had sufficient hydrogen to keep its basket partially above the waves. Waist-deep in water, Money began a long battle to remain afloat in a choppy sea as darkness fell.

He soon abandoned his basket, cutting it loose and allowing it to sink beneath him, while climbing up into the balloon hoop and clinging onto the rigging. By skilfully playing the lines, he managed to hold sufficient gas in the balloon canopy to keep it partially inflated, pulling him slowly through water almost like a kite, and giving him just enough buoyancy to stay afloat. Increasingly cold and exhausted, Money hung on grimly hour after hour as the balloon steadily dragged him further and further out to sea through the darkness. As the gas slowly escaped, he sank gradually deeper into the water, until after four hours he was up to his chest, and almost incoherent with hypothermia.

Several pleasure boats and fishing smacks had in fact set out after him, both from Yarmouth and further south from Southwold. Their crews were in sportive mood, playfully competing to find the airborne quarry. But as darkness fell they grew dispirited and bored, eventually giving up any hope of recovering him. One by one they turned to beat

---

〜 I know Long Sands quite well. I was shipwrecked off this same turbulent part of the East Anglian coast in 1991, and waited for some time in the open sea for rescue with sensations that Major Money might have recognised. All the same, my circumstances were very different. The craft I had to abandon was a yacht, not a balloon. It was just before dawn, not night time. And above all I was with two friends, not alone. Nevertheless, I was infinitely grateful to see the arrival overhead of a bright-yellow air-sea-rescue helicopter. One by one, we were winched up to safety, spinning several hundred feet into the air as we went, the choppy waves dropping away beneath our feet. I went last, and well remember those few final moments of utter isolation, waiting in the open water gazing upwards. It now strikes me that this rescue curiously reversed the conditions of Money's uncontrolled descent into the sea.

back into port, telling each other that he was either drowned or in Holland, which came to much the same thing. Agonisingly, it appears that Money had seen several of these ships. Their sails were clearly silhouetted on the western horizon behind him, dark against the dying summer light. But they were too far away, and he was now too weak even to shout. The water was colder, and the waves came up from his chest to his chin.

But one determined coastguard cutter, the *Argus*, had set out from Lowestoft. Long before a regular lifeboat service was formed, this was a professional rescue vessel, its crew skilled in the pursuit of both smugglers and mariners in distress. A balloon was a new and interesting object for them to hunt. Its skipper skilfully put the wind dead astern and, making due allowance for tides, steadily followed exactly along the balloon's last observed line of flight, with lookouts posted at his masthead. He knew that the moon was due to come up by late evening, and would illuminate the sea very well if he persisted.

Just before midnight, after Major Money had been in the water for over five hours, the pale shape of his crumpled balloon canopy was spotted on the dark waves by the crew of the *Argus*. They came gently alongside, carefully disentangled his body from the rigging, and hauled him out of the water. As he was pulled aboard, he stirred, and they realised he

was still conscious. Well-practised in revival techniques, the crew wrapped him in blankets, forced brandy down his throat, and had Major Money joking and telling his story by the time they were back in port the following dawn.[5]

Money immediately became famous throughout East Anglia. The Norfolk and Norwich Hospital received a splendid donation. He was interviewed by the local journals, and became the subject of one of the most dramatic of all the early balloon prints, a mezzotint by Paul Renaigle, entitled *The Perilous Situation of Major Money*. It showed him heroically struggling with the flapping balloon canopy, half-immersed in the water, while a ship turns away from him under a stormy sky.

Major Money remained undaunted by this experience. He later volunteered to command a French regiment at the Battle of Valmy, and for the first time saw balloons being used for observation on the battlefield. When he returned he was promoted General, and in 1803 published A *Short Treatise on the Use of Balloons in Military Operations*. This was unusual for a military manual, in that it included a number of balloon ideas set to verse:

> Great use, he thought, there might be made
> Of these machines in his own trade;
> Now o'er a fortress he might soar
> And its condition thence explore
> Or when by mountains, woods, or bog
> An enemy might lie incog
> Our friend would o'er their station hover
> Their strength, their route, and views discover;
> Then change his course, and straight impart
> Glad tidings to his chieftain's heart ...[6]

These were all to prove strangely prophetic.

5

The experience of ballooning is in a sense timeless. Man-carrying balloons are both extremely modern and extremely primitive devices. In their contemporary form, powered by stainless-steel propane-gas burners and using rip-stop nylon envelopes, they were virtually reinvented in the mid-1960s by an American, Ed Yost, experimenting in Nebraska. His ideas were quickly taken up by Don Cameron and others in Britain and France.[7] It should not be forgotten that these reinvented balloons were contemporary with the first moon landings and the earliest communication satellites.

But balloons are also ancient and symbolic devices. They have a long history, and a longer mythology, going back in various forms and dimensions thousands of years, to ancient civilisations in South America and China. There are vague accounts of man-carrying smoke balloons from the Yin dynasty of the twelfth century BC. The great scholar and Sinologist Joseph Needham suggested that Chinese of the fourth century BC used fire balloons for signalling in warfare, or perhaps for carrying love letters. There are rumours of shamanic balloon flights made by the priests of the pre-Inca civilisations. Peruvian funereal rituals involved sending corpses out over the Pacific by hot-air balloon, just as the Vikings would later send out their sacred dead by fireboat into the North Sea. The famous geometrical carvings on the Nazca plateau in southern Peru, some of them animal shapes stretching over four miles in outline, are only explicable if they were originally designed to be viewed from hundreds of feet in the air, so presumably by balloon.

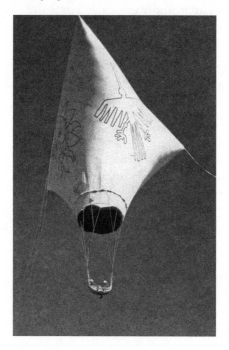

It has been suggested that the Nazca designs were made by visiting aliens, hovering in flying saucers.◊ But the modern balloonist Julian Nott successfully invented a huge smoke balloon, constructed purely from local materials, to prove that human beings could overfly and supervise the carvings even in the fifth century AD.[9]

The primitive and the sophisticated elements of ballooning are often combined. In this way the balloon may have both a practical and a symbolic function, for example when it is used as a means of escape. Among the most remarkable balloon escapes ever made was a flight across the East German border in September 1979. Its daring, and the idea of a symbolic flight from Communism to the free West, so caught people's imaginations that it was made into an adventure film by Disney, *Night Crossing* (1982), starring John Hurt and Beau Bridges.

In March 1978 two East German men, Peter Strelzyk and Günter Wetzel, living with their wives and four children at Poessneck, near the East German frontier, began working on several ideas for escaping to West Germany. Strelzyk was an aircraft mechanic and electrician with his own workshop, while Wetzel was a builder and a gifted handyman. Both were brilliant at *bricolage*, endlessly resourceful and determined. Together they hit upon the idea of secretly constructing a home-made hot-air balloon in Strelzyk's attic and workshop. There are various accounts of how they came up with this idea, but one is that Wetzel's sister-in-law gave them an illustrated magazine article about the annual Albuquerque International Balloon Fiesta, the most famous of all hot-air balloon gatherings, launched in New Mexico in 1972. After that they got all their technical information from the Poessneck public library.

◊ Admittedly, not all planetary visitors from outer space may have needed flying saucers or even balloons. In 1667, at the end of Book 2 of *Paradise Lost*, John Milton reports on Satan's epic flight from Hell up towards the shining Earth. Observing from deep space the 'emptier waste, resembling Air', Satan sees from afar our delicate planet as a small but fabulous jewel, miraculously suspended from the 'Opal Towers and battlements' of Heaven, and prophetically ready to be despoiled or ravaged (a task he later leaves to mankind):

> ... *fast by, hanging in a golden Chain*
> *This pendant World, in bigness as a star*
> *Of smallest magnitude, close by the Moon.*
> *Thither full fraught with mischievous revenge.*
> *Accursed, and in a cursed hour he hies.*[8]

Their balloon had to be very large, capable of lifting eight people to a height of at least five thousand feet, to avoid detection by frontier search-lights, and of carrying them at night over a distance of at least ten miles. They spent months surreptitiously assembling suitable materials, build-ing makeshift propane burners and testing various potential balloon fabrics – including cotton sheets, umbrella covers, waterproof-jacket linings and tenting fabric. The work was shared, but Strelzyk specialised in constructing the burners and the sheet-metal balloon platform, while Wetzel worked on the balloon canopy and rigging. He sewed all the curved balloon strips, or gores, together on a pre-war, pedal-operated sewing machine. Everything had to be bought in small quantities from different shops to avoid alerting the network of Stasi informers; they drove as far as Leipzig to cover their purchases, sometimes claiming that they represented camping or sailing clubs. Their balloon trials were carried out at night in remote areas of the Thuringian forest.

In the end, with infinite patience and ingenuity, they built three versions of their escape balloon. The first, a sixty-foot-high cotton balloon with a capacity of seventy thousand cubic feet, failed to inflate properly, due to porous fabric and a weak burner using two domestic propane cylinders. It had to be abandoned and painstakingly destroyed in April 1978. After more than a year of experiments and setbacks, they came up with a second design with a better, four-cylinder burner and tighter fabric. But during this anxious time Günter Wetzel, increasingly haunted by the risks to his family, reluctantly withdrew from the scheme, and began to consider more conventional methods of crossing the border.

The second balloon was designed to take only the three members of the Strelzyk family. Symbolically, they chose American Independence Day, 4 July 1979, for their launch. But the balloon still lacked lifting power. It flew too low, became drenched by rainclouds, and began to sink earthwards just as the border came in sight. The Strelzyks crash-landed in the bare no-man's land two hundred yards short of the actual frontier fence. By good luck they were just outside the frontier 'death zone', where the barbed wire, anti-personnel mines and automatic guns would have proved fatal. Astonishingly, the crumpled shape of the balloon was not immediately spotted by the border guards, probably because of the heavy rain. Under cover of darkness the three Strelzyks scrambled out of the wreckage, collected all the personal belongings they could carry, and somehow managed to slip back undetected to Poessneck, covering nine miles on foot before dawn. But the balloon equipment that they were forced to abandon meant that the Stasi had clues to their identity, and would soon be hot on their trail. Discovery within a matter of weeks was inevitable.

At this desperate moment, the two families joined forces again. Working around the clock, the Strelzyks and the Weltzers constructed a much bigger balloon using piecemeal sections of artificial taffetas and dress materials, hastily purchased from small shops all over East Germany. An electric engine was attached to the sewing machine, and the propane burner was redesigned. In a matter of six weeks they had a new balloon looking like a huge multicoloured quilt. When fully inflated it stood nearly ninety feet high, and had a hot-air capacity of over 140,000 cubic feet, double that of the previous balloon. Its burner was powered by four propane tanks feeding into a simple five-inch-diameter stovepipe, capable of producing a narrow, violent flame which at maximum pressure shot fifty feet into the air – within thirty feet of the inner crown of the balloon. This could in theory lift well over 1,200 pounds (544 kilograms), the equivalent of seven adults and a child plus all the balloon equipment. But everything depended on the durability of the home-made envelope, the strength and direction of the wind, and the general flying conditions (including air temperature and humidity) on the actual night of the flight.

Unable to obtain materials for a conventional wicker basket, they constructed instead an open metal platform four and a half feet square. The four propane cylinders stood in the centre of this platform, and the eight passengers carefully distributed their weight around them, having to crouch within inches of the platform's outer edge. The youngest Wetzel was held in his mother's arms. The ten guy ropes connected to the balloon were tethered to iron stanchions welded along the edges of the platform, which provided some handholds. There was also an outer guardrail made of loops of washing line, but this only came up to the adults' waists. The stovepipe burner was ignited by a household match, and at full power burnt with a tremendous roar about six feet above the passengers' heads, shooting flame high into the centre of the balloon. When this 'flame-thrower' was extinguished, they would float in absolute darkness and silence, standing virtually unprotected in the air, with no sound but the creak of the ropes against the balloon fabric, somewhere invisible above their heads. It was a magnificent, dreamlike, insane contraption. But it flew.

At 2 a.m. on the night of 16 September 1979, with a brisk eighteen-mile-per-hour breeze blowing towards West Germany, they took off from their secret base in the Thuringian forest, about six miles from the frontier. They cleared the fir trees, and with a tremendous blast from the propane burner, the balloon rose rapidly to 6,500 feet. But as it turned on its axis in the dark, they soon lost all sense of direction. Clinging together on the tiny metal platform, they peered down in silence, looking for car headlamps which would indicate roads, or the chain of lights which would mark the border.

After about twenty minutes, to their alarm, they suddenly saw searchlights springing up almost directly beneath them. They had the choice to drift downwards, steadily sinking but hoping to avoid detection in the dark, or to fire up their burner and try to climb clear. They chose to fire the burner, and with a huge sustained burst of flame, which they felt must surely be visible for miles around, rose to nearly nine thousand feet. Under either the increased heat or the air pressure, the crown of the balloon split. They began to sink again, but the balloon remained inflated, and by continuing to fire the burner until their propane ran out, they managed a crash-landing in an open field a hundred yards from a high-voltage pylon. Günter Wetzel broke his leg, but otherwise they were

all unhurt, although they had no idea on which side of the border they had arrived. Peter Strelzyk walked over and shone a torch on the 'Danger of Death' sign fixed to the base of the pylon. It belonged to a West German electricity company. They had flown to freedom – and to fame – in exactly twenty-eight minutes. 'We could have made it as far as Bayreuth,' remarked Wetzel.[10] 𝄞

6

The theme of escape, either literally from some form of imprisonment, or symbolically from the troubles of the earth itself, constantly recurs in the history of ballooning. When Dr Alexander Charles made the first ever flight by a true hydrogen balloon, two hundred years before the escape of the Strelzyks and the Wetzels, on 1 December 1783, it was the feeling of absolute and almost metaphysical freedom that overcame him.

Flying with an engineering assistant, Monsieur Robert, Dr Charles launched from the Jardin des Tuileries in central Paris, and travelled over twenty miles north-west to the country town of Nesles. His balloon was a mere thirty feet high, but was equipped with a proper wicker basket, a venting valve, and sacks of ballast to adjust its height and control its descent. His departure was witnessed by nearly half a million people, among them the American ambassador, Benjamin Franklin. After they had landed safely at Nesles, Monsieur Robert disembarked, but Dr

---

𝄞 Ideas of aerial escape have haunted many prisoners. Perhaps the most famous is the glider built in the attic of the German prisoner-of-war camp Colditz Castle in 1944. It wonderfully sustained the morale of the British PoWs during the last months of the war, but was never deployed. Afterwards, it was generally agreed that had it ever actually been launched from the high, tilting slate roof of Colditz, it would have plunged into the gorge 150 feet beneath the castle walls, instantly killing its two prospective escapees. Curiously, the much more practical method of escape by a hot-air balloon was never considered by the Colditz Escape Committee, but perhaps practicality was not the point. See Airey Neave, *They Have Their Exits*.[11] A more metaphysical version of this flight-escape fantasy is movingly told in the 1962 film *The Bird Man of Alcatraz*, with Burt Lancaster playing the real-life convict Robert Stroud. It is also a curious fact that many of the early balloonists, like Blanchard and Garnerin, had experienced periods of imprisonment and gazing up at the sky through barred windows.

Charles remained in the basket. He then achieved the first ever solo ascent, rapidly rising in the lightened balloon to a magnificent ten thousand feet. From this vantage point he saw the sun set for a second time on the same day. It was a revelation.

Dr Charles's brilliant account of this ascent was widely published in both Britain and France, and catches a euphoric tone which never quite disappears from subsequent balloon accounts. He had laid in supplies for an aerial journey of many hours – fur coats, cold chicken and champagne. But what he actually tasted was that existential substance:

Nothing will ever quite equal that moment of total hilarity that filled my whole body at the moment of take-off. I felt we were flying away from the Earth and all its troubles and persecutions for ever. It was not mere delight. It was a sort of physical rapture ... I exclaimed to my companion Monsieur Robert – 'I'm finished with the Earth. From now on our place is in the sky! ... Such utter calm. Such immensity! Such an astonishing view ... Seeing

all these wonders, what fool could wish to hold back the progress of science!'[12]

Benjamin Franklin watched the launch through a telescope from the window of his carriage. Afterwards he remarked, 'Someone asked me – what's the use of a balloon? I replied – *what's the use of a new-born baby.*'

The same sense of escaping into an utterly new world is displayed by Thomas Baldwin's *Airopaedia, or Narrative of a Balloon Excursion from Chester in 1785.* This is his account of a single flight made on 8 September 1785, flying northwards above the river Mersey, from Chester to Warrington in Lancashire. It must be one of the most remarkable books about the experience of ballooning ever written. It also included flight maps, and the first aerial drawings ever made from a balloon basket.

Baldwin was one early pioneer of the existential attitude to ballooning, in which the idea that the 'Prospect' itself – the free ascent, the magnificent views, the whole 'aerial experience' – was the real point of flight. He believed that 'previous Balloon-Voyagers have been particularly defective in their Descriptions of aerial Scenes and Prospects'. Consequently he took with him a battery of recording equipment: a variety of pens and red lead pencils, special 'Ass Skin Patent Pocketbooks', paints and brushes, drawing blocks and perspective glasses, telescopes and compasses. *Airopaedia* contained the first ever paintings of the view from a balloon basket, an analytic diagram of the corkscrew flight path projected over a land map, and a whole chapter given up simply to describing the astonishing colours and structures of cloud formations.

Baldwin also notices how the balloon responded to air currents arising from the earth beneath. His careful flight-mapping shows how it was constantly drawn downwards to follow the cool, curving airflows above the meanderings of the river. Similarly, the heady act of leaning directly over the side of the basket to paint, observe and measure makes him sensitive to shifts in shade and colour and perspective on the ground below.

One typical observation reads: 'The river Dee appeared of a red colour; the city [Chester] very diminutive; and the town [Warrington] entirely blue. The whole appeared a perfect plane, the highest buildings having no apparent height, but reduced all to the same level, and the whole terrestrial prospect appeared like a coloured map.'[13]

Baldwin also writes wonderfully well about clouds, and the prismatic effects of light. He clearly perceives a whole new world opening out

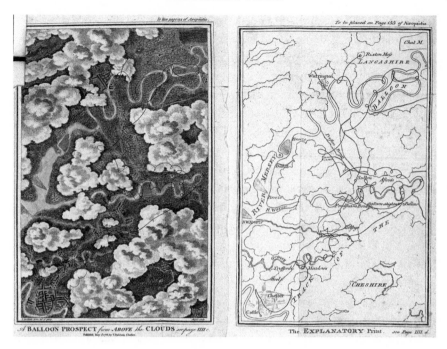

A BALLOON PROSPECT *from ABOVE the* CLOUDS *see page IIII c*

The EXPLANATORY Print   *see Page IIII d*

around him, and expresses a euphoric emotional reaction. Indeed, to keep these feelings within bounds, he writes of himself throughout his flight in the third person: 'A Tear of pure Delight flashed in his Eye! of pure and exquisite Delight and Rapture!' For him, ballooning instinctively combined both scientific discovery and aesthetic pleasure. But perhaps it should provide more? He could imagine the time when 'aerostatic ships make the Circuit of the Globe'.[14]

7

The essential mystery of ballooning – the enigmatic meaning of the original dream – was there from the start. Almost a decade after its invention by the Montgolfier brothers, with flights recorded in many nations, including Germany, Italy, Russia and America, it was still not clear, either to the Royal Society in London or the Academy of Sciences in Paris, what the true purpose or possibilities of ballooning really were. Don Paolo Andreani had flown from Milan in February 1784; Jean-Pierre Blanchard and Dr John Jeffries had traversed the Channel in January 1785; Pilâtre de

Rozier had died attempting the same crossing in the opposite direction with a composite hydrogen and hot-air balloon in June 1785 (thereby becoming the first scientific balloon martyr); Baron Lütgendorf had 'partially' flown at Augsburg in August 1786; and Blanchard had gone on to demonstrate ballooning in virtually every major city in Europe, finally crowning his international career with what he claimed was the first ever American ascent, from the city of Philadelphia in January 1793, carrying an 'aerial passport' endorsed by President George Washington, and successfully crossing the Delaware river into New Jersey.[15]

Yet all these ascents were essentially public spectacles and entertainments. 'Flight' itself remained a novel and surprisingly unexplored concept. What, in practice, could balloons actually do for mankind, except provide a hazardous journey interspersed with the fine aerial 'Prospects' that men like Dr Charles and Thomas Baldwin recorded so eloquently?

According to Barthélemy Faujas de Saint-Fond, the Parisian promoter of the Montgolfier balloons, they might, for example, provide observation platforms: for military reconnaissance, for sailors at sea, for chemists analysing the earth's upper atmosphere, or for astronomers with their telescopes. It is notable that most of these applications were based on the notion of a *tethered* balloon. In fact many of the Montgolfiers' early experiments were made with tethered aerostats, held to the ground by various ingenious forms of harness, guy ropes or winches.◊

The poet and inventor Erasmus Darwin's first practical idea of balloon power was, paradoxically, that of shifting payloads along the

---

◊ I have described the Montgolfier brothers' balloon trials of 1783 in The Age of Wonder, together with the historic first flight by Pilâtre de Rozier and the Marquis d'Arlandes from the place de la Muette across the rooftops of Paris on 21 November of that year, and Joseph Montgolfier's famous formula for a hot-air balloon: 'a Cloud in a paper bag'. But I have since discovered an unpublished letter by Philippe Lesueur, dated Paris, 22 September 1783, describing the earlier tethered ascents, the spectacular unmanned flight from the Champ de Mars, and the notorious animal ascension from Versailles on 19 September 1783 with 'a sheep, a cockerel and a duck in a wickerwork basket suspended from the Aerostatique Machine by a fifteen-foot rope'. Lesueur not only catches the first thrill and astonishment of these public experiments – 'truly the most astonishing spectacle to see such a massive object rising majestically under its own power' – but also incorporates into his letter a superb pen-and-watercolour illustration of the Versailles Montgolfier with its first animal aeronauts.[16]

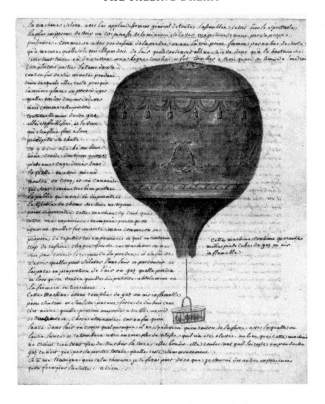

ground. He suggested to his friend Richard Edgeworth that a small hydrogen balloon might be tethered to an adapted garden wheelbarrow, and used for transporting heavy loads of manure up the steep hills of his Irish estate. This convenient aerial skip would allow one man to shift ten times his normal weight in earth, but also in bricks or wood or stones. In fact it might cause a revolution in the entire conditions of manual labour.[17]

Similarly, Joseph Banks, the President of the Royal Society, had the initial idea that balloons could increase the effectiveness of earthbound transport, by adding to its conventional horsepower. He saw the balloon as 'a counterpoise to Absolute Gravity' – that is, as a flotation device to be attached to traditional forms of coach or cart, making them lighter and easier to move over the ground. So 'a broad-wheeled wagon', normally requiring eight horses to pull it, might only need two with a Montgolfier attached. This aptly suggests how difficult it was, even for a trained scientific mind like Banks's, to imagine the true possibilities of flight in these early days.[18]

Benjamin Franklin, 'the old fox', as Banks's secretary Charles Blagden called him, was quick to suggest various menacing military applications, perhaps in a deliberate attempt to fix Banks's attention. 'Five thousand balloons capable of raising two men each' could easily transport an effective invasion army of ten thousand marines across the Channel, in the course of a single morning. The only question, Franklin implied, was which direction would the wind be blowing from?[19]

His other speculations were more light-hearted. What about a 'running Footman'? Such a man might be suspended under a small hydrogen balloon, so his body weight was reduced to 'perhaps 8 or 10 Pounds', and so made capable of running in a straight line in leaps and bounds 'across Countries as fast as the Wind, and over Hedges, Ditches & even Water ...' Or there was the balloon 'Elbow Chair', placed in a beauty spot, and winching the picturesque spectator 'a Mile high for a Guinea' to see the view.

There was also Franklin's patent balloon icebox: 'People will keep such Globes anchored in the Air, to which by Pullies they may draw up Game to be preserved in the Cool, & Water to be frozen when Ice is wanted.'[20] This contraption would surely have appealed to the twentieth-century illustrator W. Heath Robinson.

Franklin, who suffered formidably from gout, later suggested that a balloon might even be used to power a wheelchair. When he had returned from Paris to Philadelphia in autumn 1785, he began using a sedan chair lifted by four stout assistants for his daily commute from his house to the Philadelphia State Assembly Rooms. He suggested reducing the requisite manpower by 75 per cent, simply by harnessing the chair to a small hydrogen balloon, 'sufficiently large to raise me from the ground'. This would make his malady less vexatious for all concerned, by providing a 'most easy carriage', lightweight and highly manoeuvrable, 'being led by a string held by one man walking on the ground'.[21] ♨

♨ Given Franklin's weight, a hydrogen balloon of approximately ten feet in diameter, or one thousand cubic feet, would have been sufficient to perform this appealing service. One of the crucial scientific discoveries of late-eighteenth-century chemists like Henry Cavendish and Antoine Lavoisier was that hydrogen could be isolated, weighed (as counter-intuitive as that sounds), and compared against the weight of atmospheric air. It was actually named by Lavoisier.) Measurements were later

8

In 1785 Tiberius Cavallo, a Fellow of the Royal Society, put together the first British study of ballooning. His *A Treatise on the History and Practice of Aerostation* studiously adopted the French scientific term for 'lighter-than-air' flight, but moved far beyond national rivalries. He wanted to consider the phenomenon of flight from both a scientific and a philosophical point of view. He thought that ballooning held out immense possibilities, less as a transport device than as an instrument for studying the upper air and the nature of weather. This distinction between horizontal and vertical travel would have a long subsequent history.

Cavallo was a brilliant Italian physicist who had moved to London at the age of twenty-two, and had already written extensively on magnetism and electrical phenomena. Elected to the Royal Society in 1779, he quickly turned his attention to ballooning. He had some claims to be one of the first to inflate soap bubbles with hydrogen as early as 1782. Although a handsome portrait is held by the National Portrait Gallery in London, he is now largely and unjustly forgotten. Yet his study emerges as the most authoritative early treatise on the subject of ballooning in either English or French. The copy of Cavallo's book held by the British Library is personally inscribed 'To Sir Joseph Banks from the Author', in severe black ink.

---

refined, but as a rule of thumb one thousand cubic feet of air weighed seventy-five pounds. By comparison, one thousand cubic feet of hydrogen weighed only five pounds. Hydrogen was therefore approximately fourteen times lighter than air, with an equivalent lifting power. In terms of practical ballooning, this meant that a small hydrogen balloon, like Alexander Charles's of 1783, standing perhaps thirty feet high and with a capacity of approximately twenty thousand cubic feet, could lift a payload of roughly eight hundred pounds. In theory, this would be enough for two men, the basket with supplies, all the balloon equipment (fabric envelope, rigging, ropes and anchor), and many sacks of ballast. In practice, all these figures were alarmingly variable. They depended for example on the purity of the hydrogen, the temperature and humidity of the surrounding air (which changed constantly with height), and the condition of the balloon fabric. Hot air was of course even more variable. But in general a hot-air balloon needed to be three or four times as big as a hydrogen balloon to produce the equivalent lift. Pilâtre de Rozier's original Montgolfier stood ninety feet high. Even a modern hot-air balloon, powered by efficient propane burners, might require sixty thousand cubic feet – standing eighty feet high – to lift two or three men and their equipment safely. And 'safely' remains, as always, a relative term.[22]

Cavallo carefully adopted a considered and even sceptical tone, well calculated to appeal to Banks. Much had been made of Vincenzo Lunardi's historic first flight in Britain, in September 1784, when he flew from London to Hertfordshire with his pet cat. The newspapers of the day all declared Lunardi a heroic pioneer, a patriot and an animal lover, although the gothic novelist Horace Walpole – author of *The Castle of Otranto* – roundly criticised him for risking the life of the said cat. But Cavallo noted: 'Besides the Romantic observations which might be naturally suggested by the Prospect seen from that elevated situation, and by the agreeable calm he felt after the fatigue, the anxiety, and the accomplishment of his Experiment, Mr Lunardi seems to have made no particular philosophical observation, or such as may either tend to improve the subject of aerostation, or to throw light on any operation in Nature.'[23] ♔

Cavallo analysed and dismissed most claims to navigate balloons, except by the use of different air currents at different altitudes.[24] He emphasised the aeronaut's vulnerability to unpredictable atmospheric phenomena such as down-drafts, lightning strikes and ice formation. He deliberately included the first alarming account of a French balloon caught in a thunderstorm, during an ascent from Saint-Cloud in July 1784, and dragged helplessly *upwards* by a thermal:

> *Three minutes after ascending, the balloon was lost in the clouds, and the aerial voyagers lost sight of the earth, being involved in dense vapour. Here an unusual agitation of the air, somewhat like a whirlwind, in a moment turned the machine three times from the right to the left. The violent shocks, which they suffered prevented their using any of the means proposed for the direction of the balloon, and they even tore away the silk stuff of which the helm was made. Never, said they, a more dreadful scene presented itself to any eye, than that in which they were involved. An*

♔ The romantic reputation of Vincenzo Lunardi has nevertheless remained fondly in the British annals of early flight, despite reservations about his recklessly endangering the life of his cat. His use of a magnificent Union Jack design on his balloon canopy was a tactful pro-English response to French aeronautical pretentions. The stone monument recording his landing place still stands on a village green in Hertfordshire. In Alexander Korda's early black-and-white epic movie *The Conquest of the Air* (1934), Lunardi – always regarded as a typically Italian ladies' man – is played with lissom charm and a highly suggestive accent by the young Laurence Olivier.

unbounded ocean of shapeless clouds rolled one upon another beneath,
and seemed to forbid their return to earth, which was still invisible. The
agitation of the balloon became greater every moment ...[25]

Yet for all this, Cavallo was a passionate balloon enthusiast. He recorded
and analysed all the significant flights, both French and English, made
from Montgolfier's first balloon at Annonay in June 1783, to Blanchard
and Jeffries's crossing of the Channel in January 1785. He distinguished
carefully between hot-air and hydrogen balloons, and their quite different flight characteristics. He looked in detail at methods of preparing
hydrogen gas, noting that Joseph Priestley had come up with one that
used steam rather than sulphuric acid. He also examined the different
ways of constructing balloon canopies from rubber ('*cauchou*'), waxed
silk, varnished linen and taffeta.

In a longer perspective, he stressed the astonishing speed of aerial
travel over the ground – 'often between 40 and 50 miles per hour' –
combined with its incredible 'stillness and tranquillity' in most normal
conditions.[26] This he thought must eventually revolutionise our fundamental ideas of transport and communications, even if the moment had
not yet arrived. But he was less impressed by the horizontal potential of

ballooning than by its vertical one. The essence of flight lay in attaining an utterly new dimension: altitude.

He pointed out that in achieving altitudes of over two miles, balloons opened a whole new perspective on mankind's observations of the earth beneath. Man's growing impact on the surface of the planet for the first time became visible. As did the vast tracts of the earth – mountains, forests, deserts – yet to be traversed or discovered. Above all he stressed that the full potential of flight had not yet been remotely explored. The situation has perhaps some analogies with the space exploration programme, in the years following the Apollo missions.

Cavallo considered the whole range of possible balloon applications. But he finally and presciently championed its relevance to the infant science of meteorology:

> The philosophical uses to which these machines may be subservient are numerous indeed; and it may be sufficient to say, that hardly anything of what passes in the atmosphere is known with precision, and that principally for want of a method of ascending into the atmosphere. The formation of rain, of thunderstorms, of vapours, hail, snow and meteors in general, require to be attentively examined and ascertained.
>
> The action of the barometer, the refraction and temperature of air in various regions, the descent of bodies, the propagation of sound etc are subjects which all require a long series of observations and experiments, the performance of which could never have been properly expected, before the discovery of these machines. We may therefore conclude with a wish that the learned, and the encouragers of useful knowledge, may unanimously concur in endeavouring to promote the subject of aerostation, and to render it useful as possible to mankind.[27]

Cavallo's work was both a challenge and an intellectual landmark in the early history of ballooning. He was largely responsible for the historic first article on 'Aerostation', which appeared in the *Encyclopaedia Britannica*, with notable illustrations, in 1797. This was a signal date. From then on, flight was officially established as a new branch of scientific knowledge, rather than an old backwater of mythology.

9

Yet there always remains the most enduring early dream or fantasy of flying, which is a metaphysical one. The ultimate purpose is to fly as high as possible, *and then look back upon the earth and see mankind for what it really is.* This idea has persisted since the beginning, and still continues, sometimes in a satirical form, and sometimes in a visionary one.

The seventeenth-century French dramatist Cyrano de Bergerac (a fearless duellist and also an intellectual provocateur) convincingly reported a secret flight to the moon, undertaken sometime before his death in 1655. It appeared in his posthumously published work *Histoire comique des états et empires de la Lune* (*Comical History of the States and Empires of the Moon*). Cyrano's flight is powered by glass cluster balloons, filled with dew and drawn skyward as the droplets are heated by the sun and evaporate: 'I planted myself in the middle of a great many Glasses full of Dew, tied fast about me, upon which the Sun so violently darted his Rays, that the Heat, which attracted them, as it does the thickest

Clouds, carried me up so high, that at length I found myself above the middle Region of the Air. But seeing that Attraction hurried me up with so much rapidity that instead of drawing near the Moon, as I intended, she seem'd to me to be more distant than at my first setting out ...'

After an initial power failure and crash-landing, the final approach, made with the additional aid of gunpowder and a lunar force-field, is memorably disorientating: 'When according to the calculations I had made, I had travelled much more than three-quarters of the way between earth and moon, I suddenly started falling with my feet uppermost, even though I had not performed a somersault ... The earth now appeared to me as nothing but a great plate of gold overhead.'

When Cyrano eventually lands, he is captured and cross-questioned by various lunar inhabitants. One, more kindly than the others, remarks: 'Well, my son, you are finally paying the penalty for all the failings of your Earth world.'[28] Presented before the Lunar Court, he narrowly escapes being condemned to death for impiety. He has maintained the ridiculous notion that 'our earth was not merely a moon, but also an inhabited world'. He returns in sober mood, crash-landing near a volcano in Italy.[29]

Some three hundred years later, on 24 December 1968, the Apollo 8 spacecraft came round from the dark side of the moon. The astronaut Bill

Anders later recalled: 'When I looked up and saw the earth coming up on this very stark, beat-up lunar horizon, an earth that was the only colour that we could see, a very fragile-looking earth, a very delicate-looking earth, I was immediately almost overcome by the thought that here we came all this way to the moon, and yet the most significant thing we're seeing is our own planet, the earth.'

The trip produced one of the most famous colour photographs ever taken. It has become universally known as 'Earthrise'. The small, beautiful planet earth is sliding above the bleakness of the cratered moon surface, and hanging against the blackness of outer space. From this vision arose the whole modern concept of planet earth as the 'small blue dot' of life, amid a dark and mysterious universe.[30]

The dream of flight is to see the world differently.

2

Fiery Prospects

———

1

Just like space flight, early balloon flight offered military as well as metaphysical prospects. Cavallo's dream of scientific ballooning was soon displaced by prospects of a more warlike kind. Benjamin Franklin had warned Sir Joseph Banks of this possibility, and such signs of the times were recognised by many contemporaries, including Banks's clever younger sister, Sophia. She had begun to make a collection of balloon memorabilia, which she stuck into an enormous red-leather folio scrapbook especially purchased for the purpose. It would eventually run to over a hundred items.[1] One of her earliest specimens was a British cartoon, dated 16 December 1784, entitled 'The Battle of the Balloons'.

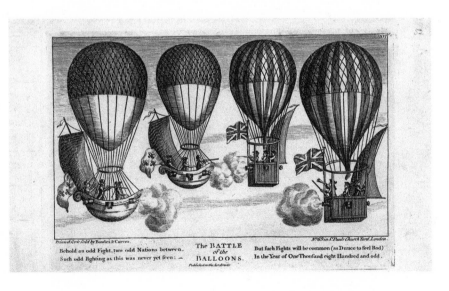

Printed first & Sold by Bowles & Carver.

Behold an odd Fight, two odd Nations between,
Such odd fighting as this was never yet seen; —

The BATTLE
of the
BALLOONS.
Published as the Act directs

No 69 in S. Pauls Church Yard London

But such Fights will be common (as Dunce to feel Rod)
In the Year of One Thousand eight Hundred and odd.

This shows four balloons, two flying the French fleur de lys and two the British Union Jack, manoeuvring for aerial combat. Their crews are armed with muskets, but also, more menacingly, with broadside cannons. Their muzzles point through portholes cut in the balloon wickerwork.[2] Here the balloon is already conceived of as a weapon of war, comparable to the navy's ships of the line.

Sophia also had an eye for many of the more eccentric examples of balloon propaganda. One set of this type was 'Mr Ensler's Wonderful Air Figures', in which balloons were constructed in various animal and mythological shapes, such as the 'Flying Horse Pegasus'. Some of these figures were intended to be provocative, like the giant '*Nymphe coiffée en ballon et habillée à la Polonaise*'. The Frenchified style of this description suggests sexual mockery. Then there was the bluffly patriotic 'Mr Prossor's Aerial Colossus', showing an enormous 'Sir John Falstaff' floating defensively above the Dover cliffs.[3] Such inventions were probably pure design fantasies, or at the most models, never actually manufactured full-size. But they suggest how balloons would become powerful forms of imaginative propaganda later, in the nineteenth century.

2

The first actual military balloon regiment, as Franklin had prophesied, was indeed French. The Corps d'Aérostiers was founded at the château of Meudon outside Paris on 29 March 1794. Less than three months later, on

---

The tradition of fantasy shapes still flourishes in modern hot-air ballooning. At the Albuquerque International Balloon Fiesta in October 2010, I saw everything from a Pepsi-Cola can to Mickey Mouse, Darth Vader and Airabelle the Cow with her Beautiful Udder. Why do giant 'air sculptures' hold such a perennial fascination? There is something dreamlike about them; they are almost 'thought-sculptures'. I have never seen one inspired by the human form – e.g. a Venus de Milo balloon; they are usually inspired by cartoon figures, publicity logos or friendly animals. Neither erotic nor monster shapes appeared popular. A humorous innocence seemed the most favoured mode, as if they had broken free from an illustrated children's book. Actually the all-black Darth Vader balloon was weirdly menacing, and no one I asked at Albuquerque liked it. 'Not such a great idea,' seemed to be the general response. Although no one said so, I had the feeling that some balloonists thought Darth Vader might bring bad luck. They preferred the Force to be with them.

26 June, the French army first made use of a military observation balloon
at the Battle of Fleurus, against an Austrian army, and again a few weeks
later at the Battle of Liège (where it was witnessed by the galloping Major
Money). The balloon, manned on both occasions by a daring young
officer, Captain Charles Coutelle, provided vital information prior to
successful cavalry charges, and both battles were won by the fledgling
French Revolutionary army.

The balloon school at Meudon was immediately expanded, and
Coutelle showered with medals and appointed its commanding officer.
He rapidly drew various lessons about military aerostation. First, that it
was difficult to inflate a balloon with hydrogen on the battlefield.
(Lavoisier was immediately coopted to invent a simpler method of generating hydrogen.) Second, that it was extremely hazardous to launch a
tethered balloon if anything more than a light breeze was blowing. Held,
kite-like and unnaturally on its cable against the force of the wind
(instead of moving tranquilly within it), the balloon canopy would often
thrash about and sometimes tear. Moreover, instead of gaining height it
would fly horizontally and low. Above all, the basket would become
highly unstable as an observation platform. Coutelle also remarked that

it was not always easy to transmit really accurate and continuous observations from an airborne basket to a ground controller. Signal flags,
scrawled messages or maps were rarely adequate. In most cases the aeronaut had simply to be winched back down, so he could deliver his appreciation verbally to a commander, in person and on the ground.
Interestingly, it proved very difficult to get any commander to go up to
see for himself.

But overall Coutelle believed that balloons promised considerable
military value. He argued that, under the right conditions, a balloon
would give an immense intelligence advantage to an army on the move,
whether defending or attacking. It provided a wholly new tactical
weapon, a 'spy in the sky' which could supply vital warning of troop
build-ups and defensive positions, as well as preparations for attack or
(equally vital) for retreat. Such observations could give a commander a
decisive initiative in the field.

More subtly, a balloon was also an extraordinary psychological
weapon. Because of its height, every individual soldier could see an enemy
balloon hovering above a battlefield. By a trick of perception, this gave the
impression to every soldier that he, in turn, *could always be seen by the
balloon.* So everything he did was being observed by the enemy. There was
no hiding place, no escape. The very presence of such a balloon above a
battlefield was peculiarly menacing and demoralising. The enemy might
certainly read an enemy soldier's intentions, and even seem to read his
thoughts. This alone made it a powerful military instrument. An Austrian
officer was reported, after his army's defeat at Fleurus, as murmuring,
'One would have supposed the French General's eyes were *in our camp.*' His
troops complained more angrily, 'How can we fight against these damned
Republicans, who remain out of reach *but see all that passes beneath.*'[4]

There was one unforeseen consequence of this. The French balloons
quickly came to be universally hated by the opposing allied armies. As a
result, they immediately attracted intense and sustained enemy fire, with
every weapon that could be mustered, from pistols and muskets to
cannon and grapeshot, directed at the observers' basket. This, concluded
Coutelle, made the military aeronaut's position both peculiarly perilous
and peculiarly glamorous.

The Corps d'Aérostiers eventually fielded four balloons, complete with
special hangar tents, winches, mobile gas-generating vessels (designed by
Lavoisier) and observation equipment. Coutelle would write a racy history

of the Meudon balloon school, with modest emphasis on both the tactical and the amorous successes of the French military aeronauts. Wilfrid de Fonvielle later observed: 'The favour of the ladies followed the balloonists wherever they went, which was not an unmixed blessing, and seems in the end to have contributed to the suppression of the corps.'[5]

With the declaration of war against Britain in 1794, many plays, poems and cartoons imagined an airborne invasion – both French and English – across the Channel. The *Anti-Jacobin* published invasion-scare cartoons featuring the French guillotine set up in Mayfair, and also extracts from a play purportedly running at the Théâtre des Variétés in Paris: *La Descente en Angleterre: Prophétie en deux actes*.[6] There were some remarkable fantasy drawings of entire French cavalry squadrons mounted on large, circular platforms sustained by enormous Montgolfiers, sailing over the white cliffs of Dover. Nevertheless, the much-feared aerial invasion of England by Napoleon's army never quite materialised.

In 1797 Napoleon triumphantly took the Corps d'Aérostiers with him to Egypt, counting on the very sight of balloons to put terror into the heart of his Arab enemies, as Hannibal's elephants had once done in Italy. On 1 August 1798 Coutelle was preparing to unload all his gear outside Alexandria from the French fleet's mooring at Aboukir Bay when Nelson

sailed in at dusk. At the ensuing three-day Battle of the Nile, half of Napoleon's ships were destroyed, and with them the entire Corps d'Aérostiers. The surviving aeronauts stayed on in Alexandria as technical advisers, like melancholy cavalry officers deprived of their horses. On his return home Napoleon disbanded the corps and the school at Meudon.

Nevertheless, rumours of a French airborne army invading Britain continued to be cultivated, and remained a powerful element in both French and British propaganda long into the nineteenth century. It was the aerial dream turned nightmare.

3

Civilian balloons and a different kind of competitive showmanship reappeared in France at the time of the Peace of Amiens in 1802. They were promoted by André-Jacques Garnerin (1770–1825), who launched his career by performing a spectacularly dangerous first parachute drop from a balloon over Paris's Parc Monceau in 1797, when he was twenty-seven. As a young man, Garnerin had fought in the French Revolutionary armies, but he had been captured and incarcerated in the Hungarian castle of Buda for three desperate years. He spent his time there designing imaginary balloons to lift him out of the prison courtyard, or parachutes with which he could leap from the castle battlements.[7] Finally released and returned to Paris, he turned his ballooning escape-fantasies into a full-time profession, and became head of the first of the famous French 'balloon families' (a role later inherited by the Godards).

Garnerin pioneered a new kind of balloon event: not merely the conventional single ascent, but a whole series of acrobatic displays, parachute drops and night-flights with fireworks. To add to the excitement, his dashing young wife Jeanne-Geneviève made the first recorded parachute jump by a woman in 1799. These shows attracted enormous crowds, and soon the Garnerins became famous throughout European capitals. Garnerin had numerous posters made of their flights, often showing him in heroic, eagle-like profile. Napoleon himself began to see that the balloon had a potential propaganda value even greater than the military one.[8]

The Peace of Amiens was barely signed when Garnerin daringly took his balloon show and parachute drops to London. His reception was surprisingly friendly, perhaps because he was joined not only by

Jeanne-Geneviève, but also by his pretty niece Lisa. Garnerin's first
London ascent, from Chelsea Gardens on 28 June 1802, attracted a huge
audience: 'Not only were Chelsea Gardens crowded, and the river covered
with boats, but even the great road from Buckingham Gate was absolutely impassable, and the carriages formed an unbroken chain from the
turnpike to Ranelagh Gate.'[9]

Fearlessly launching in a near-gale, Garnerin flew along the line of
the Thames, from the West End to the East End, directly over the City,
and then out north-eastwards over the Essex marshes. He was effectively
seen by half the population of London. Forty-five minutes later he crash-
landed in Colchester, but came back the same day in a coach, gallantly
announcing that his balloon had been torn to pieces – 'we ourselves are
all-over bruises' – but that he would fly again within the week. Indeed, he
next ascended from the old Lord's cricket ground (on the site of the
present-day Dorset Square) on 5 July. Advertising his flights with sensational engravings, Garnerin popularised night-ballooning and parachuting in England, and also the dangerous attitude that 'the balloon show
must go on' whatever the weather.

The following year, 1803, Garnerin published *Three Aerial Voyages*,
describing his London flights and including an amusing account,

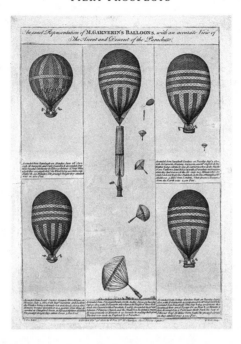

evidently intended for English readers, supposedly written by his wife's cat: 'Brought up under the care of Madame Garnerin, I may be said to have been nursed in the very bosom of aerostation, and to have breathed nothing but the pure air of oxygenated gas since the first moment of my birth. Hearing of my mistress's intended ascension, I determined to share the danger ...'[10]

Scientific ballooning was not entirely forgotten in France. In August 1804 the mathematician Jean-Baptiste Biot and the chemist Joseph Gay-Lussac made a high-altitude scientific ascent, in the one war balloon Coutelle had succeeded in bringing back from Cairo to Paris. They tested the composition of the air in the upper atmosphere, and the strength of the magnetic field, but found no significant alteration from ground level in either. Biot passed out during the descent, so Gay-Lussac went up again alone in September, climbing to 22,912 feet, a new altitude record which would stand for over half a century. Here, in the tradition of Dr Alexander Charles, along with his instrument readings he calmly recorded his breathlessness, fast respiration and pulse, inability to swallow and other symptoms, and concluded that he was very close to the limit of the breathable atmosphere. In the historical section of his classic book *Through the Air* (1873), the great American aeronaut John Wise later reflected on the courage of these early scientific ascents into the absolute unknown: 'It is impossible not to admire the intrepid coolness with which they conducted these experiments ... with the same composure and precision as if they had been quietly seated in their scientific cabinet in Paris.' Wise also raised the prophetic possibility of using such high-altitude balloons *unmanned* for weather observations: 'Balloons carrying "register" thermometers and barometers might be capable of ascending alone to altitudes between eight and twelve miles.' But such experiments would have to wait for a time of international scientific cooperation, 'when nations shall at last become satisfied with cultivating the arts of peace, instead of sanguinary, destructive and fruitless wars'.[11]

Indeed, it was the celebration balloon, used for propaganda and patriotic rather than scientific purposes, that most readily held the public's attention in France. In December 1804 Napoleon commissioned Garnerin to construct and launch a massive, decorated but unmanned balloon to celebrate his coronation as emperor in Paris. It was festooned with silk drapes, flags and banners, and carried an enormous golden imperial crown suspended from its hoop on golden chains. Having been successfully launched above Notre Dame during the coronation

GAY-LUSSAC ET BIOT A 4000 MÉTRES DE HAUTEUR (1804)

ceremony, this fantastic contraption flew southwards right across France, and amazingly crossed the Alps during the night. The following day it was spotted symbolically descending upon Rome, the imperial city, a triumph for Garnerin's craftsmanship.[12]

The huge balloon veered towards St Peter's and the Vatican Palace, then swooped down low across the Forum. But here the symbolic triumph was turned into a propaganda disaster. The enormous golden crown became hooked on the top of an ancient Roman tomb and broke off, leaving the balloon to disappear, with its banners flapping, over the Pontine Marshes. By unbelievable coincidence, or thoroughly appropriate bad luck (you can never tell with balloons), the tomb upon which Garnerin's prophetic balloon had deposited Napoleon's golden crown was that of the infamous tyrant and murderous pervert, the Emperor Nero. Napoleon's name was hooped like a deck-quoit over Nero's.♆ Once this ill-starred news was efficiently relayed back to Paris by Napoleon's diplomatic service, Garnerin and his balloons began to fall out of imperial favour.[14]

♆ Such incidents had an interesting effect on the burgeoning new genre of science fiction and fantasy. *The Surprising Adventures of Baron Munchausen* (a collection of imaginary tales about an actual historical figure) was originally published in German by Rudolf Erich Raspe in 1785, though no balloons appear in this first edition. But with the series of expanded and pirated English versions between 1809 and 1895, balloon fantasies soon abounded. 'I made a balloon of such extensive dimensions, that an account of the silk it contained would exceed all credibility,' one tall tale begins. 'Every mercer's shop and weaver's stock in London, Westminster, and Spitalfields contributed to it. With this balloon I played many tricks, such as lifting one house from its station, and placing another in its stead, without disturbing the inhabitants, who were generally asleep.' With his massive balloon, Munchausen carries Windsor Castle to St Paul's in London, makes the clock strike thirteen at midnight (for no particular reason), and then airlifts it back again before daylight, 'without waking any of the inhabitants'. Later, he uses his balloon to levitate the entire building of the Royal College of Physicians. The physicians are in session (i.e. 'feasting'), so he keeps them suspended several thousand feet in the air for 'upwards of three months'. The consequence is that, deprived of their medical services, there are absolutely no deaths in the entire population of London, while many clergymen and undertakers go bankrupt. Munchausen adds disingenuously: 'Notwithstanding these exploits, I should have kept my balloon, and its properties, a secret, if Montgolfier had not made the art of flying so public.'[13]

4

Garnerin was replaced, almost at once, by the most justly famous of all the French Revolutionary balloonists. She was a woman – the small, fearless and enigmatic Sophie Blanchard. Born at the sea port of La Rochelle in March 1778, Sophie somehow became involved with the experimental balloonist Jean-Pierre Blanchard, who had first crossed the Channel with Dr Jeffries in 1785, when Sophie was only eight. How their romance began remains a mystery, since Blanchard was already married with children, and spent much of the 1790s touring the cities of Europe and America. But it was rumoured that he first saw her when she was still a child, standing in the crowd at one of his launches, and vowed to return and marry her when she came of age in 1799.

However, the first definite record of them together is not until December 1804, when Blanchard took Sophie on her first balloon flight, above Marseille. According to him she was immediately smitten,

Madame
BLANCHARD
(MADELEINE SOPHIE ARMAND)
Aéronaute
Née vers 1778 à à Paris, 6 Juillet 1819

breaking her customary painful silence to gasp, 'Sensation incomparable!' Pictures show her to be petite and pretty, with large eyes and a dark fringe. But she was also said to be frail and 'bird-like', abnormally nervous on the ground, terrified of crowds, loud noises, horses and coach travel, and shy to the point of self-effacement. Yet all this changed completely once she was in the air. In a balloon she became confident and commanding, a natural entertainer and a provoking exhibitionist, daring to the point of recklessness.

Blanchard, who was ageing and nearly bankrupt, evidently saw the possibilities of reviving his aeronautical career with this fearless young woman, who could instinctively control a balloon, manage aerial fireworks, do acrobatics, and wear eye-catching hats and dresses to please a crowd. He married Sophie when she was twenty-six, and she became his balloon partner for several years, taking over all the arrangements as his health gradually failed. Blanchard died from a heart attack in 1810, while landing in a damaged balloon near The Hague.[15] Immediately after his death, Sophie gave her first major solo balloon display in Paris. Like Garnerin, she specialised in night ascents and firework displays, but with much greater daring and eventually recklessness. She deliberately set herself up to rival the other famous female aeronaut of the time, Garnerin's niece Lisa. Both seemed to vie for official recognition, though Lisa suffered from the waning popularity of the Garnerin name with Napoleon.

Sophie Blanchard seems to have caught the Emperor's attention during a midsummer ascent from the Champ de Mars in Paris on 24 June 1810. Soon after, she was asked to contribute to the celebration mounted by the Imperial Guard for Napoleon's marriage to the Archduchess Marie-Louise of Austria. From then on she became a fixture at the imperial court, with propaganda as well as entertainment duties. On the birth of Napoleon's son in March 1811, she took a balloon flight over Paris from the Champ de Mars and threw out leaflets proclaiming the happy event. She again performed at the official celebration of his baptism at the Château de Saint-Cloud on 23 June, with a spectacular firework display launched from her balloon. In the same year, in an ascent above Vincennes, she climbed so high to avoid being trapped in a hailstorm that she lost consciousness, and spent 14½ hours in the air as a result.

Napoleon now made Sophie's position official. He appointed her Aéronaute des Fêtes Officielles, a position especially created for her, and gave her responsibility for organising ballooning displays at all major events in Paris. It was also said that he made her his 'Chief Air Minister of Ballooning', with secret instructions to draw up plans for an aerial invasion of England. However this seems more like English counter-propaganda, as Napoleon's idea for an invasion of England had long since been displaced by the ultimately disastrous invasion of Russia, which began in the spring of 1812.

Sophie had by now developed her own peculiar free-style of balloon-ing. She abandoned her husband's large canopy and unwieldy basket, both of which were by this time much battered. To replace them she commissioned a much smaller silk balloon, capable of lifting her on a tiny, decorative silver gondola. This was shaped like a small canoe or child's cradle, curved upwards at each end but otherwise quite open. It was little more than three feet long and one foot high at the sides. One end was upholstered to form a small armchair (in which she sometimes slept), but otherwise the gondola offered astonishingly little protection. When she stood up, grasping the balloon ropes, the edge of the gondola did not reach above her knees. It was virtually like standing in a flying champagne bucket.

She also began to adopt distinctive outfits, which could be seen at a considerable distance. For this purpose her dresses were always white cotton and narrowly cut, with the fashionable English Regency style of high waist and low décolletage. Her sleeves were long, coming right down to her knuckles, presumably to keep her hands warm at high altitudes. Most important of all, she wore a series of white bonnets extravagantly plumed with coloured feathers, to increase her height and visibility. The combined effect of these dramatic clothes and the tiny silver gondola was to make her look both flamboyant and vulnerable. She also appeared terrifyingly exposed, an effect she evidently cultivated.

Sophie now began to give displays in Italy. In the summer of 1811 she took her balloon across the Alps by coach, and celebrated Napoleon's birthday, the Fête de l'Empereur on 15 August, with an ascent above Milan. She travelled on to Rome with instructions to efface the memory of the unfortunate incident of Nero's tomb. Here she ascended spectacu-larly to a height of twelve thousand feet, and stayed aloft all night in her tiny upholstered chair, claiming that she fell into a profound sleep, before landing much refreshed at dawn at Tagliacozzo. She then flew by balloon from Rome to Naples, splitting the journey in half with a stop after sixty

M. S. BLANCHARD celebre aeronauta
al momento del volo aerostatico da Lei eseguito in Milano
in presenza delle L.L. A.A. I.I. e R.R.
la sera del 10. Agosto 1811.

miles. She made a daring ascent in bad weather over the Campo Marte in
Naples to accompany the review of the troops by Napoleon's brother-in-
law Joachim Murat, the King of Naples.

In 1812, the third Aeronautical Exhibition was held at the Champ de
Mars. With all eyes on Napoleon advancing towards Moscow, it was not
a great success, so Sophie again crossed the Alps with her balloon to
make ascents at Turin. On 26 April she flew so high, and the temperature
dropped so low, that she suffered a nosebleed and icicles formed on her
hands and face.

On her return to Paris, Sophie was surprised to receive a letter on 9
June from André-Jacques Garnerin. He gracefully invited her to dinner at
the Hôtel de Colennes to discuss 'a project that might be of mutual interest'. This probably involved a proposal to make double ascents and parachute drops above the Jardin du Tivoli or the Parc Monceau with her old
rival Lisa Garnerin. It is highly unlikely that Sophie would have accepted
what she must have regarded as a demeaning and unsuitable
proposition.[16]

The defeat and subsequent exile of Napoleon seemed to pose Sophie Blanchard few problems. When the restored Bourbon king, Louis XVIII, entered Paris for his official enthronement on 4 May 1814, Sophie made a spectacular ascent from the Pont Neuf, with fireworks and Bengal lights. King Louis was so taken with her performance that he immediately dubbed her 'Official Aeronaut of the Restoration'. Evidently Sophie had no qualms about this shift of political allegiance. Her only loyalty was to ballooning.

Four more years of brilliant public displays followed, with Sophie established as queen of the fireworks night at the Tivoli and Luxembourg Gardens. Her small balloon lifted more and more complicated pyrotechnical rigs, with long booms carrying rockets and cascades, and suspended networks of Bengal lights, all of which she would skilfully ignite with extended systems of tapers and fuses. At the height of these displays, her small white figure and feathery hat would appear like some unearthly airborne creature or apparition, suspended several hundred feet overhead in the night sky, above a sea of flaming stars and coloured smoke.

Towards midnight on 6 July 1819, a hot, overcast summer evening, Sophie Blanchard, aged forty-one, began one of her regular night ascents from the Jardin de Tivoli, accompanied by an orchestra in the bandstand below. At about five hundred feet, and still climbing, she began to touch off her rockets and Bengal lights, dropped little parachute bombs of fizzing gold and silver rain, and ignited a lattice of starshells suspended on wires twenty feet below her gondola. As the gasps and applause of the crowd floated up to her through the darkness, she became aware of a different quality of light burning above her head. Looking up, she saw that the hydrogen in the mouth of her balloon had caught fire. It was amazing that it had never done so before. Many of the crowd thought it was just part of the firework display, and continued to applaud.

The flaming balloon dropped onto the roof of number 16, rue de Provence, near the present Gare Saint-Lazare. The impact largely extinguished the fire. Sophie was not severely burnt, but she was tangled in the balloon rigging. She slid down the roof and caught onto the parapet above the street. Here she hung for a moment, according to eyewitnesses, calmly calling out 'A moi, à moi!' Then she fell onto the stone cobbles beneath.

There are numerous accounts of this fiery descent from the Paris sky. They appeared in all the Paris newspapers, and also in English journals like the *Gentleman's Magazine*.[17] One of the clearest, most poignant descriptions was written by an English tourist, John Poole, who witnessed the event from his hotel room.

> *I was one of the thousands who saw (and I heard it too) the destruction of Madame Blanchard. On the evening of 6 July 1819, she ascended in a balloon from the Tivoli Garden at Paris. At a certain elevation she was to discharge some fireworks which were attached to her car. From my own windows I saw the ascent. For a few minutes the balloon was concealed by clouds. Presently it reappeared, and there was seen a momentary sheet of flame. There was a dreadful pause. In a few seconds, the poor creature, enveloped and entangled in the netting of her machine, fell with a frightful crash upon the slanting roof of a house in the Rue de Provence (not a hundred yards from where I was standing), and thence into the street, and Madame Blanchard was taken up a shattered corpse!*[18]

The death of the Royal Aeronaut profoundly changed the reputation of ballooning in France. A public subscription was raised in her honour, but it was found that Sophie Blanchard had no family, and was reported to have left fifty francs in her will 'to the eight-year-old daughter of one of

MORT DE MME BLANCHARD (1819)

her friends' (perhaps an illegitimate child?). So the two thousand francs raised was used to erect a notable balloon monument, which still exists in the 94th Division of Père Lachaise cemetery. ◊

5

Sophie Blanchard's death in 1819 effectively ended the first great wave of ballomania and the celebration of ballooning in France. Something similar happened in England with the equally shocking death of Thomas Harris five years later. Amazingly, Harris was the first English aeronaut to be killed on home ground. A glamorous young naval officer, he made a much-advertised ascent in his new balloon the *Royal George* on 24 May

---

◊ The mysterious, 'bird-like' Sophie Blanchard has attracted more romancing than any other female aeronaut. She is the subject of a novel by Linda Doon, *The Little Balloonists* (2006), in which she enchants Goethe and actually seduces Napoleon. She is also celebrated, like a local saint, at an annual event in the little Italian Alpine town of Montebruno, where she once landed in her balloon. She has become adopted as a feminist heroine, 'the first professional female aeronaut', and the film-maker Jen Sachs is making an animated cartoon of her life, *The Fantastic Flights of Sophie Blanchard*.

1824. As part of his publicity, he took with him a dazzlingly pretty eighteen-year-old cockney girl, known to the newspapers only as 'Miss Stocks', who was generally assumed to be his mistress. Miss Stocks and the balloon, which had cost Harris a thousand guineas to construct, had both been exhibited at the Royal Tennis Court in Great Windmill Street, and stirred much excitement and comment.[19]

The balloon had a new kind of duplex release valve, which Harris said would allow him and Miss Stocks to make a perfectly controlled landing. One valve was housed inside the other at the top of the balloon. The smaller, inner valve was the conventional safety mechanism, as invented forty years previously by Alexander Charles, designed to release excess gas pressure during flight, or to commence a controlled descent. The larger outer valve was a radical solution to the problem of keeping the balloon safely on the ground once it had landed. When the larger valve line was pulled, it would deflate the entire balloon in a matter of seconds (the equivalent of the 'rip panel' in a modern hot-air balloon). This, claimed Harris, would prevent the terrible bouncing and dragging across fields which had caused so many injuries, and so much damage to crops and property (especially chimneys and rooftops) which had undermined the general popularity of balloonists.

Harris circulated a campaigning pamphlet saying that he was trying to save the declining art of ballooning in England. 'The Science of Aerostation has lately fallen into decay, and has become the subject of Ridicule,' he lamented. This decline was caused by the 'total want' of serious technical inventions by recent aeronauts, who had been content (like that Frenchman Garnerin) to exploit frivolous novelties like parachutes and fireworks. The *Royal George*, with its new system of valves and its beautiful young passenger, would show the way ahead. In the event it showed something quite else.

Harris took his balloon and Miss Stocks from the West End to the East End to generate further interest, and launched successfully from the large courtyard of the popular Eagle Tavern, in City Road. It was noted that Lieutenant Harris wore his best blue naval uniform, and Miss Stocks a charming dress, much as if they were a honeymoon couple, which perhaps they were. The change in venue was probably made because the wind was blowing south-westwards that day. It took the balloon back across London and the river Thames, an excellent display route, and then on into Kent and towards Croydon. All went well in the basket, champagne was drunk, and Harris then attempted his first perfect display landing at Dobbins Hill, just outside Croydon. However, this did not quite go to plan.

Distracted either by Miss Stocks or by his new duplex valve, he forgot to hang out his grapnel line in time, and was forced to throw out ballast to avoid colliding with some nearby trees. This was by no means a disastrous error, but it evidently rather flustered Harris. The balloon rose several hundred feet in the air, and was carried on over Beddington Park, on the other side of Croydon. Here Harris evidently prepared for a second attempt at a landing.

The swift sequence of events that followed has remained a matter of dispute ever since. For no accountable reason, the *Royal George* suddenly began to descend from several hundred feet 'with fearful velocity'. As it dropped, it was claimed by witnesses that some kind of struggle was briefly observed in the basket. The partially deflated balloon plummeted 'with frightful rapidity' into a large oak tree in the park, tore through the light spring foliage (it was only May) and dashed its passengers to the ground. 'They were shortly afterwards discovered, buried beneath a monumental pile of silk and network.' Both of them were outside the basket, but while Thomas Harris was dead, Miss Stocks was alive and conscious, though quite unable to give a coherent account of what had occurred during the last few seconds.

What had destroyed the balloon seemed obvious to the coroner. The new large gas valve – 'the preposterous aperture' – had been released prematurely. The coroner ascribed this to Harris's fatal error in pulling the wrong valve line while still in the air. Later aeronauts, like Charles Green, analysed the sequence of events more subtly. Green suggested that Harris's only error was to have tied the larger valve line to a point in the basket, precisely to keep it safe and out of the way, especially with Miss Stocks aboard. However it had pulled itself taut – 'a longitudinal extension of the apparatus' – when the balloon contracted, thereby unexpectedly and fatally releasing the deflation valve by itself. This was a kindly, if ingenuous, explanation, which did not quite square with the evidence that Miss Stocks gave afterwards: 'Miss Stocks declares that she distinctly heard the peculiar sound which always accompanies the shutting of the valve, *as soon as Mr Harris had let go the line.*' This sounds as if Harris had indeed pulled the line himself, and realising his error, had let it go too late. But of course, Miss Stocks could have meant 'as soon as Mr Harris *had untied* the line'.[20]

The much larger question, and the one that made 'le mort de Harris' a *cause célèbre* in France, was a more human mystery. Why did Miss Stocks survive when Lieutenant Harris died? The disposition of the bodies gave little clue, both presumably having been thrown from the basket on impact. But the suggestion almost inevitably arose that the gallant Harris had somehow saved the beautiful Miss Stocks.

British commentators were brusque about this mystery 'that has hitherto clouded the event', and gave romantic explanations short shrift. Most probably, a branch of the oak tree, 'projecting horizontally', had protected Miss Stocks (though unfortunately not Harris). Just possibly Miss Stocks 'had fainted, and fallen forward ... upon the body of Mr Harris'. Thereby he had unintentionally protected her from 'the first violence' of the impact. Anything else pandered to 'false and scurrilous reports'.[21]

French journalists were more liberal in their interpretations. Lieutenant Harris and Miss Stocks could have been distracted 'in many ways', so that the wrong valve line was pulled. Lieutenant Harris would surely have tried to protect Miss Stocks with his body during the terrifying descent, 'if not before'. But most likely of all, and the real reason for

MORT DE HARRIS (1824)

her survival, was that Lieutenant Harris 'in a spirit of admirable chivalry' had leapt from the balloon basket before the moment of impact, in a quixotic attempt to reduce the speed of her fall.[22] It was a gesture of true English gallantry from a true English naval officer. The romantic image of Lieutenant Harris leaping from the balloon to save his beloved became an iconic one in France, and featured alongside images of the death of Pilâtre de Rozier and Sophie Blanchard in a famous series of French balloon cards. They marked the end of an era.

The growing scepticism about the future of ballooning was summed up by the satirical artist George Cruikshank. In his brilliant coloured cartoon of 1825 entitled 'Balloon Projects', he depicted a row of gaudily striped balloons tethered down the length of St James's, like a rank of hackney cabs waiting for hire. A fashionable couple are about to climb gingerly into one of them (with pink plush upholstery), to embark on a 'one shilling' flight from Mayfair to the City. Each balloon car is manned by a suitably villainous driver, one of whom is shouting to another, 'I say, Tom, give my balloon a feed o' gas, will you!'

Above them the sky is filled with a mass of grotesquely shaped balloons, one in the form of a hogshead of beer, another with a weather-cock on top, and a third on fire with its passengers leaping out. Behind them all the buildings are advertising balloon companies and businesses,

which is probably the real point of Cruikshank's satire. These dubious establishments include 'The Balloon Life Assurance Company', 'The Bubble Office', 'The Office of the Honourable Company of Moon Rakers', and perhaps most ingeniously, 'The Balloon Eating House – Bubble & Squeak Every Day'.[23]

The historical painter Benjamin Robert Haydon, friend of John Keats and Charles Lamb, who kept a close if disillusioned eye on London freaks and fashions from his studio near Baker Street, noted grimly in his diary for 6 June 1825: 'No man can have a just estimation of the insignificance of his species, unless he has been up in an air-balloon.'

Of course, not everyone was so disillusioned. A remarkable balloon is featured in Jane Loudon's extraordinary science fiction novel *The Mummy! A Tale of the Twenty-First Century* (1827). It was written when Loudon was aged twenty, 'a strange, wild novel', as she herself called it, full of totally unexpected technical inventions, such as 'steam percussion bridges', heated streets, mobile homes on rails, smokeless chemical fuels, electric hats (for ladies), and most spectacularly a full-size balloon made from a nugget of highly concentrated Indiarubber. Small enough to be stored in a desk drawer, such a 'portable' balloon once inflated is large enough carry three people from Britain to Egypt.[24]

Loudon's description of an aerial balloon carnival is brilliantly prophetic of modern balloon fiestas, especially the more psychedelic American ones: 'The air was thronged with balloons, and the crowd increased at every moment. These aerial machines, loaded with spectators till they were in danger of breaking down, glittered in the sun, and presented every possible variety of shape and colour. In fact, every balloon in London or the vicinity had been put into requisition, and enormous sums paid, in some cases, merely for the privilege of hanging to the cords which attached to the cars, whilst the innumerable multitudes that thus loaded the air, amused themselves by scattering flowers upon the heads of those who rode beneath.' The fantastic balloon shapes include, for more sportive individuals, 'aerial horses, inflated with inflammable gas', and for the more languid, 'aerial sledges' which can be flown from a supine position.[25]

3

Airy Kingdoms

———

1

By the end of the 1820s the early pioneering days were over, and the dream of some universal form of global air transport by navigable balloons had faded across Europe.

Indeed, quite another form of revolutionary transport system was starting to emerge, in the shape of the railway network. The opening of the twenty-six miles of the Stockton to Darlington line in 1825 heralded an era of universal railway building. Five years later, the first genuine passenger service opened between Manchester and Liverpool. From then on, the heavy engineering of the Victorian steam engine, establishing powerful new notions of speed, reliability and a regular 'time-table' came to dominate the whole concept of travel.

These weighty considerations were, of course, the exact opposite of what a hydrogen balloon could provide. Victorian ballooning would eventually become the antithesis of the Victorian railway. It would be seen as poetic as against prosaic; as natural rather than man-made. The Victorian railway would mean iron, steam, noise, power and speed, as Turner envisaged in his painting. It would bring 'railway time', and a form of mass transport which was both a vital means and a literary symbol of industrialisation. By contrast, ballooning would come to be seen as essentially bucolic, even pastoral. It was silent, decorative, exclusive, and refreshingly unreliable: a means to mysterious adventure rather than a mode of mundane travel.[1] ⍟

⍟ Mary Shelley introduces just such a mysterious, poetic balloon into *The Last Man* (1825), the science fiction novel intended to follow up her success with *Frankenstein*.

So, at this very time of the railway boom, a new kind of flying was starting to capture people's imagination. It was marked by the emergence of what might be called the 'recreational' balloon. These were increasingly large, sophisticated and well-equipped aerostats, designed to take several paying passengers, and with luck to make a profit for their owners. Though they were commercial propositions, they retained an ineffable romance. They often had 'royal' in their names, in deference to the new young Queen Victoria, who came to the throne in 1837.

They were recreational balloons in several senses, constructed and flown by a new breed of balloon businessmen and entertainers, aerial entrepreneurs who regarded themselves as skilled professionals as well as artists of the air. Crucial to their commercial success was the discovery that coal gas, cheap and reliable and easily available from the urban 'mains', could be substituted for expensive and unstable hydrogen. Their most memorable flights would also be over urban landscapes.

Drifting silently above London or Paris, or any of the industrialised centres of northern Europe, they granted their passengers a new and instructive kind of panorama. They revealed the extraordinary and largely unsuspected metamorphosis of these cities, with their new industries and their hugely increased populations. What they found spread out

---

The passage is striking for its sense of movement, light and airy freedom, in a novel which is otherwise notably dark and claustrophobic. Her hero Lionel, faced with a global cholera pandemic which will destroy all mankind, is escaping from London to the Highlands of Scotland:

> Everything favoured my journey. The balloon rose about half a mile from the earth, and with a favourable wind it hurried through the air, its feathered vanes cleaving the unopposing atmosphere. Notwithstanding the melancholy object of my journey, my spirits were exhilarated by reviving hope, by the swift motion of the airy pinnace, and the balmy visitation of the sunny air. The pilot hardly moved the plumed steerage, and the slender mechanism of the wings, wide unfurled, gave forth a murmuring noise soothing to the sense. Plain and hill, stream and corn-field, were discernible below, while we unimpeded sped on swift and secure, as a wild swan in his spring-tide flight.[2]

Yet this balloon is already an anachronism. Mary Shelley thinks of it as a navigable sailing ship, or even a large feathered bird, rather than a practical aerostat. She herself had never been up in one, but she had launched many model fire balloons with Percy Shelley in the early days of their love affair. She knew his sonnet 'To a Balloon Laden with Knowledge', always associated his spirit with air and fire, and even gave him a model balloon as a present to celebrate his twenty-fourth birthday in 1816.[3]

below them had never been seen before in history. It was both magnificent and monstrous: an endless panorama of factories, slums, churches, railway lines, smoke, smog, gas lighting, boulevards, parks, wharfs, and the continuous amazing exhalation of city sounds and smells. Above all, what they saw from 'the angel's perspective' was the evident and growing divisions between wealth and poverty, between West End and East End, between blaze and glimmer.

2

The most celebrated British balloonist of this second generation was Charles Green (1785–1870). He was famed for his 526 successful ascents, and his absolute sang-froid in emergency. An annual Charles Green Silver Salver is awarded to this day by the British Balloon Association for the most impressive technical flight of the year. Green's portrait in the National Portrait Gallery presents him like a plain, good natured farmer. He sits stiffly at a window, with his famous striped *Royal Nassau* balloon hanging in the sky above his right shoulder. His image is rendered with the solidity and simplicity of an English pub sign. Yet his character was far from straightforward: a curious mixture of earth and air, of ballast and inflammable gas, of the worldly and the visionary.

Everyone agreed that the stocky and rubicund Green was jovial and easy-going on the ground. Yet he became strangely fierce and commanding in the air, a veritable martinet. This kind of Janus-like character can sometimes be found among amateur yachtsmen – twinkling and expansive in the club bar, but fiery authoritarians at the helm. It has perhaps something to do with the loneliness and stress of command. In ordinary company, a contemporary journalist found Green 'garrulous, and delighting all with his intelligence, his enterprise, his enthusiasm and his courtesy'. But once installed in the aerial kingdom of his balloon basket, he became 'taciturn, and almost irritable', rarely speaking, but always ready to 'roar out' commands at crucial moments of manoeuvring or landing. Apart from his balloon logbooks, he left few, if any, written accounts of his own flights. Yet he inspired others to do so, and many of the classic descriptions of Victorian ballooning, especially over London, by John Poole (1838), Albert Smith (1847) and Henry Mayhew (1852), were written after memorable voyages with Charles Green.[4]

Green came from a family of fruit merchants, based in Goswell Street on the edge of the City of London. This was, and still is, a bustling district of small businesses and shops, modest houses and working cafés. It is where Dickens's *Pickwick Papers* begins, with Mr Pickwick leaning out of his lodging windows onto Goswell Street and greeting 'that punctual servant of all work, the sun' on a cheerful morning in 1827. How or when Green became fascinated by ballooning remains a mystery. Fruit and veg do not seem an obvious source of aerial longings. He would only say that ballooning started as a hobby, after he had been trying to improve the gas lighting in the family shop. Possibly the idea for using mains coal gas to inflate a balloon came to him during these experiments.[5]

Born in 1785, Green was working full-time in his father's shop from the age of fifteen, and little is known about his earliest flights until 1821, when he was approaching forty. It would seem that he began ballooning when he was already a successful businessman, perhaps as a kind of sport, and was determined to find a way around the huge expense of hydrogen. He made his first ascent using coal gas, from the newly installed Piccadilly mains, launching from Green Park in London on 19 July 1821. But Green must already have had some reputation as a balloonist, for the launch was part of the public celebrations for the coronation of George IV, the balloon was decorated with the royal arms, and the site at Green Park was close to St James's Palace. This was not usually a place of popular entertainment, and the ascent must have been officially sanctioned and even subsidised.

Green gradually established himself as part of a new generation of professional balloonists. They could hire out their services, and their balloons, to any employer or institution who wanted to celebrate a special occasion or mark a notable event. But initially Green still had to put on novelty shows like Garnerin, and he is recorded as ascending on the back of a horse harnessed to his balloon hoop from the Eagle Tavern, City Road, in August 1828. This was the same tavern from which Lieutenant Harris and Miss Stocks had launched their fatal flight four years previously.

A friend gave a whimsical description of Green in these early fairground days, going aloft on the back of a pony, while feeding him by hand with beans. He added a brief biography of the pony's career with Green: 'Finally having experienced more *ups and downs* than any horse, perhaps, that ever existed, he quitted a life of public service, and was buried in the garden of his master at Highgate where he now reposes.'[6]

What made Green exceptional, quite apart from his flying skills, was his ability as a businessman. He drastically reduced the cost of commercial ballooning by negotiating an agreement with the London Gas Light and Coke Company to purchase coal gas to power his balloons. The concept of using 'pit-coal' gas for ballooning was not new, and had been recommended by Tiberius Cavallo in his *Aerostation* study as long ago as 1785. But it took Victorian technology and enterprise to apply it. The London Gas Company had itself only been founded in 1812, but within five years it had laid nearly thirty miles of gas mains, and a decade later had more than enough capacity to inflate Green's balloons.

Green had calculated, using experiments with small models, that household coal gas was about half as effective as hydrogen, but less than a third of the cost. A seventy-thousand-cubic-foot balloon would cost £250 to inflate with hydrogen, but only about £80 with coal gas. It was also much quicker and safer to use, as it was delivered under pressure by the urban gas mains, which also supplied street lamps and public buildings.[7]

By 1835 Green had made over two hundred ascents, and established himself as the leading professional balloonist in Britain. He had also invented a new piece of balloon technology: the trail rope. This was a simple, self-regulating ballast device, which allowed a gas balloon to adjust its own height when flying at altitudes below five hundred feet. Made of heavy manila cordage, the trail rope was winched out of the basket and simply left to drag along the ground several hundred feet below. Whenever the balloon dropped closer to the ground, more trail rope – and hence more ballast weight – was transferred from the balloon basket to the earth. Thus lightened, the balloon would rise again to a new point of balance. It was a very neat form of homeostatic device, typical of Green's supremely practical turn of mind. He never attempted to patent it, and it was soon employed by aeronauts across Europe.

Of course the trail rope could not be used over cities, or any kind of built-up or industrialised area. But apart from London and the sea ports, the cathedral cities of the shires and the manufacturing towns of the Midlands, England was still very largely a rural landscape. So Green employed his trail rope with what now seems amazing insouciance across the whole countryside: dragging it crashing through lines of trees and hedgerows, hissing across fields of crops or cattle, and not infrequently lifting the odd tile or slab of stonework from church roofs or isolated

barns. This was a land still without barbed wire, let alone telegraph lines
or electrical pylons, and the trail rope suggests a lost age of open park-
lands, isolated villages and largely unpopulated countryside.ᵠ If he
descended on a gentleman's estate, Green could expect immediate help
from the estate labourers, and usually a warm welcome and much hospitality from the squire at the great house. The balloon was in its own way
an old-fashioned expression of pastoral values and pleasures.

But it was also a commercial proposition. In 1836 Green negotiated
an agreement with Frederick Gye and Richard Hughes, proprietors of the
immensely popular Vauxhall Pleasure Gardens, on the banks of the
Thames, to supply balloon events as part of the garden's attractions to
Londoners, alongside brass bands, food stalls and horse-riding. Balloons
were also a superb new means of publicity. For the next twenty years
Vauxhall became associated with spectacular balloon launches, and various balloon dramas, until it was finally closed in 1859.

Part of Green's contract provided him with the finances to construct
to his own specification a huge eighty-foot-high coal-gas balloon with a
seventy-thousand-cubic-foot capacity. It cost £2,000, an unheard-of sum,
and was known initially as the *Royal Vauxhall*. Its livery was a

---

ᵠ Telegraph and power lines are still the greatest single hazard to balloonists. When
I was in Wellington, New Zealand, in the year 2000, I remember the beautiful hot-air
balloons going up on summer mornings and flying low above the tranquil, open farmland, which seemed like a balloonist's pastoral paradise from a hundred years ago. Yet
I was puzzled by the fantastic array of telephone and power lines which seemed to
loop down almost every country lane and crossroad. In January 2012 a hot-air balloon
carrying ten passengers and a pilot inexplicably struck one of these ten-metre-high
power lines while landing in the completely open farming district of Carterton, north
of Wellington. It is just possible that the pilot was trying to avoid a pony. The impact
was slight, but the basket was snagged and began to drag along the cables, still thirty
feet in the air. Then the power lines shorted and arced, the basket started to catch
fire, and the two youngest passengers jumped out to save themselves. Suddenly
relieved of their weight, the balloon broke free of the cables, shot up to five hundred
feet, and burnt like a Roman candle before dropping into the next field. All eleven
passengers, including the two who jumped, were killed. The ones who had remained
in the basket, including the expert pilot, who had over a thousand hours of flying
experience, were so badly burnt that it took several days to identify their individual
bodies. Tragedies like this are, thankfully, extremely rare; yet their dark shadow lies
behind even the most sunlit moments of ballooning. A balloon flight is never safely
over until the basket is on the ground, and the canopy back in the bag.[8]

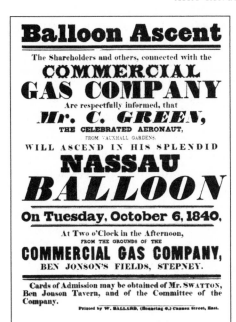

red-and-white vertical candy stripe, with a Latin inscription from Ovid's *Metamorphoses* emblazoned grandly round the circumference: '*Caelum certe patet, ibimus illi*' – 'Surely the sky lies open, let us go that way!' This was taken as signifying an appropriate pedagogical gesture on the part

---

 This striking motto has sometimes been rendered in later aviation history as simply 'Reach for the Sky', as in the title of Paul Brickhill's great biography (1954) of the legless fighter pilot Douglas Bader. Certainly the idea of overcoming or escaping earthly difficulties (such as gravity, poverty, imprisonment or having no legs) by soaring upwards into the heavens is deeply embedded in its meaning. This is also, as Green instinctively saw, central to the aspirational metaphysics of ballooning. Thus the full quotation comes from the Icarus section in Book VIII of the *Metamorphoses*. Daedalus is determined to escape the island of Crete, where he has been imprisoned by Minos. '*Terras licet*' inquit '*et undas obstruat; et caelum certe patet; ibimus illac*'. '"Though he [Minos] may barricade the earth and the waves," he said, "surely the sky at least stands open; let us go that way."' This strangely inspirational motto continues to turn up in thought-provoking places. It was adopted by the Junior Classical League of America when meeting at the University of Richmond, Virginia, in 1984. The League, of course, regards the Classics as under siege. Moreover, Richmond, as we shall soon see, was the besieged capital of the rebel South during the American Civil War, and no stranger to balloons.

of Green or the Vauxhall proprietors: balloon travel was not mere entertainment, it would open the mind as well as the skies.

The *Royal Vauxhall* was capacious, with a nine-foot oval wickerwork
basket, capable of carrying at least nine passengers, as well as a massive
grappling iron and a huge hand-cranked winch to raise and lower the
famous manila trail rope. To begin with it supplied paying passengers
with short, tethered ascents to a hundred feet, so they could admire a
view that stretched right up the Thames to the Houses of Parliament and
St Paul's. Equally, of course, the balloon could itself be seen for miles
around London, a hugely effective piece of advertising.

After various preliminary trials, Green pulled off the most spectacular advertising coup for the Gardens in November 1836, when he set out
on a legendary overnight flight from London to the Continent. The
balloon team consisted of Green himself, Robert Hollond, a wealthy MP
who partly financed the flight, and Monck Mason, an Irish musician and
balloon enthusiast (suitably, he played the flute) who talked and wrote
with great fluency, and undertook to record the whole event.

The departure was well publicised, especially the spectacular list of
supplies that were intended to victual the three-man crew for up to three
weeks. These included forty pounds of ham, beef and tongue; forty-five
pounds of cooked game and preserves; forty pounds of bread, sugar and
biscuits; and not least sixteen pints each of sherry, port and brandy,
together with several dozen bottles of champagne. By dividing these
figures by sixty, it is possible to estimate the *minimum* that each of these
three Victorian gentleman was expected to consume *per diem*. For example: one and a half pounds of meat, half a pound of biscuits, a pint of
fortified wines, and several glasses of champagne. Indeed, Mason made
many jokes about the '*high* flavour and *exalted* merits' of their supplies.

Much of the weight of these provisions (two hundred pounds) would
eventually of course become expendable organic ballast, although the
exact arrangements for disposing of this were not advertised. To this was
added four hundredweight of the actual sand ballast, hung in sacks
round the outer edge of the basket.[9] Overall, the weight and worldly
solidity of the crew's creature comforts were seen not as luxuries, but as
a guarantee of good preparation and serious intent. In a gas balloon the
amount of ballast defined the potential for remaining airborne.

Of course, food supplies were not their only baggage. There was heavy
clothing including cloaks and fur hats; carpet bags for personal items;

repair equipment and maps; 'speaking trumpets, barometers, telescopes, lamps, wine jars and spirit flasks'; the mighty trail rope; hundreds of extra yards of rope and cordage; Bengal flares; and a patent safety lamp, designed on the principle of the Davy miner's lamp. Finally there was Green's particular delight, a patent portable coffee-brewer. Ingeniously, it worked by 'slaking' a supply of quick lime with water in a metal canister, thereby producing 'chemical heat' (calcium hydroxide) without any open flame. Moreover, it could be emptied and replenished as required.[10]

Most of these articles were suspended above the crew's heads, around the wooden hoop of the balloon, in a carousel of swinging sacks and nets and bags, producing the effect of some fabulous airborne hardware store. The total payload or lifting capacity of the eighty-foot *Vauxhall* – including the men, the equipment, the supplies and the sand ballast – was just under three thousand pounds, the equivalent of about fifteen robust men (or a modern rugby team with their boots on).

3

Watched by an enormous crowd, they launched from the Cremorne section of the Vauxhall Gardens at 1.30 p.m. on 1 November 1836, with approximately three hours of daylight in hand. They sailed rapidly eastwards across London, down the Thames, over Rochester, diagonally across north Kent towards Canterbury and the North Foreland. This line of flight would take them over the Goodwin Sands, out across the North Sea towards the Baltic, and possibly even Scandinavia. It was much too far north for Green's liking.

Green immediately impressed his crew with a quietly confident demonstration of balloon navigation. If they gained height, he announced, they would turn south. He briskly ordered Mason to release half a sack of ballast, and they watched silently as the whole horizon appeared to revolve beneath them, turning slowly and 'majestically' northwards at Green's command. At first confused, Mason gradually realised that the *Vauxhall* had entered an upper airstream and was flying due south towards Dover. 'Nothing could exceed the beauty of this manoeuvre,' he thought.[11]

They sailed over the first twinkling lights of Dover port at exactly 4.48 p.m., 'almost vertically over the Castle'.[12] It was precisely the point

from which Blanchard and Jeffries had begun their historic flight almost
fifty-one years before. They crossed the Channel just before dusk, overfly-
ing Calais at three thousand feet, and then dropping to recover the easterly airstream. As the last light failed, Green calculated their average land
speed so far at twenty-five miles per hour, with the probability that it
would increase over the flat expanses of Flanders and Belgium. They were
now on a compass course of approximately 100 degrees, a fraction south
of due east, headed in the general direction of Brussels, Liège, Cologne,
Frankfurt, Prague, Moscow ... So the balloon disappeared into the gathering penumbra of Continental Europe.

Their next act was to sit down to a huge meal of cold meats and wine,
spread on the central work bench of their basket, and accompanied with
'other liquors'. Mason noted that the champagne was unmanageable at
any altitude, as due to the lower pressure it simply shot frothing out of
the bottle, revealing what he called its 'natural tendency to *flying*'.[13]
Perhaps under the influence of these refreshments, the landscapes of
northern France seen after dusk, with isolated points of candlelight
'burning late' in the villages below, seemed infinitely romantic and
mysterious. Equally, the bigger towns, now lit by gaslight, glowed on the
horizon like unearthly sources of energy and activity. Their reflected radi-
ance bloomed yellow and purple in the thickened atmosphere above the
balloon: a first indication of urban pollution.

By midnight they had been airborne for nearly twelve hours, already
something of a record. They were now flying towards the flaring lights of
some huge industrial complex, distinctly set on the banks of a large river
running north and south. They identified this as the river Meuse, and
realised that they must have long since passed south of Brussels. The
town was too big to be either Charleroi or Maastricht, and the only possibility remained Liège. They were astonished that such an ancient city
should be surrounded by so much modern industry. Approached by air,
and at night, this became dramatically evident.

Nestling in the valley of the river Meuse, with its historic churches
and ancient markets, Liège was once the tranquil centre of the traditional textile trade of northern France. But as part of the newly independent Belgium it been transformed into one of the largest centres of
heavy industry in the coalmining belt of northern Europe. With a population of nearly a hundred thousand labourers, it supported a growing
number of huge ironworks and foundries, which were worked in shifts

without ceasing twenty-four hours a day. Its commercial port was the third largest river port in Europe, with direct river and canal connections to Antwerp, Rotterdam and Aachen. Massive supplies of coal, iron ore and other raw materials were constantly shipped in by barge, while a steady stream of metal goods, guns and engineering parts were hurried away south-eastwards into France and south-westwards into Germany.

The balloonists, silently approaching through the night and flying very low, were transfixed by the unearthly glare of the fiery foundries moving swiftly towards them out of the darkness. The ancient centre of the city itself, set peacefully round the great oxbow curve of the Meuse, 'the theatres and squares, the markets and public buildings', slid quietly beneath them. But the surrounding districts 'appeared to blaze with innumerable fires ... to the full extent of all our visible horizon'.[14]

As they floated over this industrial inferno, they were gradually overwhelmed by the thunderous machine noise, the choking industrial smells, and the haunting sound of men below still working on the night shifts. There was disembodied shouting, coughing, swearing, metallic banging and sometimes, weirdly, sharp echoing bursts of laughter. They were being granted a unique, nightmare vision of the new industrial future, a world of ever-extending ironworks where every street was 'marked out by its particular line of fires'. It was what Mason called the strangest glimpse of a 'Cyclopean region'.

Mason's disturbing account of passing over Liège later became famous in aeronautical circles. The French aeronaut Camille Flammarion recalled it as he made the same night flight over the symbolic city thirty years later. But for Flammarion, the experience was subtly different. He pointed enthusiastically to the industrial lights as they approached: 'See, *mon ami*! See how beautiful this is! Do not dream of days gone by ...' The vision of manmade power and productivity deeply impressed him: 'The Belgian towns, lit up by gaslight and the flames issuing from the smoky summits of the blast furnaces seemed to us silent aerial navigators the most dazzling spectacle. The deep sound of the Meuse, as it flowed along its course, was orchestrated by the sharper noises from the workshops, whose mysterious flames and dark smoke rose in the distance all around us.'[15]

But the English balloonists were shaken. Uncharacteristically distracted, Green mistakenly let their expensive coffee-brewing device fall overboard. Mason recorded this unusual error dispassionately. Having opened it over the side of the basket to shake out the expended materials, 'Mr Green unfortunately let it slip from his hand'.[16] Even less in character, Green then started to play a curious trick on the unsuspecting foundry workers below.

He lit a blazing white Bengal light, and lowered it on a rope until it was skimming 'nearly over their heads'. He then urged Mason to shout down through the speaking trumpet 'alternately in French and German', as if some supernatural power was visiting them from on high. The ironworkers were being visited by the gods of the air.

Mason complacently imagined how this aeronautical trick must have 'struck terror' into even the boldest hearts and wisest heads of the 'honest artizans' beneath: 'Catching alone the rays of the light that preceded from the artificial fire-work that was suspended close beneath us, the balloon, the only part of the machine visible to them, presented the aspect of a huge ball of fire, slowly and steadily traversing the sky, at such a distance as to preclude the possibility of it being mistaken for any of the ordinary productions of Nature ...'[17]

As the Bengal light went out, they completed this supernatural effect by emptying half a bag of ballast sand directly onto the upturned faces a hundred feet below. Then the balloon sailed silently and invisibly away, leaving behind the puzzled tribe of Belgian foundry workers staring

uncomprehendingly upwards, as these mysterious superior intelligences disappeared. 'Lost in astonishment, and drawn together by their mutual fears,' Mason concluded, 'they stood no doubt looking up at the object of their terrors.'[18] So the gods were also treating the ironworkers as if they were some primitive tribe.

Mason's account of the voyage, *Aeronautica*, has several illustrations, views and cloudscapes, among them a very strange, dramatic one entitled 'Balloon over Liège at Night', taken from an imaginary point outside the basket looking across at the crew. Their faces are weirdly illuminated by the Davy lamp hung from the balloon hoop. The curving river Meuse, and the blazing foundries, are visible in the darkness below.

After midnight it was the crew's own turn to be alarmed. Gradually all human lights on the ground disappeared. The moonless night seemed to close in around them, encircling them completely, even from below. It was an increasingly disturbing sensation. 'The sky seemed almost black with the intensity of night ... the stars shone like sparks of the whitest silver scattered upon the jetty dome around us. Occasionally faint flashes of lightning would for an instant illuminate the horizon ... Not a single object of terrestrial nature could anywhere be distinguished; an unfathomable abyss of "darkness visible" seemed to encompass us on every side.'[19]

What was so frightening and disorientating was that the darkness seemed increasingly *solid*. Gone was the classic balloon feel of airy vistas, glowing luminosity and huge benign openness. The night was thickening into an alien substance. It was menacing and claustrophobic, entrapping and imprisoning them. Mason records no conversation with Green or Hollond at this time, but afterwards tried to describe what were clearly shared sensations: 'A black, plunging chasm was around us on all sides, and as we tried to penetrate this mysterious gulf, we could not prevent the idea coming into our heads that *we were cutting a path through an immense block of black marble* by which we were enveloped, and which, a solid mass a few inches away from us seemed to melt as we drew near, *so that it might allow us to penetrate even further into its cold and dark embrace.*'[20]

The idea of the men being thrust into or entombed in 'an immense block of black marble', and held there forever in its 'cold and dark embrace', is strangely unsettling. Is it a shivering anticipation of the Victorian horror of being buried alive; or even of some nightmare of sexual entrapment? The passage is curiously reminiscent of some of the later horror stories of Edgar Allan Poe, such as 'The Tell-Tale Heart'. In fact it is highly likely that Poe read this very description as soon as it was published, for it turns out that he was following the accounts of Green's balloon adventures very closely from the other side of the Atlantic.

4

The year before Green's epic flight, Edgar Allan Poe had written one of the earliest of his fantasy stories, 'The Unparalleled Adventure of One Hans Pfaall', published in the mass-circulation newspaper the New York Sun in June 1835. It is a highly technical and perversely well-imagined account of a successful ascent to the moon – in a home-made balloon of 'extraordinary dimensions', containing forty thousand cubic feet of gas.[21]

Pfaall's lengthy preparations are given in great detail, his equipment including a specialised telescope, barometer, thermometer, speaking trumpet, 'etc etc etc', but also a bell, a stick of sealing wax, tins of pemmican, 'a pair of pigeons and a cat'. Immediately upon launching, an explosion leaves him hanging upside down from a rope beneath the balloon basket. This proves to be a typically Poe-like state of horrific suspension ('I wondered ... at the horrible blackness of my finger-nails') which would often be repeated in later stories.

Ingeniously recovering himself by hooking his belt buckle to the rim of the basket, Hans describes how his balloon, ascending 'with a velocity prodigiously accelerating', rapidly overtakes the record height achieved by 'Messieurs Gay-Lussac and Biot'. He soon crosses 'the definite limit to the atmosphere'. On the way he has another Poe-like vision into the centre of a stormcloud: 'My hair stood on end, while I gazed afar down within the yawning abysses, letting imagination descend and stalk about in the strange vaulted halls, and ruddy gulfs, and red ghastly chasms of hideous and unfathomable fire.'

Hans succeeds in breaking out of the earth's gravitational field, and uses a patent 'air-condenser' to breathe. But his ears ache and his nose bleeds. During nineteen days and nights, he observes the steadily retreating surface of the planet, gradually reduced to a curving globe of gleaming blue oceans and white polar ice caps: 'The view of the earth, at this period of my ascension, was beautiful indeed ... a boundless sheet of

PENGUIN ⟨O⟩ CLASSICS

The Science Fiction of Edgar Allan Poe

unruffled ocean ... the entire Atlantic coasts of France and Spain ... of individual edifices not a trace could be discovered, the proudest cities of mankind had utterly faded away from the face of the earth ...'

He eventually floats upwards into the moon's gravitational sphere, and begins to drift into lunar orbit. At this point the balloon turns round and begins a rapid descent towards the lunar surface. After landing, the balloonist is surrounded by an aggressive mob of small, ugly-looking creatures, 'grinning in a ludicrous manner, and eyeing me and my balloon askance, with their arms set akimbo'.

After desultory greetings and unsatisfactory conversations, Hans turns from them 'in contempt', and lifts his eyes longingly above the lunar horizon. The version of 'earthrise' which follows is one of the most hauntingly poetic passages in the entire story. 'Gazing upwards at the earth so lately left, and left perhaps forever, [Hans] beheld it like a huge, dull, copper shield, about two degrees in diameter, fixed immovably in the heavens overhead, and tipped on one of its edges with a crescent border of the most brilliant gold.'

Poe's final, delicate irony is that the moon creatures do not believe the earth is inhabited. They think Hans Pfaall is a great and inveterate liar. And when, after a five-year lunar sojourn, he somehow contrives to get a message taken back across space by 'an inhabitant of the Moon' to the earth, addressed to the 'States' College of Astronomers in the city of Rotterdam', they in turn dismiss Hans as a 'drunken villain', and his missive as 'a hoax'. Poe's readers are sardonically asked to draw their own conclusions.

Within less than a decade, Poe would return to the subject of balloons and amazing flights. This first pioneering tale, written when he was only twenty-six, is evidently inspired by Cyrano de Bergerac's Histoire comique. But its technical originality and brilliance, its mixture of scientific realism and metaphysical terrors, suggests the wholly new dimension of science fiction.[22]

5

Perhaps to counter just such metaphysical terrors, somewhere towards 1 a.m. Green allowed the balloon to sink until the long trail rope, though invisible, was again reassuringly in touch with the ground. They minimised the flame of their overhead safety lamp, and gradually 'the intensity of the darkness yielded' and they could pick out very faint shapes below – vast, shadowy stretches of forest looming out against snow, and the dull gleaming curve of an enormous river which they calculated must be the Rhine. These shapes were an extraordinary relief, familiar forms which, as Mason wrote, 'acknowledged the laws of the material world'.[23]

Yet flying so close to the earth in what was evidently a landscape of steeply wooded valleys was a risk, and they had frequent moments of alarm. On one occasion around 3 a.m. a thin, luminous shape, like a watchtower or a spire, suddenly seemed to be approaching them at terrifying speed and at exactly their height. For several agonising moments all three leaned out of the basket, desperately trying to see what the obstacle was, and how they could possibly avoid it. Finally Green realised that it was a section of their own stay rope, hanging down from the crown of the balloon, not more than twenty-five feet outside the basket. But, caught by the reduced light from their low-burning lamp, it gave the alarming illusion of a distant object hovering directly in their path. Once again the night had deceived them.[24]

The night had also become increasingly cold, and their thermometer now dropped well below freezing. Even the coffee – deprived of its lime heater – was frozen solid in its canister. They held it just above the lamp to thaw it back into liquid form. The crew were tense, and morale was a little low. They drank brandy and talked of the great polar navigator Captain William Parry, and his heroic attempts to discover the North-West Passage through the frozen wastes of the Arctic Circle, and to reach the North Pole (as it turned out, a prophetic conversation).[25]

At 3.30 a.m. Green decided to climb back up to a safer height, and look for the first welcome indications of dawn. He discharged a little ballast, but to his surprise the balloon seemed to gather momentum as it climbed, and very shortly their barometer indicated a height of twelve thousand feet, far higher than he had intended, more than two miles up. Once again they were surrounded by total enveloping blackness and complete silence, except now there were a few scattered stars high above. As they were gazing up at these, something really terrifying happened. There was a sharp cracking sound from the balloon canopy overhead, a sudden jerk on the hoop, and then the whole basket began to drop away beneath their feet.[26]

Mason vividly described his sensations of horror. His narrative suddenly leaps into the present tense:

> At this moment, while all around is impenetrable darkness and stillness most profound, an unusual explosion issues from the machine above, followed instantaneously by a violent rustling of the silk, and all the signs which may be supposed to accompany the bursting of a balloon ... In an instant the car, as if suddenly detached from its hold, becomes subjected to a violent concussion, and appears at once to be in the act of sinking with all its contents into the dark abyss below. A second and a third explosion follow in quick succession ...[27]

Rigid with terror, clinging to the basket's edge, Mason knew that nothing could now avert his death. Then, with equal suddenness, everything about the balloon reverted to normal. The basket became steady, the balloon canopy smooth and silent above them; all was just as tranquil and reassuring as before. Mason stood gazing blankly at Hollond, both men still clinging to the edge of the basket, pale with shock.

Green reassured his shaken passengers about what had happened. It was all quite normal, he told them with a smile, and could be explained by simple physics. While they had been flying near the ground and in increasingly cold air, the canopy of the balloon had gradually shrunk and folded in on itself, as its volume of hydrogen contracted. But as it was night, no one (except Green) had observed this. Then during their rapid ascent the balloon entered regions of lower pressure, and the hydrogen rapidly expanded again. This forced the canopy to reinflate more swiftly than usual. The loose folds of silk, concertina-ed or 'corrugated' together, and partially stuck by ice, did not immediately open. Only when sufficient hydrogen pressure had built up to snap them forcefully apart did the balloon resume its full shape in a series of sharp, violent unfolding movements.

Moreover, chuckled Green, the terrifying jerks on the hoop were actually the basket being pulled *upwards* as the balloon expanded. The sensation of falling was strictly speaking an illusion: they were actually 'springing up' rather than dropping down. All was well. But perhaps they should all have some more brandy?

How far this account reassured his passengers is not clear; nor even how far Green himself had been taken by surprise. It must have occurred to him that the unexpected rapidity of their ascent was in fact extremely perilous, as one of the frozen folds of silk could easily have ruptured before it was forced apart by the pressure of the hydrogen. He later emphasised to Mason that he had 'frequently experienced the like effects from a rapid ascent'.[28]

Altogether it was a huge relief when the November dawn slowly began to lighten the sky. The ground below seemed strangely smooth and luminous, and they gradually realised that they were again passing over 'large tracts of snow'. The bitter cold had increased: they could see the plumes of their own breath, and the glistening ice that had formed on the lower canopy of the balloon. The question of their exact location now became pressing. According to their compass they had been travelling steadily due east for most of the night. Green made a quick dead-reckoning calculation, and concluded that, based on the speed with which they had reached Liège, it was possible they had travelled up to two thousand miles from England. This would put them somewhere over 'the boundless

ASCENT OF THE
GREAT NASSAU BALLOON
FROM THE MONTPELLIER GARDENS, CHELTENHAM.

planes of Poland, or the barren and inhospitable Steppes of Russia', an alarming prospect. In fact, as Mason later admitted, this was an unduly 'extravagant' estimate, over 110mph, largely inspired by the long period of darkness, disorientation and terror they had experienced.[29] When they eventually landed at 7.30 a.m., descending inelegantly into a stand of snow-covered fir trees (their sand ballast had frozen solid and could not be properly released), they found that they were still in north Germany. Local foresters, tactfully recruited by means of the balloon's copious stores of brandy, led them in triumph to the little country town of Weilburg. They were thirty miles north-west of Frankfurt, in the Duchy of Nassau.

Yet their achievement was spectacular. They had travelled 480 miles in eighteen hours.[30] This was a long-distance record for a balloon flight, at an average speed of just over twenty-six miles per hour, roughly the same rate as when they set out. But they had covered an astonishing eastwards trajectory, on a line that ran roughly through Calais and Brussels, to Liège, Coblenz, and almost as far as Frankfurt. The *Vauxhall* had survived in good order, and was immediately rechristened the *Royal Nassau*, after their landing site. News of the flight caused an international sensation. On the way back Green spent some time in France, and flew his famous balloon from several sites around Paris and at Montpellier Spa. His international reputation was made.

6

The flight inspired something like a renewed balloon craze. Crowds of tourists and foreign visitors flocked to the Vauxhall Gardens. Numerous articles, editorials and poems were published in the press. The fashionable painter John Hollins produced a striking composite portrait entitled *A Consultation Prior to the Aerial Voyage to Weilburgh*, which is now in the National Portrait Gallery. The balloonists and their financial backers (including Hollins himself) are shown gallantly grouped around a large planning table, with maps and sheets of calculations, like generals working out a military campaign. Green, seated at the right, gazes purposefully across the table at Robert Hollond MP, seated on the left, while Monck Mason, their historian, stands between them apparently lost in thought. The *Royal Vauxhall* – now the *Royal Nassau* – can be seen outside through a window, tethered like an impatient warhorse.

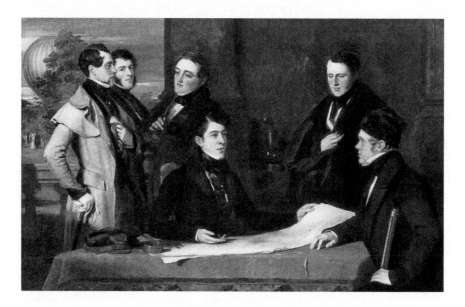

If the flight was heroic, it also had – like all balloon flights – its comic aspects. It had flown over several countries; but mostly at night, when nothing could really be seen. It had achieved a distance record, certainly; but without the balloon ever being capable of steering towards any destination. It had revolutionised long-distance travel, but without making it any more practical. Thomas Hood, famous for his Chartist ballad of the working man, 'The Song of the Shirt', wrote several humorous poems in praise of ballooning, including 'The Flying Visit'. But he outdid himself with a bubbling, mock-heroic party piece, 'Ode to Messrs Green, Hollond and Monck on their late Balloon Adventure'. It opens with a high, jocular, punning invocation in a 'champagne style' that he had invented especially for the occasion:

> O lofty-minded men!
> Almost beyond the pitch of my goose pen
> And most inflated words!
> Delicate Ariels! Etherials! Birds
> Of passage! Fliers! Angels without wings!
> Fortunate rivals of Icarian darings!
> Kites without strings! ...[31]

Hood, with his gaseous puns and mocking emphasis on the amount of food and drink consumed by the aeronauts under the stars, tended to treat the whole expedition as an enormous prank. But Monck Mason was serious. After publishing several articles, his carefully completed account of the voyage appeared two years later in book form as *Aeronautica* (1838). He made the story especially memorable by his haunting description of night flying.

Mason later added a hundred-page appendix on the general experience of flying in balloons: the euphoric sensations of ascent, the diminution of the people below, views of clouds and sunlight, the spherical appearance of the earth beneath, and the particular panoramic and sharply 'resolved' view of cities, rivers, railways. He is surprisingly perceptive about the strangely 'delineated' sounds heard from below, and the way they create an entire sound 'landscape': the rural world of farm dogs, cattle, sheep-bells; but also sawing, hammering and agricultural flails; sportsmen's guns or the 're-iterated percussion' of mill wheels.

Some of this is weakened by Mason's orotund pseudo-sublime manner, by which he attempts to give ballooning a kind of contemplative gravity, the exact opposite of Hood's mad 'levity'. But there are many fine existential passages: on the sense of extreme solitude and silence in 'the immense vacuity'; on the sublime appearance of cloudscapes and the 'Prussian blue' zenith at high altitudes; and on the uneasy feeling of 'intruding' on God's territory, 'the especial domains of the Almighty'.[32] Above all, he attempts to evoke the strange and beautiful other world of the kingdom of the air:

> Above and all around him extends a firmament dyed in purple of the intensest hue, and from the apparent regularity of the horizontal plane on which its rests, bearing the resemblance of a large inverted bowl of dark blue porcelain, standing upon a rich mosaic floor or tessellated pavement. In the zenith of this mighty hemisphere – floating in solitary magnificence – unconnected with the material world by any visible tie – alone – and to all appearances motionless, hangs the buoyant mass by which he is upheld ...[33]

Throughout such passages there is a curious mixture of scientific terminology – horizontal plane, zenith, buoyant mass – with the rhetoric of Victorian poetry and sublimity; even on occasion of Victorian prayers or hymns. This seems to reflect a philosophical, or even theological, problem later expressed by many Victorian balloonists. To what extent is the upper sky, where Prussian blue deepens into black, 'the space beyond the limits of our atmosphere',[34] a scientific zone or a celestial one, or both? What unknown powers or energies lurk in the terrific 'black and fathomless abyss' – an abyss paradoxically overhead? What monsters or deities does the upper deep contain? ᵠ

To give his book further weight, Mason added six other appendices to later additions. Appendix B consisted of a short biography of Charles

---

ᵠ The ambiguity also appears in the use of the word 'heavens' or 'heaven'. Is God up there in space? The implied question is posed, in a similar way to which it was already posed by Lyell's geology, and would soon be by Darwin's theory of evolution: Is God back there in time? It has also been suggested that Monck Mason's highly coloured prose anticipates the way Victorian ballooning would celebrate an unexpectedly 'feminised' version of Romantic sublimity in the air. The flight of the Victorian balloonist could be seen to represent a curious Freudian 'regression' into a passive, childlike, dreamlike, oceanic or 'infantile' state. The aeronaut is suspended 'above reality' and the harshness of the Victorian industrial world of labour, regularity, masculine effort and control. See Elaine Freedgood, Victorian Writing About Risk.[35]

Green, with accounts of an earlier test flight made with Green from Vauxhall to Chelmsford on 4 October 1836, and of Green's part in the fatal Cocking parachute experiment of 1838, in which the over-confident inventor Robert Cocking leapt to his untimely death above the Thames Estuary, and almost killed his pilot Green into the bargain. Appendix C was an alphabetical checklist of all known European aeronauts between 1783 and 1836, with longer individual notes on the early pioneering figures like Blanchard, Lunardi, Sadler, Gay-Lussac and Garnerin. Appendix D, 'On the Mechanical Direction of the Balloon', investigated the old question of navigating a balloon. Appendix E considered Green's use of the guide rope and other 'equilibrium' devices. Appendix F reflected on the limitations of bird flight (but with special praise for the gliding capacities of the South American condor 'above the lofty peaks of the Andes'). And Appendix G indulgently reprinted further verses in praise of the *Nassau* flight.

Thanks to Green, ballooning had once again caught the imagination of writers, but its significance was interpreted in increasingly various ways. In 1837 Thomas Carlyle used the image of balloons in his introductory chapter, 'The Paper Age', in *The French Revolution*, to express the political risks and hopes of the time: 'Beautiful invention, mounting heavenward – so beautifully, so unguidably! Emblem of our Age, of Hope itself.'

In 1838 John Poole (who had once seen Sophie Blanchard fall to her death in Paris) gave a more satirical but shrewdly perceptive account of a flight with Green at night over the East End of London. It seems to him very different from Paris, with its boulevards, parks and cafés. The sinister, garish lights of gin 'palaces', taverns, apothecaries and brothels – alternately twinkling 'blue, green, purple and crimson' – are used to explore the notion of the hidden city of poverty, sickness and crime. On landing near Hackney Marshes, the balloon is surrounded by a threatening mob, which has pursued it all the way 'from Stepney, Limehouse and Poplar'. Prophetically, it was as if the balloon had trespassed into an African jungle, and stirred up an unfriendly horde of howling 'natives'. Assaulted by 'their yells, their savage imprecations, curses both loud and deep, their threats to destroy the balloon', Poole, Green and his burly crew just manage to pack their equipment onto a cart, and beat a strategic retreat to the local Eagle and Child public house. Here they hole up until one in the morning, when it is safe to slip back through the silent streets to the West End and 'civilization'.[36]

In his poem 'Locksley Hall' (1842), with its own visions of social disturbance and upheaval, the thirty-three-year-old Alfred Tennyson imagined the aeronauts not merely as romantic adventurers, but also as busy commercial traders. They descend in flocks through the evening skies, to settle upon distant marketplaces around the globe. They are part Homeric travellers in the tradition of Jason and the Argonauts; but also partly hungry commercial travellers, with just a hint of a cloud of locusts descending upon an innocent land at dusk:

> For I dipt into the future, far as human eye could see,
> Saw the Vision of the world, and all the wonders that
>   would be;
> Saw the heavens fill with commerce, argosies of magic
>   sails,
> Pilots of the purple twilight, dropping down with costly
>   bales ...

The poem was originally drafted in 1835. But Tennyson also foresaw, like Franklin before him and H.G. Wells afterwards, balloons producing the terror of aerial warfare:

> Heard the heavens fill with shouting, and there rained
>   a ghastly dew
> From the nation's aerial navies, grappling in the
>   central blue.[37]

Charles Green had established himself as much more than a balloon showman, or the publicity agent of the Vauxhall Gardens.

He had resurrected the old dream of ballooning, but adapted it to the coming Victorian age. Bronze medals were even cast in his honour.

In his Preface to the second edition of *Aeronautica*, Mason suggested that Green's ambitions were turning towards an Atlantic crossing. Green apparently took a quite nonchalant view of the huge distances and meteorological challenges this would involve: 'In his view, the Atlantic is no more than a simple canal: three days might suffice to effect a passage. The very circumference of the globe is not beyond the scope of his expectations: in fifteen days and fifteen nights, transported by the trade winds, he does not despair to accomplish in his progress the great circle of the earth itself. Who can now fix a limit to his career?'[38]

This was heady talk, and made good journalistic copy. But Mason was not a successful balloon pilot himself, merely a successful balloon passenger, and had perhaps had his head turned by all the excitement and publicity. In the same Preface he cheerfully advocated the use of a trailing guide rope 'above fifteen thousand feet in length'. He saw no problem in this monster appendage dragging across 'trees, houses, rivers, mountains, valleys, precipices and plains' with what he described as 'equal security and indifference'.[39]

7

Two years after the publication of *Aeronautica*, in 1840, Green issued his own proposals to fly the Atlantic. He claimed to have identified a prevailing west-to-east wind current in the upper atmosphere, which meant that he would start the crossing from America. 'Under whatever circumstances I made my ascent, however contrary the direction of the wind below, I uniformly found that at a certain elevation, varying occasionally but always within 10,000 feet of the earth, *a current from west to east, or rather from the north of west, invariably prevailed.*'

He also explained that a two-thousand-foot guide rope, fitted with canvas sea drags and copper floats, would be enough to stabilise an eighty-thousand-cubic-foot balloon and keep it airborne, without expending additional ballast, for 'a period of three months'. He said he was only awaiting a generous sponsor to undertake the trans-Atlantic flight immediately.[40] In the end, the astute Green could find no financial

backer, refused to depart without one, and the Atlantic attempt was
never made.

But it was made in fiction. Green's proposals inspired a further brilliant invention by Poe, published in the *New York Sun* in 1844. This time
it was a news story hoax. 'The Atlantic Balloon' coolly presents an
extraordinarily detailed and convincing account of Green and Monck
Mason crossing the Atlantic from England in seventy-three hours. Much
of the story is drawn from the well-publicised flight of the *Royal Nassau*.
As the third member of the balloon crew, instead of Robert Hollond MP,
Poe mischievously added his rival, the popular British thriller writer
Harrison Ainsworth.

Poe's story broke on Saturday, 13 April 1844, when the *New York Sun*
announced that it would be issuing an 'Extra' containing a detailed
account of a transatlantic crossing by a balloon, the 'flying machine'
Victoria. There was also a postscript in the morning edition of the *Sun*,
with an appropriate accumulation of exclamation marks: 'By Express.
Astounding intelligence by private express from Charleston via Norfolk!
– The Atlantic Ocean crossed in three days!! – Arrival at Sullivan's Island
of a steering balloon invented by Mr Monck Mason!!!'

The Extra created an immediate sensation. According to Poe's own
account, a large crowd gathered in the square surrounding the *New York
Sun* to wait for it, and when it appeared at two in the afternoon, it sold
out immediately. The account consists of an introductory section and a
journal kept by Monck Mason, to which Mr Ainsworth added a daily postscript. The introduction details the invention of the balloon by Mason
(rather than Green), who adapted an Archimedean screw for the purpose
of propelling a dirigible balloon through the air, inflated with more than
forty thousand cubic feet of coal gas.

In contrast to the newspaper announcement, Poe's own 'reportage'
remains cool and apparently factual. The plain and straightforward
narrative works on several levels. First, it genuinely explores the technical, scientific challenge of crossing the Atlantic, which was already
beginning to obsess American aeronauts like John Wise. Next, it quietly
touches on a vein of social satire, a mockery of scientific presumption
and hubris which would become characteristic of the later science fiction
genre. Finally, as with so many of Poe's stories, it is a psychological study,
an exploration of collective delusion, a group 'suspension of disbelief'.
Here Coleridge's famous term takes on a new, strangely literal

meaning.[41] The desire to be dazzled by scientific wonders may be associated with a conscious willingness to be bamboozled or hoaxed.

Needless to say, it is also a brilliant exploitation of the growing newspaper tradition of the 'scoop' – and the fake scoop. American editors were shrewdly realising that their readers did not mind occasionally being taken for a ride, especially such an airborne one. This fruitful connection between balloons and newspapers was ready to expand.

4
Angel's Eye

―――

1

Throughout the 1840s and 1850s, dramatic ballooning stories gained increasing notice in the popular press, both in Britain and America. With the arrival of new illustrated journals, such as the *Illustrated London News*, founded in 1847, it was soon clear that they also offered superb opportunities for picture stories. The sheer size and glamour of a balloon, especially when contrasted with human crowds and cityscapes, were natural material for full-page and even double-page balloon 'spreads'.

Few pieces of mid-Victorian aeronautical journalism could match Henry Mayhew's long and rapturous account, 'A Balloon Flight over

ASCENT OF MR. GREEN'S BALLOON, ON MONDAY LAST.

London', which appeared in the *Illustrated London News* for 18 September 1852.

Much of Mayhew's previous writing life had prepared him for this extraordinary essay. He was one of the greatest journalists of the age, whose interests spanned everything from the fine arts to social reform. He also wrote poetry, plays, operas and would go on to produce hugely successful accounts of the early lives of two scientists: *Young Humphry Davy* (1855) and *Young Benjamin Franklin* (1861). His most famous work, *London Labour and the London Poor,* had been published in instalments throughout 1851, deliberately timed to coincide with the Great Exhibition, as a sobering correction to its Victorian triumphalism.

After spending much of his twenties knocking about Paris, freelancing alongside his friends William Thackeray and Douglas Jerrold, Mayhew returned to London full of ideas for a new kind of streetwise journalism. He was much taken with the irreverent and satirical style of the French magazine *Le Charivari,* to which the pioneering French aerial photographer Félix Nadar and many others contributed. In 1841 he helped the journalist Mark Lemon launch a quite new kind of humorous British periodical. It became *Punch,* with its mixture of witty essays and clever but good-natured satirical cartoons. It had an immediate success, but Mayhew and Lemon soon parted company, though remaining on excellent terms. Lemon continued at *Punch,* becoming a comfortable fixture in London literary clubland, and eventually one of Dickens's most trusted editors. Meanwhile Mayhew struck out on his own, gradually developing a new kind of investigative journalism. He went far beyond the gentle, sardonic scope of *Punch,* contributing edgy, groundbreaking pieces to the *Morning Chronicle.* His special subject was London, and the underside of city life. Mayhew's London was the city that few middle-class readers ever glimpsed: London from beneath.

For the next decade Mayhew produced hundreds of vivid, detailed reports of life in the backstreets and the rookeries, and especially on the marginal trades and skills that sustained the poorest men and women – and not least the children – of the capital. Among his celebrated and scandalous subjects were street vendors, costermongers, milkmaids, ratcatchers, mudlarks, crossing sweepers, fire eaters, prostitutes, pickpockets and dustmen. Each of his accounts was written with the clipped shape and high polish of a short story. They were buttressed by statistics, glinting with minute visual details, and brought to life with inimitable passages of dialogue.

Often these develop into simple but disturbing sequences of question and answer. 'I make all kinds of eyes,' the eye-manufacturer says, 'both dolls' eyes and human eyes; birds' eyes are mostly manufactured in Birmingham, and as you say, sir, bulls' eyes at the confectioner's ... A great many eyes go abroad with the dolls ... The annual increase in dolls goes on at an alarming rate. As you say, sir, the yearly rate of mortality must be very high, to be sure, but still that's nothing to the rate in which they are brought into the world ... I also make human eyes. Here are two cases, in the one I have black and hazel, in the other blue and grey. Here you see are the ladies' eyes ... There's more sparkle and brilliance about them than the gentlemen's ... There is a lady customer of mine who has been married three years to her husband, and I believe he doesn't know she has a false eye to this day.'[1] Such material, with its mixture of the mundane and the gothic, its small revelations of human eccentricity and affection, would clearly influence the later and darker novels of Dickens.

When he had amassed about half a million words of material, Mayhew began to edit and reorganise the pieces into the form of his grim masterwork *London Labour and the London Poor*. Once the work was completed, he cast around for a suitable way to celebrate. It struck him that an airy overview of the great city, in whose backstreets and dark corners he had spent so many years almost buried, would be suitable. So he accepted an invitation to take a flight in one of Charles Green's balloons.

Officially this was to be one of Green's frequently-advertised 'Last Ascents' from Vauxhall Gardens. For Mayhew, the flight was also to be, in a sense, the culmination and farewell to much of his previous journalism. But it was also a celebration and a release from it. Having seen London from the darkest and most labyrinthine street level, he now wished to sail into the clear air above it. He wanted to see his huge and 'monstrous' city at last in the grand perspective, or – as he put it with poignant irony – from 'an angel's view'. What's more, he would write about it for the leading current-affairs weekly in Britain, the *Illustrated London News*.

In case the 'angel' approach seemed rather presumptuous, he began by explaining that he was naturally 'a coward – constitutionally and habitually timid'. As it did for most of his readers, the idea of flying in a

balloon frankly appalled him: 'I do not hesitate to confess it'. The best he could say was that he was motivated by 'idle curiosity, as the world calls it'. Having made this apparently modest disclaimer, Mayhew immediately admitted to the most heroic previous adventures:

> I had seen the great metropolis under almost every aspect. I had dived into holes and corners hidden from the honest and well-to-do portion of the Cockney community. I had visited Jacob's Island (the plague-spot) in the height of the cholera ... I had sought out the haunts of beggars and thieves ... I had seen the world of London below the surface, as it were, and I had a craving to contemplate it from far above it.

Even if balloon flight turned out be more terrible than anything he had previously experienced, he was determined to try it. What he hoped to see from Mr Green's balloon was a new vision of the city. Supposing it would be something both familiar yet apocalyptic, Mayhew prepared himself to behold

that vast bricken mass of churches and hospitals, banks and prisons,
palaces and workhouses, docks and refuges for the destitute, parks and
squares, and courts and alleys, which make up London – all blent into one
immense black spot – to look down upon the whole as the birds of the air
look down upon it, and see it dwindled into a mere rubbish heap, to
contemplate from afar that strange conglomeration of vice, avarice, and
low cunning, of noble aspirations and humble heroism, and to grasp it in
the eye, in all its incongruous integrity, at one single glance – to take, as
it were, an angel's view of that huge town where, perhaps, there is more
virtue and more iniquity, more wealth and more want, brought together
into one dense focus than in any other part of the earth.[2]

One of Mayhew's most powerful images was of London miniaturised and
made safe, like some huge child's toy. He, the weary and hardbitten
observer of the streets, was somehow lifted clear and transformed by his
airborne vantage point. The balloon conferred on him a kind of innocence, a kind of grace:

To hear the hubbub of the restless sea of life and emotion below, and hear
it, like the ocean in a shell, whispering of the incessant strugglings and
chafings of the distant tide – to swing in the air high above all the petty
jealousies and heart-burnings, small ambitions and vain parade of 'polite'
society [– was to] feel, for once, tranquil as a babe in a cot.

All this gave him a strange sensation, something close to a religious experience, a celestial transfiguration. It was as if, he wrote,

you are hardly of the earth, earthy, as Jacob-like, you mount the aerial
ladder, and half lose sight of the 'great commercial world' beneath, where
men are regarded as mere counters to play with, and where to do your
neighbour as your neighbour would do you constitutes the first principle
in the religion of trade – to feel yourself floating through the endless realms
of space, and drinking in the pure thin air of the skies, as you go sailing
along almost among the stars, free as 'the lark at heaven's gate', and enjoy-
ing, for a brief half hour, at least, a foretaste of that Elysian destiny which
is the ultimate hope of all.

Mayhew was a master of tone and phrase. These last sentences start carefully, with an evangelical earnestness, the language of Charles Kingsley
and moral uplift, but slowly elide into something more sentimental and

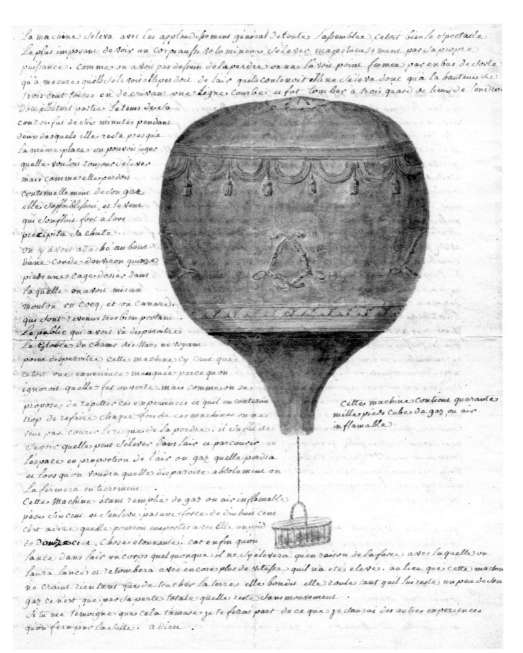

Philippe Lesueur, letter from Paris, 22 September 1783, an eyewitness account of the early sixty-foot Montgolfier balloon that was launched at Versailles with a sheep, a cockerel and a duck in the basket, all of which lived to tell the tale (and did so in several pamphlets).

'The Perilous Situation of Major Money', 1785. John Money's descent
into the sea twenty miles off Lowestoft while gallantly attempting
to raise funds for the new Norfolk and Norwich Hospital.

Dr Jacques Alexandre Charles, the first successful pilot of a
hydrogen balloon, Paris, December 1783. An official portrait
by Bailly, commissioned nearly forty years later by the
Institut Royal de France, 1820.

Jacques Garnerin, the first great French
balloon showman after the Montgolfier
brothers, in heroic profile, Paris 1802.
The modest Latin device reads:
'Praise the Intrepid Aeronaut who
dares to take to the Air'.

Sophie Blanchard, the great French female
aeronaut, drawn by Jules Porreau at the
height of her fame, Paris, 1815.

Sophie Blanchard in her tiny silver gondola above Milan in August 1811,
to celebrate Napoleon's arrival in the city and carrying his Imperial standard.

COLLECTION 476   2ème Série (No 8)   ROMANET & Cie IMP. EDIT. PARIS

MORT DE HARRIS (1824)

'Mort de Harris', Croydon, 1824. The gallant death of Lieutenant Harris, and
the mysterious survival of his beautiful passenger Miss Stocks. From a popular
French balloon collecting-card series, issued by Romanet & Cie, Paris, 1895.

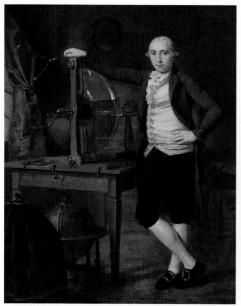

Tiberius Cavallo, physicist and Fellow of the Royal Society, the first natural philosopher to inflate soap bubbles with hydrogen gas, and the first serious historian of ballooning in English. Portrait by unknown artist, c. 1790.

Charles Green, amateur portrait, 1835. Painted in a 'tavern sign' style, when the great British balloonist was still largely known as a 'novelty' showman.

The *Nassau* balloon team, 1836, painted by John Hollins. Left to right: Sir William Melbourne James (Lord Justice of Appeal), John Hollins (artist), Walter Prideaux (lawyer), Robert Hollond MP (seated), Monck Mason, Charles Green.

The *Nassau* balloon at night over the industrial foundries of Liège,
Belgium, 1836. This was the 480-mile long-distance trip that brought
Charles Green an international reputation in Europe and especially America.

Charles Green, epic British balloon pilot, painted by
John Hollins, mezzotint by G. T. Payne, 1838.
The long instrument is Green's treasured Italian
mercury barometer, which served as a precision
altimeter on more than five hundred flights.

WRITERS WHO TOOK TO THE AIR

Mary Shelley

Percy Shelley

Edgar Allan Poe

Jane Loudon

Charles Dickens

Henry Mayhew

populist. Sailing among the stars, singing 'at heaven's gate', dreaming of the Elysian fields, was really a subtle reversion to the imagery of popular Victorian songs and street ballads.

Mayhew was so pleased with this piece that he included an edited version of it in his later book *The Criminal Prisons of London* (1862). His view of London evidently influenced Gustave Doré, as well as Dickens's novels *Bleak House* and *Hard Times*.

2

Charles Dickens was strangely intrigued by balloons. He witnessed many launches in Vauxhall Gardens, and wrote about them several times. He knew Charles Green, and observed his aeronautical calm, his skill with a crowd, and his waving of his white top hat, with a professional admiration. Yet his reactions were far more complicated than Mayhew's.

It is surprising that the great master of human exotica, and the writer who enshrined the English stagecoach in imaginative literature (notably in *The Pickwick Papers*), never actually ventured to set foot in a balloon basket himself. There is no record of Dickens ever leaving *terra firma*, except in his dreams. As a result, unlike Mayhew, all Dickens's balloon observations are made from the ground. Of course he may simply have had a quite reasonable fear of heights; or there may have been more mysterious influences at work. He may even have regarded balloon flying as immoral – as a sort of suicidal surrender of self-command.

Dickens gives a surprisingly sarcastic account of one of Green's early balloon launches in a light-hearted piece of reportage entitled 'Vauxhall Gardens by Day', later collected in *Sketches by Boz* (1836). He seems particularly exercised by the gullibility of the crowd, its surrender to the meaningless novelty overhead.

> *The gardens disgorged their multitudes, boys ran up and down screaming 'bal-loon;' and in all the crowded thoroughfares people rushed out of their shops into the middle of the road, and having stared up in the air at two little black objects till they almost dislocated their necks, walked slowly in again, perfectly satisfied.*

There is a sense here that ballooning is an art of illusion, almost a conjuring trick played upon a credulous audience. Yet it is also a symbol of

novelty and popular excitement, and Dickens's illustrator Phiz used it to
witty effect in his cover drawing for the collected *Boz* essays. This ambiguous impression is sharpened in Dickens's comments on the subsequent
newspaper coverage, making a series of sly digs at an interview that
Green gave to over-enthusiastic reporters.

> *The next day there was a grand account of the ascent in the morning*
> *papers, and the public were informed how it was the finest day but four in*
> *Mr. Green's remembrance; how they retained sight of the earth till they lost*
> *it behind the clouds; and how the reflection of the balloon on the undulat-*
> *ing masses of vapour was gorgeously picturesque; together with a little*
> *science about the refraction of the sun's rays, and some mysterious hints*
> *respecting atmospheric heat and eddying currents of air.*[3]

The key note here is one of bathos. Balloon science is all gas, self-inflation,
and altogether much ado about nothing. The tone is similar to that of
Dickens's celebrated satire on the newly formed British Association for
the Advancement of Science, which he memorably attacked under the
mocking title of 'The Mudfog Association for the Advancement of
Everything' (1837).

Later, as editor of the weekly journal *Household Words* (1850–59),
Dickens recognised the drama and popularity of balloons, and

commissioned several articles on the subject. These included some short pieces of straight reportage, such as 'Over the Water', 'A Royal Balloon' and 'A Royal Pilot-Balloon'.[4] But by far the longest was a well-researched but inescapably comic treatment of the entire history of aerostation, dwelling in loving detail on its most satisfactory catastrophes. It was simply entitled 'Ballooning'.

The piece seems to have been triggered by the extraordinary gallery of aerostats on display at the Great Exhibition of 1851. Whether Dickens regarded these as an expression of imperial hubris, or simply as a display of scientific absurdity, he deliberately commissioned a hostile feature. His chosen reporter for the task was Richard Hengist Horne, a literary adventurer and poet who had travelled in Mexico and Canada, and would soon emigrate to Australia. Although he had once been the schoolfellow of John Keats, Horne's aeronautical credentials were not evident. His previous works included a verse drama, *Prometheus the Firebringer*, and he had had a long, passionate friendship with Elizabeth Barrett Browning. But his droll essay clearly met with Dickens's editorial approval, as it ran to sixteen columns and was given the lead in *Household Words* No. 33 on Saturday, 25 October 1851.

Horne kicked off with the deadpan observation that travelling through 'the sublime highways of the air' was not entirely natural. Man was never intended to be 'lord of the clouds'. The urge to fly might have existed from 'time immemorial', yet among balloonists it seemed to take on a morally questionable form. 'Eccentric ambition, daring, vanity, and the love of excitement and novelty' inspired them quite as much as 'the love of science and of making new discoveries'.

Horne then embarked on a relentlessly mocking history of man's disastrous attempts to become airborne: 'We do not allude to the Icarus of old, or any fabulous or remote aspirants, but to modern times.' These attempts included 'a flying monk of Malmesbury' who became 'impudent and jocose' on the subject of tail-feathers; a flapping French marquis who crash-landed in the Seine and broke his leg against 'one of the floating machines of the Parisian laundresses'; and a citizen of thirteenth-century Bologna who was persecuted by the Inquisition because he failed to drown when his flying machine landed in the river Reno, thus of course inadvertently proving he was a witch.

After summarising the various adventures of Charles Green, Horne turned to the fantastical collection of balloons on display in the

Aeronautical Hall of the Great Exhibition. He enumerates them without comment: 'One has the appearance of a huge Dutch vegetable marrow ... another a silver fish with revolving fins ... a huge inflated bonnet ... a large firework case ... the skeleton of some fabulous bird'. Dickens had clearly given Horne *carte blanche*, and the article continued in this supercilious vein to the end.[5]

Why should Dickens have felt so hostile towards ballooning? He was always ready to poke fun at scientific pretentions, but his mockery seemed to go deeper than this. It is clear that he despised balloons as a form of mass entertainment. He felt that they exploited both the credulity of the public and the courage of the balloon 'artist'. But he may also have feared them at some less conscious level. A clue appears in an extraordinary essay he wrote on the subject of 'Nightmares'. Here he makes a strange and startling comparison between the expectant crowd at a Vauxhall balloon launch and the similarly expectant spectators at a public hanging outside a London prison.

This intimate essay, which Dickens nevertheless published in *Household Words*, opens with the author lying awake in the dark, insomniac, unable to settle his thoughts, and besieged by obsessive and even perverse images. He tries to distract himself:

> *The balloon ascents of this last season. They will do to think about, while I lie awake, as well as anything else. I must hold them tight though, for I feel them sliding away, and in their stead are the Mannings, husband and wife, hanging on the top of Horse-monger Lane Jail. In connexion with which dismal spectacle, I recall this curious fantasy of the mind. That, having beheld that execution, and having left those two forms dangling on the top of the entrance gateway – the man's, a limp, loose suit of clothes as if the man had gone out of them; the woman's, a fine shape, so elaborately corseted and artfully dressed, that it was quite unchanged in its trim appearance as it slowly swung from side to side – I never could, by my uttermost efforts, for some weeks, present the outside of that prison to myself (which the terrible impression I had received continually obliged me to do) without presenting it with the two figures still hanging in the morning air ...*

Here the essay breaks off in horror. Then Dickens tries again with balloons:

The balloon ascents of last season. Let me reckon them up. There were the horse, the bull, the parachute, – and the tumbler hanging on – chiefly by his toes, I believe – below the car. Very wrong, indeed, and decidedly to be stopped. But, in connexion with these and similar dangerous exhibitions, it strikes me that that portion of the public whom they entertain, is unjustly reproached. Their pleasure is in the difficulty overcome. They are a public of great faith, and are quite confident that the gentleman will not fall off the horse, or the lady off the bull or out of the parachute, and that the tumbler has a firm hold with his toes. They do not go to see the adventurer vanquished, but triumphant.[6]

For Dickens, the balloon basket and the public scaffold seemed intimately, even vertiginously, linked. They both hold out the idea of humiliation, exposure and death: the horrific promise of a fatal fall. The novelty ascents arranged by Green and others – the man on a horse, the woman on a bull (surely a Dickens invention?) – make this even worse by trivialising the terror. Worst of all is the lone 'tumbler', hanging over the abyss 'chiefly by his toes'.

And here perhaps lies a possible explanation. It is with this solitary acrobat, totally exposed above the crowd, that Dickens the solitary writer surely identifies. Both ballooning and writing are 'dangerous exhibitions'. The writer, like the balloonist, hopes to be 'triumphant' in front of his audience, the 'public of great faith'. But he – or she – may fail, 'vanquished' despite all their skill, and drop to their long death as from a scaffold. Ballooning haunted Dickens because it reminded him of the permanent, secret terror of successful writing, the ultimate exposure of the popular entertainer, and the public fall from grace.

It is no coincidence that Dickens also slipped a balloon, almost unnoticed, into the famous, grim opening of *Bleak House* (1853). He wrote: 'Fog in the Essex marshes, fog on the Kentish heights ... Fog in the eyes and throats of ancient Greenwich pensioners ... Chance people on the bridges peering over the parapets, into a nether sky of fog, with fog all round them *as if they were up in a balloon and hanging in misty clouds*.' Here the balloon has become again an image of helplessness and doom. The very word 'hanging' has an uneasy echo of Dickens' scaffold nightmare.

3

But usually the Victorian balloon had far more progressive associations. Scientific ascents also took place from Vauxhall Gardens, manned by serious gentlemen in top hats, prepared to observe and measure and speculate.

The use of the aerial panorama was even encouraged as a tool of sociological investigation. By studying the city from above, with the objective 'angel's eye', it was possible to reveal much about its social structure, its balance of commercial and private dwellings, and especially (as both Poole and Mayhew had remarked) its savage contrasts of rich and poor. One indirect result of this was the famous 'poverty maps' compiled by the philanthropist Charles Booth in the 1880s.

These, with their colour codings and careful urban annotations, adapted the balloon overview as a technical device for compiling and storing new kinds of information. Here the 'angel's eye' has become both analytical and philanthropic. The balloon perspective has become an expression of the Victorian social conscience. The 'panoptic' view leads potentially to both planning and improvement. It is 'godlike' in a new, secular way. It is an instrument of social justice, even moral redemption.

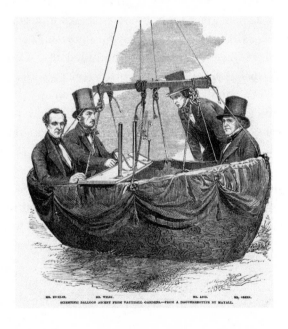

MR. HICKLIN.   MR. WELSH.   MR. ADIE.   MR. GREEN.
SCIENTIFIC BALLOON ASCENT FROM VAUXHALL GARDENS.—FROM A DAGUERREOTYPE BY MAYALL.

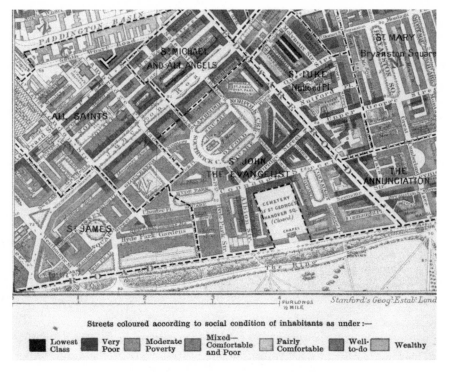

Streets coloured according to social condition of inhabitants as under :—

Lowest Class   Very Poor   Moderate Poverty   Mixed—Comfortable and Poor   Fairly Comfortable   Well-to-do   Wealthy

Another, more commercial, use of the panoramic 'airborne' view was to develop a new kind of tourist's or visitor's guide to the great cities. These were particularly successful with the main landmarks and thoroughfares of London and Paris. They invited the newcomer to overfly the great metropolis in imagination, to float calmly above its streets and squares, and to link one district with another. Thus journeys could be planned in a new way, and the visitor could achieve a new kind of 'orientation'. They may even have helped people to think about the layout of big cities in a different way, no longer as a series of fixed localities or distinct villages, but as a flowing, dynamic urban environment actively linked by a moving network of cabs, horse-drawn omnibuses and trams; and later by underground railways and motor cars. Indeed, the first section of the London Underground (part of the Metropolitan and City Line) opened as early as 1863.

Panoramic, fold-out maps began to be published in the 1850s and 1860s, forerunners of the famous A to Z guides. One of the most successful was published by Appleyard & Hetling in 1854, 'In a Case for the Pocket', priced one shilling and sixpence. It was comprehensively

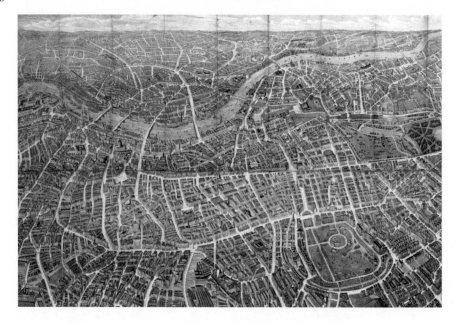

entitled *A Balloon View of London Taken by Daguerreotype Process, Exhibiting Eight Square Miles, Showing all Railway Stations, Public Buildings, Parks, Palaces, Squares, Streets etc ... Forming a Complete Street Guide.*[7] In fact the 'daguerreotype' claim was certainly misleading, as there is no record of a genuine aerial photograph of a city before 1858–59 (Paris and Boston were to be the first). But the combination of balloon and photography evidently had a fashionable, up-to-the-minute appeal.

The 'angel's eye' might also be used to celebrate or commemorate particular events. One of the most memorable of these was the airborne view of the Great Fire of Newcastle, which broke out at one o'clock in the morning of 6 October 1854. Starting with a horrific explosion in a chemical warehouse in Gateshead containing hundreds of tons of sulphur, naphtha, brimstone, and arsenic, the flames leaped across the river Tyne into Newcastle and burned for two days, causing over a million pounds' worth of damage, and terrible loss of life.

An image of this catastrophe was presented by the *Illustrated London News* on 14 October, like an action photograph taken from a balloon. From an imaginary viewpoint some five hundred feet above Gateshead, it gives a startling panorama of houses, bridges, churches, quaysides, ships and factories, looking across the Tyne towards the great railway viaduct running through the centre of Newcastle. The pale autumnal

tone of the print, predominately blue and white, is clearly the wan, aching light of dawn. But the picture is also realistically coloured and animated with leaping flames, wind-torn smoke and tiny fleeing figures, as if it was being observed in real time. (To the modern observer there is an unmistakable resonance with the hurrying, peopled cityscapes of L.S. Lowry.) It achieves a kind of mythic quality, a vision of the industrial city devoured by fire, the icon of a modern secular version of hell. Or perhaps more accurately 'cleansed' by fire, and thereby becoming a kind of purgatory.

The picture was published above a vivid piece of reportage, which itself achieved the extraordinary effect of an all-seeing eye.

> *The streets in the neighbourhood of the explosion presented a most melancholy spectacle. Men, women, and children in their night dresses might be seen rushing from their abodes in search of shelter, they knew not whither. In Gateshead particularly the scene was most distressing – mothers were vainly trying to return for a child, forgotten in the suddenness of escape – and children were searching for their parents. The quay on the Newcastle side of the river was literally strewed with burning staves and rafters, covered with sulphur, and burning like matches. Adults and children, confused by the awful catastrophe, went staggering to and fro as if*

intoxicated, uttering lamentable and piercing cries. At one time the whole
town seemed to be devoted to the rampant agency of fire ... The shop fronts
and windows upon the Quayside, the Sandhill, the Side, and all the neigh-
bouring streets, were almost universally blown out, and the gas lights, for
a square mile around the spot, were extinguished in a moment, adding a
weird and horrible confusion to the scene. The streets rapidly filled with
the entire population of the lower parts of Newcastle, hundreds of them in
their night clothes, and seriously injured. The blood-begrimed counte-
nances of many, and the shrieks, wailing, and lamentations to be heard
on every side, commingling with the voices of others devoutly calling upon
the Lord to have mercy upon them, made up a scene which has been seldom
paralleled.[8]

The fire's impact was so great that a national disaster fund was launched
to relieve the destitute citizens, and the first contribution was made by
Queen Victoria herself. The *Illustrated London News* reported a striking
example of Victorian philanthropy: 'The public sympathy for the numerous poor families, who were rendered destitute by this terrible catastrophe, was displayed in the most marked manner throughout the kingdom. Upwards of £11,000 were subscribed for their relief. No less than
eight hundred families applied for assistance from the funds ...' Money
was also given to institutions like the Newcastle Infirmary and the
Gateshead Dispensary. The image of the burning northern industrial city,
with its displaced citizens wandering the streets like lost souls in purgatory, struck very deep. It was said that Queen Victoria, in an unprece-
dented departure from royal protocol, ordered that the royal train on the
way to Balmoral should halt on the famous High Level Bridge above
Gateshead so she could look down at the devastated city and weep.

Balloons were also used to celebrate colonial cities, and inspire
imperial links, notably in Australia. In 1858 the British balloon the
Australian made some startling flights over Melbourne and Sydney.
There was a late-summer-night ascent in March from Cremorne
Gardens, Melbourne, in which a basketful of local dignitaries sailed
over the Botanical Gardens in bright moonlight, with a magical sight
of the festival fireworks far below. But, attempting to land at Battam's
Swamp, they found themselves in a working-class district, and the
balloon basket was seized by a violent crowd. Amid vocal democratic

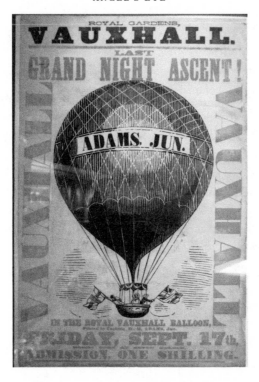

objections to such 'superior' transport, the distinguished guests were forced to escape by jettisoning champagne bottles, picnic hampers, several bags of sand ballast, and finally throwing off a few hardy objectors still clinging to the sides of the basket. Unlike America, ballooning in Australia remained an essentially urban entertainment. There is no record of any practical attempts to explore the Australian interior by balloon at this time. Burke and Wills, starting on their epic journey from Melbourne to the Gulf of Carpentaria in August 1860, stuck firmly and fatally to the ground.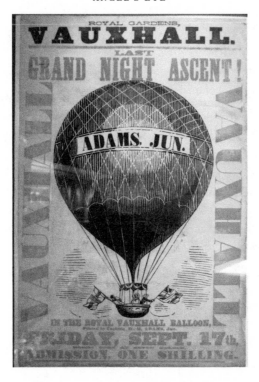

Vauxhall Gardens finally closed after its 'Last Night for Ever' on 26 July 1859. Many reasons were given for this. The proprietors blamed the magistrates who continually banned their most popular attractions as

---

However, hot-air ballooning was firmly established as an Australian sport by the late twentieth century. In 2008 I made a flight over the capital city, Canberra, crossing low over Lake Burley Griffin at dawn, and attempting to land on the trim lawns of the National Parliament building, until waved away by a genial security officer who threatened to give us a parking ticket.

either too dangerous, or too disruptive to the newly respectable neigh-
bourhood of Kennington. Ballooning and fireworks displays were particularly blamed for this. But other factors certainly played a part: the
gardens had become run-down and tawdry, and were considered old-
fashioned; the railway, which ran past the main entrance, had made
travel further afield much easier and cheaper; seaside towns, with their
Vauxhall-like piers, were becoming fashionable; and, finally, the site itself
was too valuable as property, and the blandishments of developers eventually persuaded the proprietors to sell up.[9]

At about this time Charles Green, after more than five hundred
successful ascents and now in his seventies, also went into retirement.
He purchased an elegant little house on a hillside above the Holloway
Road, North London, and named it 'Aerial Villa'. But he kept a weather
eye on the horizon.

5

Wild West Wind

———

1

For American balloons the horizon was just opening up. From the 1840s, long before the establishment of the Union Pacific railroad in 1869, a generation of small-time fairground aeronauts and showmen had begun to dream of achieving the ultimate airborne feat and publicity coup. It was of course the big one, the epic: a single non-stop balloon flight three thousand miles right across America.

American balloonists, unlike their British counterparts, had a vision of their nation's untamed nature, the wilderness and vastness – the endless great prairies, forests and lakes. Their long and daring attempts at trans-America flights were always made from west to east because of the prevailing winds. They were also haunted by the idea of crossing the ultimate wilderness, the three and a half thousand miles of the Atlantic Ocean.

This was at a time when most long-distance transport in America was still by horse, wagon or stagecoach, or else by boat slowly along one of the great rivers like the Ohio or the Mississippi. Railroad-building had only begun in 1830, with the Baltimore and Ohio Line, and by 1840 there was still only about 2,500 miles of track in the whole country, almost all of it confined to local lines on the eastern seaboard, between Charleston and Boston. The great cities of the mid-west, like Chicago and Cincinnati, were served primarily by paddle steamers or Wells Fargo stagecoaches until the 1850s, and serious railroad-building westwards did not begin until after the passing of the Railroad Act of 1862.

When Charles Dickens went to America in 1842, although he was rapturously received, hospitably cared for and most generously financed,

his five-month 'national tour' remained largely along the east coast, visit-
ing Boston, New York, and Washington. He got as far north as the
Niagara Falls, no further west than Kentucky, and no further south than
Missouri and a thoroughly unpleasant ride in a steam paddleboat down
the Mississippi. Subsequently he complained about most of it in his
American Notes (1842).

No one was sure where a trans-American balloon flight should start
from, or in which direction it should go. But it had huge symbolic power
as an idea. There was no real equivalent challenge in Europe. Such a
flight would celebrate the land as one vast, rolling entity. It would in
a sense both discover America, and knit it together. It would also be a
potential money-spinner.

No aeronaut was crazy enough to suggest starting in California, still
part of Mexico and not yet part of the Union. The thought of attempting
to fly across the Rockies was simply suicidal. A balloon was rumoured to
have crossed the Alleghenies in West Virginia, but large parts of the mid-
western states were still settler and cowboy country, with the barest township amenities. In practice any launch site required at least three things:
a local source of coal gas or the means to produce hydrogen gas; a local
newspaper that could whip up interest and funding; and a telegraph link
which could carry the news and generate publicity. It would also help to
have a populace who were wealthy enough and educated enough (or at
least sufficiently gullible) to subscribe good money.

2

The experience of the great professional French aeronaut Eugène Godard,
who began his first American tour in 1854, suggests the possibilities of
American ballooning up to that date. Godard, then aged twenty-seven,
and already a regular star of the Paris Hippodrome, arrived in New York
with a complete and glittering aeronautical roadshow. His suite of five
balloons included his flagship, an impressive 106,000-cubic-foot aerostat,
gloriously decorated and diplomatically named the *Transatlantic*. He
hired a publicity agent to advertise them as 'The Most Beautiful Balloons
in America'. He himself was billed as 'Member of the Parisian Academy
of Arts and Sciences', and 'Chief Aeronaut to the Austrian Army', both
largely invented designations.

His crew consisted of his fearless wife, and a small team doubling up as aerial trapeze artists, spangled female acrobats, and daredevil parachutists. The aerial horseriding, particularly relished on the American plains, was performed with appropriate sang-froid by Madame Godard, while the comic rope or 'lasso' act (with breathtaking slips and catches) was executed with Gallic style by Monsieur Godard himself. Their first tour went as far south as New Orleans, then up the Mississippi Valley, via St Louis and on to Cincinnati, then known as 'the Metropolis of the West', though really the capital of the mid-west. Here, significantly, the *Cincinnati Gazette* appointed Mr J.C. Bellman as its first official 'Balloon Editor'. Godard found he could earn good money not merely by charging a twenty-five-cent admission fee to his launches, but from taking on board paying passengers, who could enjoy the heady novelty of wining and dining at several thousand feet.[1]

The newspaper link was the crucial one. Balloons supplied wonderful copy, combining opportunities for lyrical descriptive writing with dramatic incidents and the satisfactory suspense of near-disasters. In fact, a complete disaster was the best copy of all. Bellman accompanied Godard on several of his showpiece ascents, including one long-distance flight to Hamilton, Ohio, which produced a memorable article with much emphasis laid on 'alfresco repasts of cold duck and turkey' at fifteen thousand feet, and the tossing overboard of empty champagne bottles. Actually Godard refilled them with water, so Bellman could time their explosive impact after a fall of exactly three minutes twenty-five seconds, a piece of 'science' that somehow fascinated his readers.[2]

On a later flight they encountered a prairie storm, and crash-landed in a tree near Caesar's Creek, Waynesville, fifty miles from Cincinnati. One of the paying passengers broke three ribs, and Bellman was badly cut and bruised, but this produced even better journalistic copy. Later Godard's family show went north to Boston (where he earned $3,000 for a single ascent); west again to Columbus, Ohio; and south to New Orleans. He was even rumoured to have made an ascent from Cuba.

But the cost of replacing broken equipment was high, and by the end of 1857 Godard was virtually penniless, and considering joining the New Orleans Minstrel Show. However, before he finally left America in 1858, he took part in a widely advertised balloon derby against an American balloon, the *Leviathan*, piloted by 'Professor' John Steiner. Forty thousand people paid to attend the launch. A thrilling collision between the two

balloons occurred at fifteen thousand feet above Cincinnati, a kind of
aerial joust with both pilots behaving with chivalric gallantry. The
balloons somehow survived, and flew on for over two hundred miles
beyond Dayton, Ohio. Eventually (and perhaps tactfully) Godard lost the
race, but he had recovered his reputation and largely recouped his
fortunes.[3] Back in Paris by 1859, with the glamour of his American tour
to add lustre to his name, he had soon established the most celebrated
balloon-family dynasty in France, sharing his legendary status with his
younger siblings Eugénie, Auguste, Jules and Louis.

As the popularity of aerial shows, parachute stunts and balloon races
(not to mention the spangled French-style artistes) spread throughout
the mid-west in the 1850s, the unique American challenge of the truly
long-distance flight also began to emerge.[4] Thanks in part to Godard,
Cincinnati was now established as an ideal jumping-off point for such
attempts. Its geographical position seemed ideal. A west wind blowing
out of the prairies of Kansas or Iowa would carry a balloon virtually due
east to Washington, a distance of four hundred miles. Admittedly the
Allegheny Mountains lay in between, and the Atlantic seaboard beyond.
Equally, if the eastward wind trajectory turned north, it would swing a
balloon in an ever increasing arc towards New York (560 miles), then
Buffalo on the Great Lakes (six hundred miles), or even to Montreal,
Canada (980 miles). On the other hand, if the wind turned southwards,
the arc would swing towards Richmond, Virginia (five hundred miles),
then Charleston, South Carolina (630 miles), and eventually Florida (810
miles).

3

From the shifting population of local American balloonists and barnstormers, three men were to make their mark by the late 1850s in a way
that would soon make the Godard-style circus look flashy and old-fashioned. They were a different breed from such itinerant showmen: men of
book-learning, business and scientific aspirations. They could lecture and
write, as well as fly. They often adopted the courtesy title of 'Professor',
and wore bow ties even when in a balloon basket.

All three also had names that looked suitably memorable on a publicity poster: John Wise, John LaMountain and Professor Thaddeus S.C.

Lowe. These became the great triumvirate of the Golden Age of American ballooning; and naturally, they became the most celebrated rivals too. Their inspiration was the long-distance European flights of Jean-Pierre Blanchard and Charles Green. But they were even more haunted by the entirely fictitious Atlantic crossing of Poe's 'great balloon hoax'.

Born in Lancaster, Pennsylvania, in 1808, John Wise, as his name usefully suggested, was the oldest and most experienced of the three. In fact he had invented the name. He was from German immigrant stock, originally Johan Weiss. His father was a musical instrument-maker, and he himself was brought up with Lutheran sobriety – encouraged to study music and mathematics, to be bilingual in German and English, and to work hard. An early desire to study theology was transformed into a fascination with the visible cosmos: 'I would spend hours in the night lying upon a straw-heap looking at the stars and moon, and the arrival of a comet gave me rapturous joy. It was this kind of natural bent that first led me to indulge in aerial projects.' This began, as it had with Franklin, with kites, then tissue-paper parachutes, then small Montgolfier fire balloons.[5]

He was apprenticed for five years to a cabinetmaker in Philadelphia, and began to specialise in the delicate craft of piano-making. In his spare time he read scientific journals, studied 'pneumatics and hydrostatics', and continued to dream of flying.[6]

Philadelphia still gloried in the name of Franklin, and still proudly remembered the symbolic flight of Jean-Pierre Blanchard, from the yard of the Walnut Street prison, in 1794. Supported by his father, Wise made his first six ascents in a series of small home-made muslin balloons in 1835–36. Then a silk balloon, unhappily called the *Meteor*, exploded while deflating, throwing him ten feet in the air, severely burning his hands and face and blinding him for several days. It also set fire to the clothes of several bystanders, though strangely there were no legal ramifications, possibly because the flight had been funded by public subscription. Wise was soon back in the air, with a balloon more cautiously named the *Experiment*. By 1837 he had reached a useful deal with the Philadelphia Gas Works Company, and was learning to cultivate the local press, one of the most influential on the east coast. By his thirties he was an acknowledged figure in the town, and had made launches from many of its squares and parks.[7]

He had a gift for evangelising on the subject of ballooning, and giving good, quotable interviews. 'Ballooning is about half a century *ahead* of

the age,' he would announce. Balloons would soon make the much-vaunted railroads and steamships look old-fashioned, uncomfortable and above all slow. 'Our children will travel to any part of the globe, without the inconvenience of smoke, sparks and seasickness, – *and at the rate of one hundred miles per hour.*'[8]

Wise was studying the infant science of meteorology, and making technical innovations too. After several rough landings in which he was dragged across fields and through hedges, unable to deflate his balloon swiftly enough, he came up with the idea for a 'rip panel'. This was a strip section of the balloon gore, sewn separately into the top of the envelope, which could be instantly torn away by pulling a red-painted 'rip-cord', thus rapidly releasing the hydrogen and deflating the balloon in seconds. Many friends thought this 'safety' device was in fact suicidally risky, open to all kinds of technical failure and human error. But Wise first used it successfully on 27 April 1839, and it was soon universally adopted, the first serious balloon invention since Charles Green's trail rope.[9]

In the mid-1840s, John Wise's exploits in his silk balloon the *Hercules* were the subject of a full-page illustrated article in the *Philadelphia Inquirer*. Ballooning was hailed as a serious act of scientific demonstration, as well as a respectable entertainment. The article praised Mr Wise's 'orderly' launch from the city centre, accompanied by three passengers, one of whom was his wife. This was observed by an enthusiastic but well-behaved crowd, a large proportion of whom were 'females accompanied by children'. There was no drunkenness, and no riot. The launch was saluted not by guns, but by a brass band. The article concluded in an exalted and patriotic manner, containing a witty reference to Milton's *Paradise Lost*:

> Our Philadelphia friends have generally paid much attention to the subject of aerial navigation, and the Allegheny Mountains were crossed in this manner as early as the summer of 1837. It is said that Lucifer himself is 'the Prince of the Air', but we shall not be at all surprised to see his dominions invaded by some enterprising Yankee in a profitable style of travel ... Dr Franklin's paper kite led to the discovery of some very important first principles of science which have since benefited the whole world. Therefore we say to our scientific ballooning friends – Go on and prosper! Or let them take the motto of New York, and cry out – 'Excelsior!'. Our humble endeavour will be to aid in the publicity or illustration of all such flights of true genius.

The article was run alongside a literary essay by William Hazlitt on the same page. Ballooning was becoming a proper part of American culture.[10]

John Wise's long-term business plan was to establish the first pan-American aerial service, carrying people and mail back and forth across the continent, and eventually across the Atlantic to Europe.[11] But he also had a visionary, almost religious belief in ballooning itself, in its existential value. Ballooning was good for the body, but also for the soul. Its advantages were both physical and metaphysical. As Wise wrote during one of his high flights across the grand prairies around Lafayette, Indiana: 'I feel rejoiced – invigorated – extremely happy! God is all around me – *Astra Castra, Numen Lumen* [the Stars my camp, the Deity my lamp]. The manifestations around me make me rejoice in God's handiwork. Glorious reverie! ... With me it never fails to produce *exhilaration* ... The mind is illuminated.'[12]

But this was not all. The glorious reverie also brought measurable physiological improvements. The upper air was hygienic, tonic: ballooning was a kind of aerial health cruise. In fact, ballooning produced a high: 'The blood begins to course more freely when up a mile or two with a balloon – the gastric juices pour into the stomach more rapidly – the liver, the kidneys, and heart work with expanded action in a highly calorified atmosphere – the brain receives and gives more exalted inspirations – the

whole animal and mental system becomes intensely quickened ...' A two-
hour balloon trip 'on a fine summer's day' was worth more than an entire
fortnight's sea cruise across the Atlantic 'from New York to Madeira'.[13]

4

John Wise believed that there was a permanent west–east air current
blowing right across the entire American continent. It had perhaps been
ordained by God. It was destined to open up the whole vast land, and
then the Atlantic Ocean itself, to balloons. Above all it offered the possibility of high speeds over enormous distances, a truly American-style
revolution in communications. In 1850 he published A *System of
Aeronautics* to present this case. In 1853 he petitioned Congress for public
funding, but was turned down. In spring 1859 he became involved with a
private scheme to construct a specially equipped balloon to make a series
of pioneering long-distance west–east flights. It was ambitious in size –
120,000 cubic feet, standing 120 feet high – and being constructed of the
finest Chinese silk, it cost the very considerable sum of $30,000. Wise
intended to finance this as a conventional business venture, with capital
from private investors. Together with a number of enthusiasts, he formed
the Trans-Atlantic Balloon Corporation.[14]

Wise decided on a start point three hundred miles west of Cincinnati, at the booming frontier city of St Louis, Missouri. His declared aim was epic: to fly due eastwards from St Louis to New York, a distance of nearly a thousand miles. If necessary he would skirt the Great Lakes, along the Canadian border. No European ballooning could remotely match this ambition. The voyage was intended to demonstrate that a west–east crossing of the three thousand miles of the Atlantic Ocean was also perfectly feasible. Accordingly, the Balloon Corporation hung a fully rigged sixteen-foot lifeboat beneath the basket, and christened their balloon the *Atlantic*.

At this time the competing idea of a railroad network right across America also seemed a viable, if still visionary, commercial proposition, although the Union Pacific railway would not be completed for another decade. Nevertheless, the rate of railway building had increased exponentially. By 1860 over thirty thousand miles of track had been laid, mostly in the north-eastern states, and the railroad clearly offered a potential national transport system at some time in the near future.[15] The old Victorian rivalry between the iron horse and the silken cloud was being played out once more.

After some negotiations, Wise assembled a three-man crew: his young protégé, the twenty-nine-year-old New York balloonist John LaMountain; his chief business investor, Mr Oliver A. Gager from Vermont; and his favourite journalist on the *St Louis Republican*, William Hyde, whose job was to produce the dazzling publicity.

Hyde immediately launched into a fine enumeration of the creature comforts the crew would enjoy: 'Cold chicken, tongue, potted meats, sandwiches ... champagne, sherry, sparkling catawba, claret, madeira, brandy and port ... a plentiful supply of coats, shawls, blankets, and fur gloves ... a pail of iced lemonade ... bundles of the principal St Louis newspapers ... business cards and perhaps other articles which have escaped me.'[16] He also noted that a large official mailbag from the United States Express Company was stowed away on board. This vital piece of cargo represented Wise's key business idea of establishing the first rapid mail service from the central plains to the eastern seaboard.

Balloon mail was a real possibility in America at this date, in a way it had never been in Europe. Communications between an ever-expanding western frontier and a relatively stable, prosperous eastern seaboard, were vital in both business and human terms, but still badly served. The

Transcontinental Telegraph was not established until after the Civil War, in 1871. Wells Fargo, whose overland mail service only began in 1857, was soon in uncertain competition with Pony Express and other rival mail-coach services. The vast distances that had to be crossed, and the unreliable state of the roads and railways, made air transport a genuine alternative on the eastern or 'homebound' routes. Wise hoped to establish his Balloon Corporation as a business monopoly in air mail, which could pay for his pioneer flight many times over, and make a famous return for his investors. A family letter, a *billet-doux* or a business document from St Louis might reach New York or Washington by balloon in twenty-four hours, not four or five days.[17]

They began their epic American trip at 6.45 on the midsummer evening of 1 July 1859. The 'impetuous state' of the inflated balloon did not allow them to wait until the symbolic 4 July for their launch. The west wind would take them on a curving north-easterly trajectory of hundreds of miles, across many different kinds of landscape – 'woods, roads, prairies, farms, railroads, streams, towns' – over Missouri, Illinois, Indiana, Ohio, the Great Lakes, and into upper New York State, to the very edge of the Canadian border.[18] Both of the passengers, Gager and Hyde, and the two aeronauts, Wise and LaMountain, would write their own highly contentious accounts of the journey. These were dashed off as competing articles for various St Louis and New York newspapers, and rarely agreed with each other.

No manuscripts, or actual balloon journals, have survived from the journey. The earliest accounts of it exist in the form of newspaper articles, evidently much edited and expanded after the event. So even these 'first-hand' witness accounts conform to the literary and newspaper conventions of the time. *Frank Leslie's Illustrated Paper* claimed to publish extracts from 'Mr LaMountain's Balloon Journal', though this was probably merely an attempt to claim authenticity.[19] William Hyde, being a professional journalist, would certainly have kept a reporter's notebook. This is also suggested by the stylistic difference between the earlier and later parts of his story, which change from deadpan reportage to melodramatics worthy of Poe. Wise himself drily observed Hyde 'sitting in the boat with pencil and paper in hand, but whether to make notes of his voyage or write his last will and testament I could not tell'.[20]

Wise himself apparently kept a balloon logbook, which he used in a series of letters to the *New York Tribune*.[21] But his magnificent, hyperbolic travelogue only appeared fourteen years later, as part of his book *Through the Air* (1873). ⊕ It is a work of conscious art, combining precise technical details with florid passages of aerial description and – most remarkably of all – memorable snatches of dialogue, often shouted from the basket above to the boat below. Running to fifty pages, it forms one of the most memorable accounts of an American balloon voyage of the entire nineteenth century.

Wise opens his account of the journey with a wide-screen panorama, which captures the paradoxical sense of the whole vast American continent on the move below them, while the balloon hangs steady and detached in the dusk. The mighty land slips rapidly westwards into the setting sun. It was a new version of American pastoral: the stretching, unrolling vision of a rich, powerful, various land.

> In a few minutes after we started we were crossing the great father of American waters – the Mississippi. For many miles up and down we scanned its tortuous course of turbid water. Its tributaries – the Missouri and the Illinois – added interest to the magnificent view ... The city of St Louis, covering a large area of territory, appeared to be gradually contracting in its circumferential lines, and finally hid itself under a dark mantle of smoke ... The fruitful fields of Illinois were now passing rapidly underneath us, seemingly bound for a more western empire ... The plantations and farm-houses appeared to be travelling at the rate of 50 miles per hour ... The feeble shimmer of the new moon was now mantling the earth beneath in a mellow light, and the western horizon was painted gold and purple.[22]

⊕ The epic journey appears as Chapters XLI and XLII, 'St Louis, Missouri to New York State', with several dramatic engravings, including the first aerial view of Niagara Falls. Something of the evangelical flavour of Wise's account can be gleaned from the full title and epigraph of his book: *Through the Air: A Narrative of Forty Years' Experience as an Aeronaut. A History of the various Attempts in the Art of flying by artificial means from the earliest Period down to the Present Time. With an Account of the Author's Most Important Air-Voyages and Many thrilling Adventures and Hairbreadth Escapes.* 'Stand still, and consider the wondrous Works of God. Dost thou know when God disposed them, and caused the light of his Cloud to shine? Dost thou know the balancing of the Clouds, the wondrous works of Him which is perfect in knowledge?' (Job, xxxvii) There was always something of the Lutheran preacher about John Wise, with the balloon basket as his airborne pulpit.

By nightfall they had moved beyond the Mississippi, and were headed into Indiana. They rose to eighteen thousand feet, where it was starlit, beautiful and icy cold. Wise briefly fell asleep, and dreamed of 'interplanetary balloon voyages'. The air current, as predicted, was steadily eastwards, but the electrical atmosphere presaged a storm, and the balloon canopy was lit up by the strange, unearthly phosphorescence of St Elmo's fire, so bright that Wise could read his pocket watch by it. Somewhat unnerved by this – would the hydrogen explode? – they drank brandy, valved gas, and dropped a little lower. Wise noted that by calling down through the dark and 'talking to the dogs', he could estimate the nature of the countryside beneath. A single bark indicated a landscape of lonely, isolated farms; a chorus indicated villages, townships or railheads. Complete silence spoke eloquently of the great open prairies of America.[23]

By dawn the next day they were headed over Fort Wayne, Indiana, and towards Lake Erie. At 10 a.m. they were flying lower, and beginning to turn in a more northerly direction, so the intended target of New York City had to be recalibrated as Boston. Approaching Lake Erie, the land was more populated. As more and more people came out from their houses, waving and calling up to them, Wise realised that he was witnessing a historic phenomenon. *The balloon was expected.* News of their flight had preceded them by telegraph from St Louis. The telegraph lines, following the new railroad network spreading westwards, had anticipated him. The people of Buffalo, Black Rock and Rochester already knew he was coming. In fact, though he did not appreciate it at the time, telegraph combined with railroad had already outdistanced the balloon as a system of communication.[24] Nevertheless, the balloon was about to do something extraordinary: overfly the Niagara Falls at a height of about two miles.

A striking feature appears in the different impressions the four balloonists had of this historic crossing. The enormous cascade of Niagara, situated between Lake Erie and Lake Ontario, drops 165 feet and delivers over four million cubic feet of water every minute. Already established as one of the crowning showpieces of the American wilderness, it was visited by thousands of sightseers, honeymoon couples and celebrities every year. For most of them, it was an image of hypnotic primal power, divine or otherwise. Harriet Beecher Stowe, 'maddened' by its beauty, wanted to throw herself over it in 1834. Charles Dickens, on that first American tour of 1842, gazing in wonder at the 'vague immensity', exclaimed, 'It would be hard for a man to stand nearer God than he does there!'[25] Thomas Carlyle would use the mighty, unstoppable cascade of 'bright green waters' in his essay 'Shooting Niagara – and After?' (1867) as an immediately recognisable metaphor for cataclysmic social change, the 'Niagara leap of completed Democracy'. So Niagara was already a powerful symbolic presence in the British as well as the American psyche. For John Wise and his crew, flying over it, the first human beings ever to do so, it should have been a moment of national epiphany.

Yet William Hyde, for one, remained unexpectedly cool in his impressions:

> We had reached a height of more than a mile, the barometer marking 23.6 inches. At 12 o'clock we were nearly between the Falls and Buffalo ... The famous Falls were quite insignificant, seen from our altitude. There was, to us, a descent of about two feet, and the water seemed to be perfectly motionless. The spray gave the whole appearance as of ice, and there was nothing grand or sublime about it. Passing the western terminus of the Erie Canal, the balloon was borne on directly towards Lake Ontario ...[26]

This laconic description demonstrates the effect of 'miniaturisation' and map-like flattening which was by now well known to European balloonists. But it seems strange for an American journalist, on his first ever flight in a balloon, to have found the mighty Falls so prosaic, so small, so lacking in the anticipated 'sublimity'. In short, so completely un-newsworthy.

Yet perhaps this was partly a stylistic pose. The balloon's altitude allowed all three of Wise's fellow travellers to adopt a superior attitude, as he wryly remarked: 'Niagara Falls was deemed rather a tame sight by my companions ... and a bottle of Heidsieck champagne that was uncorked in honour of the world-renowned cataract made more of a commotion and a livelier spray than did the Falls, to all appearances. Mr. Gager observed that it was "no great shakes, after all". Mr Hyde thought it looked "frozen up". Mr. LaMountain said it would do for a "clever little mill-dam" – convenient water power.'[27]

By contrast, Wise's own description was enthusiastic, a characteristic mixture of the pedagogic and the pious. His companions needed to 'listen and observe' more closely. Niagara demonstrated how Nature had been cutting through the rocks over great aeons of geological time. The huge clouds of spray, though apparently 'frozen' to the viewer, were in

reality combining with the atmosphere to form their own distinctive and dynamic weather system. They also produced several 'beautiful miniature rainbows', of miraculous and 'faery-like' appearance. Everything showed Nature's extraordinary mixture of power, delicacy, order and design. Nature was sublime – and God a sublime Engineer.

> *Do you see what a wonderful cloud manufactory this Niagara is? Cloud upon cloud is rising up from its evaporized water. See how orderly they take up their line of march eastwards, as they rise up, perhaps to carry their treasured moisture to some distant parching land. It is a sublime spectacle this – a laboratory of Nature – an irrigating engine. Nothing is formed in vain.*
>
> *And now listen to its music. It is not a roaring, thundering, dashing, tumultuous sound, but a music of sweetest cadence. Like an Aeolian harp it sends up its vibrations. If it is not the music of the spheres, it is at least the rhythmic language of Motion, wherein we perceive that noble proverb illustrated that 'Order is Heaven's first Law'.*[28]

Yet Nature's power, embodied in the west wind, was soon to show itself in quite a different manner to the balloonists. Any complacency was about to be violently swept away. By midday Wise realised that they were 'riding the advance wave of a coming great storm', a summer hurricane boiling up behind them from the prairies of the south-west. As yet his three inexperienced companions had no idea of this, 'nor had they the slightest conjecture of what was revolving in my mind'. Wise could already see the thickening, 'milk-like' atmosphere above, and the signs of the gathering wind in the shaking trees, the bending fences and the whitened water far below them. Of course the *Atlantic* remained perfectly steady in the air, but Wise calculated that they were now travelling at ninety miles an hour over the ground, a truly terrifying speed.[29]

Ironically, the others were now enjoying themselves immensely, 'engaged in cheerful conversation' as they hung unconcerned over the basket, enchanted by the animated vision of America that rose up gleaming beneath them.

> *The grand panorama of the two Great Lakes, with innumerable cities and towns in full view, the railroads and canals on which were trailing sundry snake-like lines of moving trains, with all the concomitants of a thickly-populated district, and silvery lines of tortuous watercourses, interspersed with golden patches of grain-fields, garnished with the music of steam-whistles, ringing of bells, firing of guns, shouting from a thousand throats; indeed, the country below had become thoroughly alive for many miles to the flitting of our airship among the clouds.*[30]

It was only when Gager noticed Wise's unaccustomed silence – 'Professor, what in the world makes you look so contemplative?' – that the reality of their position began to dawn on the others. William Hyde's cool, journalistic style alters noticeably from this moment on.

> *A terrible storm was raging beneath us, the trees waving and the mad waves dashing against the shore of Erie in an awfully tempestuous manner. But above the careering whirlpools and thundering breakers swam the proud Atlantic, not a cord displaced, not a breadth of silk disturbed, soaring aloft with her expectant crew and gaily heading for the salt crests which bound our vast Republic.*[31]

Up till now, there had always been the possibility of landing at any point they chose. Gager had suggested that Wise drop off himself, Hyde and the mailbag at Rochester, where they could pick up the railroad for New York.[32] Wise agreed to try this, but pointed out that not only was the west wind increasing savagely, but they were being driven northwards,

towards a second vast stretch of water. Lake Ontario is not the biggest of
the Great Lakes, but it is huge all the same, covering an area of over seven
thousand square miles, and stretching some two hundred miles. Being
deep (in places nearly a thousand feet, compared to Lake Erie's more
modest two hundred feet), it is subject to massive building waves and
vicious storms. Wise was desperate not to take his crew over it in these
'tornado' conditions. 'Crossing the second lake', he shouted to Hyde,
would be 'sheer recklessness and hardihood'. He sent all three crew
members down into the lifeboat, ready to leap out immediately with the
mail, if he could get a grappling iron to hold for a moment.

But despite valving much gas, Wise failed to achieve even the briefest
touchdown on the wind-blasted foreshore. Caught in the now screaming
north-easterly air current, and flying far too low, they were swept towards
a last line of trees before Lake Ontario. 'LaMountain cried up to me at
this moment, with emphatic voice, "Professor what's to be done?" –
"Throw everything overboard you can lay your hands on," was my reply,
"or we shall be torn to pieces if we strike the ground."'[33]

With no ballast left, LaMountain threw out the lifeboat's brass
'propeller gearing', thus effectively disabling it. At last the *Atlantic* rose,
cleared the trees, and shot out 'like a bullet' over the huge expanse of
seething lead-grey water. 'O! how terrible it was, foaming, moaning, and
howling,' thought Wise to himself. In minutes all land disappeared from
view. 'I guess we are gone,' remarked Hyde simply. LaMountain, possibly
with the stylistic help of his newspaper editor, subsequently recalled:
'Above, the clouds were black as ink; around the winds were howling as if
alive with demons; and below the water, capped with foam ... swept up in
swells fifteen feet high, that ran in every conceivable direction.'[34]

The balloon had now lost too much gas, and was sinking again. Their
situation was, in Wise's words, 'dismal enough'. The great waters of
America were going to swallow them up. They hit the wave tops several
times, and Wise called everyone back up into the basket. There was a
fierce dispute as to what they should do next. Wise thought they should
take to the lifeboat and cut away the balloon. LaMountain thought the
opposite: they should stay in the balloon, and cut away the lifeboat as so
much useless ballast. It was a terrible dilemma. In these storm conditions it was not clear whether the boat would actually float, or the balloon
would actually fly. Both moves would be irrevocable, and either might
prove fatal.

While Wise hesitated, LaMountain, impetuous and determined to show his mettle, seized a hatchet and clambered back down into the swaying and half-swamped craft below. He began cutting the seats and fittings away, plank by plank, getting rid of as much wooden ballast as possible but leaving the canvas sides so it might still float. The waves continually engulfed him as he worked. At one point Wise thought he saw his head whirling away in the spray, and shouted, 'LaMountain's gone!' Then a muffled voice called back, 'No I ain't. It's only my hat!' He was lying in the bottom of the boat with his arms around the remaining cross-seat.[35]

Gradually they threw out everything that could be spared as ballast, every instrument, every bottle of champagne and brandy, even Wise's own personal valise with all his clothing and a 'treasured silver cigar case presented by a friend'.[36] Lastly they threw out the vital United States Express Company mailbag itself. The *Atlantic* kept hitting the wave tops, and then staggering back into the air. Its crew hung on grimly in the upper basket for two more hours, praying for landfall. Wise and LaMountain stared ahead, keeping a lookout, and occasionally shouting grimly at each other.

There were odd incidents. When Gager tried to open the last bottle of champagne with Wise's special penknife, he was warned sharply by

Wise 'not to ruin' the edge of the blade. Gager beamed with relief. 'Do you expect ever to have use for this knife again?' he asked. 'Certainly I do,' came the reply. 'Then,' said Gager, 'you don't calculate we'll die very soon.'

When they saw a steamer, the *Young America*, butting through the waves below, Wise proposed to ditch in its path and swim for it, but the others refused. So instead, as they shot past, he ordered them to give the vessel three cheers. Coming out on deck, 'the ladies waved their handkerchiefs and the men began to hurrah'. Wise thought: 'Little did they dream we were sailing with death warrants in our hands.'[37]

After two and a half hours, the balloon finally made landfall on the north-eastern shores of Lake Ontario, near the township of Henderson, in the remote forested regions of upper New York State, just short of the Canadian border. It was still travelling at over sixty miles an hour, and Wise assumed they would all die the moment it hit the trees. The iron grapnel snapped immediately he threw it out – 'Leviathan tied to a fish hook' – and the balloon dashed along through the treetops 'like a maddened elephant through a jungle'.

In the event it was the big elm trees that saved them, gradually shredding the balloon envelope as it went along, slowing it down, and finally catching the balloon basket in the fork of a stout branch, while the remaining silk tore itself to pieces above the crew's heads. The four battered occupants, who had not spoken a word to each other during this final ordeal, were amazed to find themselves uninjured, and able to descend fifty feet to the ground with ropes.[38]

Wise calculated that they had covered 'over sixteen degrees of eastern longitude' across America, a formidable achievement. Allowing for the steadily northern curve which the west wind gave to their eastward course, he reckoned they had actually travelled 'about twelve hundred miles' since St Louis, 'the greatest balloon voyage that was ever made'. In fact the official figure was later set as 809 miles in twenty hours and forty minutes, nevertheless 'establishing a world distance record that would stand until September 1910'.[39] But because of the loss of the mail, the flight had no commercial value whatsoever, though it appears that the mailbag did eventually float ashore at Oswego, on the southern side of Lake Ontario.

All the same, Wise chose to end his account on a light-hearted note. A small party of homesteaders had come out from Henderson, and stood round the aeronauts as they attempted to assemble the strewn wreckage of the *Atlantic*. 'An elderly lady with spectacles made the remark that she was *really surprised* and astonished to see so sensible-looking a party as we appeared, to ride in such an *outlandish-looking vehicle*. She anxiously enquired where we came from; and when told from St Louis, she wanted to know how far that was from there, and when informed it was over a thousand miles, she looked very suspiciously over the top of her spectacles, and said: "*That will do now*."'[40]

William Hyde had a scoop that made his career, and John LaMountain instantly established his name among the most daring of the younger aeronauts. But he also soon became known as an irascible troublemaker and publicity-seeker. He wrote articles accusing Wise of incompetence, indecision and cowardice during the flight, and wildly claimed that he alone would 'cross the Atlantic in October'.

Finally, he publicly challenged Wise to a suicidal trans-American balloon race, three thousand miles across the continent, riding the west wind from the Pacific to the Atlantic. 'If Mr Wise considers the matter of sufficient importance to test our relative capacities – *scientifically*

considered – in a trial trip from San Francisco to the Atlantic seaboard,
with balloons of equal size – he knows my address.'[41]

In fact the possibility of this madcap scheme actually coming to fru-
ition cooled overheated heads and partly reconciled the two men.
Amazingly, the wreckage of the *Atlantic* was retrieved, and Oliver A.
Gager, undeterred, paid for LaMountain to construct a new, smaller
balloon from its remains. With this, LaMountain undertook a series of
shorter but equally perilous flights, one of which landed him deep in the
Canadian wilderness, north of Ottawa. It was mainly memorable for the
three weeks it took him to hike back to civilisation on foot. Meanwhile
Wise quietly returned to St Louis, and continued his propaganda for
balloon mail, long-distance flights, and the discovery of spiritual hygiene
in the upper air. But nothing could change the fact that the great dream
of crossing the Atlantic had eluded them all.

5

In many ways the most ambitious of the American aeronautical 'Professors' was Thaddeus Sobieski Coulincourt Lowe (1832–1913). The unlikely extravagance of his name led many people to believed he was a Polish burlesque artist. In fact he was a formidably serious and well-organised man from New Hampshire, and his doting mother had named him after the hero of the Scottish romance novel she had been reading, Jane Porter's *Thaddeus of Warsaw*. Six feet tall, headstrong, unmistakably glamorous and with imposing moustaches, Lowe believed passionately in ballooning as a new science, and to the end of his life he would make characteristically bullish statements about it: 'Every invention, every innovation in the history of the world, has been laughed at. Columbus was renounced as a faker; Morse was called a crank; Franklin a fool; Charles Darwin ridiculed for years. It seems to be the fate of every man or woman who discovers a new fact, to be made the subject of attacks of the most violent nature, without rhyme or reason.'[42]

Like Wise, like LaMountain, Lowe harboured the big dream of flying eastwards across America and then the Atlantic. He said this had come to him like a revelation in childhood: 'I remember as a boy lying on my back under the trees, I had often seen through the leaves overhead, the clouds moving in different directions, the higher ones going East and the lower ones West, and it occurred to me that once in this upper current, I could sail across the Atlantic and land on the continent of Europe.'[43]

In his autobiography, *My Balloons in Peace and War*, Lowe describes his balloon apprenticeship as part of a lifelong vocation to fulfil a uniquely American destiny. At the age of twenty he read John Wise's *A System of Aeronautics*, which confirmed his dream of riding on the continuous high air current, or permanent west wind, blowing eastwards across America, and just waiting to carry a balloon to glory. But he saw this vocation as essentially scientific, rather than semi-religious, as Wise did: 'My fondness for science found expression in many ways. I saved every dollar I made – read all the scientific books I could obtain – courted the society of men who knew something – and tried to store my mind with knowledge that would be of service to me along the line of scientific investigation.'[44] One of the men whose society he courted most assiduously would be a genuine professor, the great American physicist Joseph Henry (1797–1878).

From the first, Lowe had extraordinary single-mindedness and drive. He found work early as technical assistant to an itinerant chemistry lecturer, but within a year was studying and lecturing on the chemistry of gases for himself, using a 'portable laboratory'. Increasingly these lectures included hydrogen-balloon demonstrations. So he was much struck when a pretty nineteen-year-old French actress named Léontine Gaschon appeared in the front row of his lecture audience, and expressed lively interest in the hydrogen balloons flown by her countrymen like Eugène Godard. A week later, on St Valentine's Day 1855, Lowe married Léontine, and they went on to fly many balloons and have ten children together.[45]

Lowe's first recorded flight, in a home-made balloon, took place the year following his marriage, in 1856, when he was twenty-four. Within another two years reports of his ascents were appearing in the newspapers, and he celebrated the laying of the first Atlantic telegraph cable in 1858 with a series of well-publicised flights from Ottawa and Portland, Maine. The idea of the Atlantic cable of course served to sharpen business interest in the idea of transatlantic balloon flight, and Lowe was soon in correspondence with John Wise's Trans-Atlantic Balloon Corporation in St Louis, on a possible collaboration.

It is nonetheless astonishing that by July 1859, the very month of Wise's *Atlantic* flight over Lake Ontario, Lowe had made his own separate bid for national fame, and somehow organised the financing of a rival balloon, the *City of New York*. This was far bigger than Wise's *Atlantic* (a 'mere' 120,000 cubic feet), with a capacity of 725,000 cubic feet, stood two hundred feet high, and carried a dangling three-ton steam-powered life-boat. In autumn 1859 this monster was grandly installed in the grounds of the Crystal Palace, at Forty-Second Street and Sixth Avenue, in the heart of New York, where it attracted huge crowds of sightseers, and wide but sceptical press coverage.

Indeed, the *City of New York* had distinct teething problems, notably the fact that it could not get off the ground. Having been ceremoniously connected to the New York gas mains through an enormous gasometer, the balloon was still less than one-third inflated after fourteen days of constant supply. This effort threatened to black out the drawing-room lamps of the entire city, making the ambitious enterprise increasingly unpopular. It did not help that the fall weather was freezing. Lowe arrived to supervise the inflation himself, wrapped in an enormous fur coat, look-ing and acting like 'a Russian bear', according to press reports. He also renamed his balloon the *Great Western*, in an attempt to twin it with Isambard Kingdom Brunel's iron *Great Eastern*, the largest ship ever constructed, which was shortly due to arrive in New York harbour after its first transatlantic crossing. Neither strategy had the desired effect. The balloon was now leaking slightly faster than it was filling, and the stench of gas filled the Crystal Palace park and drifted across Forty-Second Street. This first attempt to inflate the *Great Western* was finally abandoned, and it never got off the ground in New York.

Undaunted, Lowe took it the following spring to Philadelphia, at the invitation of the Benjamin Franklin Institute and the Philadelphia Gas Works, together with encouragement from the *Philadelphia Inquirer*.[46] These three patrons – the learned institute, the gas company and the newspaper – were representative of the three wealthy American lobbies which still expressed a strong if speculative interest in supporting a transatlantic balloon. The scientific establishment sought new technolo-gies. The commercial world sought new sources of investment. And the press sought a steady supply of new human-interest scoops. There would soon be a fourth interested party: the US Army. What it sought was also something new and suddenly urgent – military intelligence.

But Lowe had overplayed his hand with the *Great Western*. Possibly misled by his wealthy New York backers – he was still only twenty-eight – what he had built was essentially a brilliant publicity machine, rather than a safe flying machine. In June he did achieve a brief, lumbering test flight into New Jersey (with a terrified journalist from the *Inquirer* aboard). But on 8 September 1860, during yet another attempt at inflation, the monster simply exploded. Nothing was left to prove it had ever existed but a few fragments of coloured balloon-cloth blown across the rooftops of Philadelphia (and few mocking cartoons).[47]

But it was now that Thaddeus Lowe demonstrated the inner buoyancy of his own character. Refusing to accept defeat, he shrugged off the disaster and immediately set to work on a quite new approach to his Atlantic balloon scheme, turning away from commerce and back to pure science. He at once wrote to the greatest acknowledged expert on chemistry, electromagnetism and meteorology in America: Professor Joseph Henry, first Director of the Smithsonian Institution, Washington, DC.

Professor Henry received innumerable applications from scientific cranks, but he took Lowe seriously. In a decisive exchange of letters, Henry confirmed the 'meteorological phenomena' of the American West Wind. A high, permanent, 'prevailing easterly air current' had been fully established by ten years of 'continuous observation' at the Smithsonian. He therefore gave his opinion that a balloon of 'sufficient size and impermeability to gas' could indeed travel on this wind across the Atlantic. His one proviso was shrewd and strictly practical: Lowe should undertake what we would now call research and development. He should take much

smaller balloons, fly them eastwards over land (not over the ocean), gaining practical experience 'accumulated by voyages over the interior', and
thus ensure that the whole project should be 'thoroughly tested' before
he finally set out.[48] Professor Henry added one paradoxical but masterly
suggestion: Lowe should only take off on days when the wind at ground
level was demonstrably blowing the wrong way, to the west. Observers
would then see with their own eyes that once he had gained sufficient
altitude, he would invariably find the predicted easterly air current, and
the balloon would triumphantly turn back and head for the distant
Atlantic.

Accordingly in spring 1861, Lowe took a much more modest balloon,
the thirty-three-thousand-cubic-foot Enterprise, to the early home of
experimental ballooning, Cincinnati, Ohio.[49] He was scheduled to take
off in the second week of April. But then the balloon went up in a quite
different way.

6

Spies in the Sky

———

1

On 15 April 1861 the newly elected President Abraham Lincoln declared an 'insurrection' in America's Southern states, and sent out a call for seventy-five thousand troops to volunteer – but only for a brief three-month period.[1] His aim was rapidly to crush what was then seen as a local and ill-organised farmers' 'rebellion' in the deep South, largely in the six cotton-growing slave states – South Carolina, Mississippi, Florida, Alabama, Georgia, Louisiana – which had formed a loose armed Confederacy. The 'rebellion' was initially a scattered and desultory affair, having begun with the storming of Fort Sumpter in Charleston harbour, South Carolina. But the seceding states, joined by Texas, soon founded a new Southern capital in distant, rural Montgomery, Alabama, over a thousand miles from Washington, and brazenly elected a new Congress there.

With a huge preponderance of men, weaponry and industrial materials, the North or Union appeared to have an unassailable advantage, and intended to bring the South to heel with a brief and punishing blockade. However, by the end of May, four more slave states – Tennessee, Arkansas, North Carolina and part of Virginia – had joined the Confederacy. Moreover, several brilliant generals had rallied to its ranks: notably Robert E. Lee and 'Stonewall' Jackson – and the spirit of Southern patriotism, the spirit of 'Dixie', had caught light. With startling confidence the rebels dashingly moved their capital north from Alabama to Richmond, Virginia, a mere ninety-eight miles from Washington, and soon had some two hundred thousand troops in the field.[2] They also repudiated the anti-slavery rhetoric of the North, and began to create the powerful mythology of a graceful, productive, agrarian South, shamefully set upon by a brutal, bullying, mechanised North. In a curious way, balloons would eventually contribute to both sides of this propaganda warfare. From then on the conflict escalated into a full-blown civil war, which would endure for four long and terrible years. By the end, Lincoln would have been forced to raise over a million troops.

The rebels' provocative initiative in occupying Richmond in April 1861 effectively dictated the first eighteen months of the conflict. Other early campaigns were fought in Missouri, the Shenandoah Valley, and distant Tennessee; and the rebel city of New Orleans in Louisiana fell in April 1862.[3] But for the Union the early watchword became 'On to Richmond!' In consequence, most of the early fighting was restricted to the state of Virginia, stretching between the Blue Ridge Mountains and the Atlantic Ocean.

It was an area divided by four great rivers running south-eastwards (like the fingers of a spread hand) into Chesapeake Bay – the Potomac, the Rappahannock, the York and the James. Moving between these, the conflict was highly mobile and violent, yet strangely indecisive. It involved a series of back-and-forth manoeuvrings, alternately to subdue or to defend the rebel stronghold. The campaign opened with the First Battle of Bull Run, just west of Washington, in July 1861, and ended with the Seven Days Battle just north of Richmond in July 1862.[4]

Balloons were present at most of these bloody encounters, and especially in the latter phase, fought along the narrow strip of land between the James and the York rivers, known as the Virginia Peninsula. Balloon operations were restricted, even marginal, but in some cases were crucial in supplying military intelligence. Yet their value was always disputed. Remarkably, they played virtually no part in the war after the Battle of Fredericksburg (on the Rappahannock river) in December 1862.[5] But for the aeronauts, all Atlantic schemes were hastily put aside from April 1861 onwards. Instead, the great rivalry between Lowe, Wise and LaMountain shifted to the struggle to become appointed as the official aeronaut of the Union army. In this endeavour, patriotism, rather than profit, seems genuinely to have motivated them.[6]

Lowe had already experienced the growing ill-feeling between North and South at first hand. On 19 April 1861, just as the drums of war had begun to sound in Washington, he had launched his small, businesslike balloon the *Enterprise* from a vacant lot in the heart of Cincinnati. This

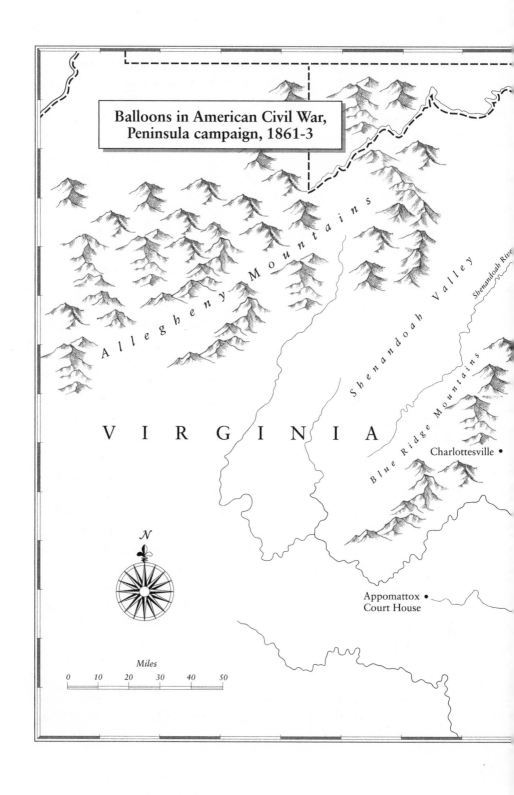

Balloons in American Civil War,
Peninsula campaign, 1861-3

Allegheny Mountains

Shenandoah Valley

Shenandoah River

Blue Ridge Mountains

V I R G I N I A

Charlottesville •

Appomattox •
Court House

𝒩

Miles

0   10   20   30   40   50

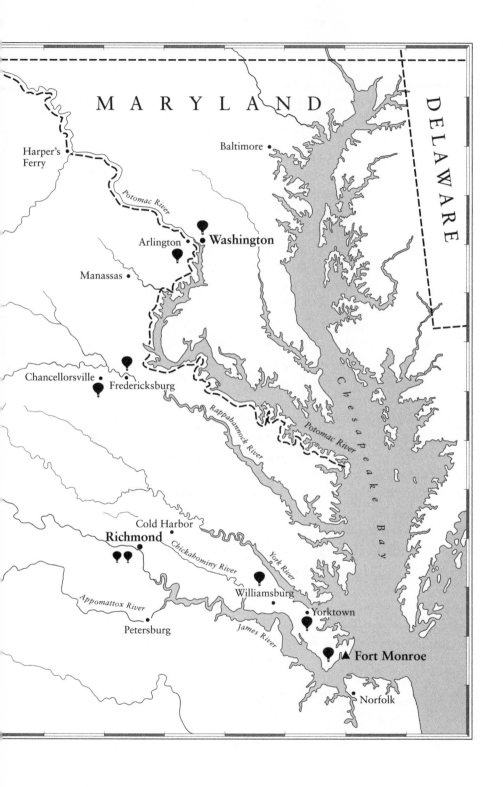

MARYLAND

DELAWARE

Harper's Ferry

Baltimore

Potomac River

Arlington   Washington

Manassas

Chancellorsville   Fredericksburg

Rappahannock River

Potomac River

Chesapeake Bay

Cold Harbor

Richmond

Chickahominy River

York River

Appomattox River

Williamsburg

Yorktown

Petersburg

James River

▲ Fort Monroe

Norfolk

flight had originally been conceived, on Professor's Henry advice, as a
scientific investigation of the easterly air current, but at some point it
had become a demonstrative effort to impress the Union government.

Accordingly, Lowe's ambitious plan was to fly five hundred miles due
eastwards over the Allegheny Mountains, and to land in Washington,
ideally perhaps on the front lawn of Lincoln's White House. Here he
might offer his services to the Union cause, and outflank his rival aeronauts. In the event he met a rebel breeze, and ended up much further
south, having skirted Kentucky and Tennessee, and finally touching
down near Unionville in the heart of the seceded state of South Carolina.
Nevertheless, he had flown 650 miles solo in just over nine hours, a feat
that compares well with the Wise–LaMountain flight of 1859.[7]

On landing, Lowe found the Civil War already declared and decidedly
in progress. The local cotton farmers were not impressed by his flying
skills or his Yankee accent. On the contrary, he was arrested as a spy for
supposedly carrying despatches from the Union North, blatantly piled in
the corner of his balloon basket. With some difficulty, due to local illit-
eracy, he was able to demonstrate that these despatches were actually a
special balloon edition of the *Cincinnati Daily Commerce*, and thereby
escape being lynched.[8]

The *Enterprise* having been extricated undamaged, Lowe and his
balloon were packed off unceremoniously on a coach back westwards.
Once safe in Kentucky, which had not seceded to the South, he switched
to the railroad and hurried north to Washington, with his balloon and
basket in packing cases. Here the news was that the Union Army of the
Potomac was preparing to invade rebel Virginia, and was already skir-
mishing across the Potomac river near Arlington. Its new commander,
General George C. McClellan, like a good Yankee, was in principle sympathetic to advanced technology. Lowe consulted with Professor Henry at
the Smithsonian, and came up with a revolutionary new idea. Provided
it was securely tethered, the *Enterprise* could carry up telegraph equipment and a wire, and send direct aerial observations to a commander on
the ground. He demanded to demonstrate this to President Lincoln
himself as a matter of acute urgency.

It says a great deal about Lowe that, amidst all the administrative
chaos of a newly declared war, he achieved exactly this. On Sunday, 16
June 1861, Lowe ascended in his balloon some five hundred feet above
Constitution Mall, Washington, with a telegraph key and an enthusiastic

Morse operator. The telegraph wire was strapped to the tether line and winch, and then run directly across the lawn and into a service room in the White House. Lowe transmitted the following message:

> Balloon Enterprise. Washington, D.C. 16 June 1861
>> To President United States:
>> This point of observation commands an area nearly fifty miles in diameter. The city with its girdle of encampments presents a superb scene. I have pleasure in sending you this first dispatch ever telegraphed from an aerial station and in acknowledging indebtedness to your encouragement for the opportunity of demonstrating the availability of the science of aeronautics in the service of the country. T.S.C. Lowe[9]

This historic message was seen by Lincoln himself, who called Lowe in for discussions that same evening. On 21 June Professor Henry, having observed the demonstration from the roof of the red-brick Smithsonian building on the other side of the Mall, wrote a decisive letter of support to Lincoln's Secretary for War, Simon Cameron: 'From experiments made here by Professor Lowe, for the first time in history, it is conclusively proved that telegrams can be sent with ease and certainty between the balloon and the headquarters of the Commanding Officer.'[10]

In this dramatic fashion Lowe succeeded in persuading Lincoln to allow him to form an official Military Aeronautics Corps within the Union army. But administrative things moved slowly. A month later, on 21 July, Lincoln had to write a second order in his own hand urging that 'Professor Lowe and his balloon' should receive the fullest cooperation from General Winfield Scott, the slow-moving C-in-C of the Union forces.[11] In August the Aeronautics Corps was finally placed under the direct control of General George McClellan, the much younger and more dynamic commander of the Army of the Potomac, who was soon to command the entire Union army in the Virginia theatre. McClellan quickly became one of Lowe's warmest supporters, and would make his first balloon ascent with Lowe on 7 September 1861. He believed that balloon observation would be vital to the new, highly mobile form of infantry warfare.

Lowe had finally received Union funds to build further balloons in August 1861. His fleet eventually consisted of no fewer than eight military aerostats: the Union, the Intrepid, the Constitution, the United States, the Washington, the Eagle, the Excelsior, and the original Enterprise. The new balloons were small and functional, ranging in size from fifteen thousand to thirty-two thousand cubic feet, though each could carry enough tether and telegraph cable to climb to five thousand feet.

To save weight, their observation baskets were unbelievably small, not much bigger than modern tea-chests, approximately two feet square and two feet deep. For a man sitting, the sides would barely reach to his armpits; for a man standing, they would not even reach to his knees. By way of reassurance, they were painted on the outside with the Union stars and stripes on a white background; though it was said that this provided a better target for Confederate sharpshooters to take a bead on. Only the fearless Sophie Blanchard had ever flown in baskets as small as this. But then, she had never been fired at while doing so.

The first balloon designed specifically for military use, the *Union*, was ready for action on 28 August 1861. On 24 September, Lowe ascended in it to more than a thousand feet near Arlington, across the Potomac River from Washington, and began telegraphing intelligence on the Confederate troops located at Falls Church, Virginia, more than three miles away. Union guns were then calibrated and fired accurately on

these enemy dispositions without actually being able to see them. This was an ominous first in the history of warfare, by which destruction could be delivered to a distant and invisible enemy.

2

Meanwhile, rival aeronaut John LaMountain was also attempting to provide balloon services for the Union. He too wrote to Lincoln in spring 1861, but having no influential backers like Professor Henry, he did not receive a reply. However, from July 1861, using his heavily repaired and battered balloon *Atlantic* (still financed by the faithful Oliver Gager), he made several reckless demonstration ascents at Fort Monroe, on the strategic southern tip of the Virginia Peninsula. These were at the invitation of the local Union commander, General Benjamin Butler, and provided information on the Confederate troops massed along the James River. Though employed as a civilian, and not officially belonging to the Army of the Potomac, LaMountain could claim to have made the first aerial reconnaissance of the Civil War.[12]

The remarkable thing about these ascents is that many were free
flights, without tether ropes or telegraph cables, and made directly over
enemy lines. LaMountain eloquently explained why he took such risks:

> *Typical ascensions, with balloon attached to the earth by cords, do not
> allow the attainment of an altitude sufficient to expose a considerable
> view ... To the eye of the [free] aeronaut, who can by knowledge of his art
> ... sail directly over points impenetrable by pickets or scouts, secrets of the
> most important character are clearly revealed. The country lies spread
> before him like a well-made map, with all its varieties of hill and valley,
> river and defile, distinctly defined, and every fort, encampment, or rifle-pit
> within range of many miles, manifest to observation.*[13]

The flamboyant style of the aeronautical 'art' employed by LaMountain
was typical of him. Irascible and arrogant he may have been, but no one
could doubt his skill and impetuous courage. On several occasions he
took his balloon at low level right across the Confederate lines at
Hampton on the James River, and later at Alexandria on the Potomac,
sowing dismay and fury among the enemy troops, who shouted and shot
at him, but without effect. The Confederates thought he was doomed to
come down behind their own lines, where he would undoubtedly be
captured and summarily shot as a spy.

But at this point, LaMountain would drop ballast and soar up to
eight thousand feet.[14] Here he entered, as he knew he would (or at least
certainly hoped he would), the top layer of an air 'box'. This is a fixed air
pattern – fixed, at any rate, at certain seasons and times of day – in which
the upper current is exactly reversed in direction from the lower. ℣ So

---

℣ The air 'box' is still used by modern hot-air balloonists. There is, for example, a
famous north–south 'box' above Albuquerque, New Mexico, which is used to stunning
effect during the famous mass ascents at the Albuquerque International Balloon
Fiesta. I have taken off at dawn with three hundred other balloons flying south, and
then an hour later, after modestly landing on a downtown baseball field, looked up to
see fleets of balloons thousands of feet overhead, steadily returning to their birth-
place like some miraculous airborne shoal of glinting salmon returning to their original spawning grounds. The effect of a balloon successfully returning to its launch point
after a flight of several hours, a thing that should be logically impossible, is curiously
moving and heartwarming. If I were American, I would say it was like your favourite
baseball team scoring a home run.

LaMountain sailed mockingly back over the entire Confederate army, and, valving fast, brought the *Atlantic* safely back virtually to its point of departure in the Union rearguard, and delivered his report.

On at least one occasion he was nearly shot by a German brigade of Union troops as he landed, an early example of 'friendly fire'. 'An infuriated crowd of officers and men were intent on destroying the balloon and myself ... One bullet passed rather unpleasantly close to my head,' as he remarked laconically. These flights, both free and tethered, caused a sensation among the opposing armies. The *Scientific American* remarked on LaMountain's reckless courage, and the *New York Times* reported that he had been able to view the whole Confederate encampment right up the east side of the James River, and later all the rebel manoeuvrings on the west side of the Potomac. A new era in aerial observation had begun.[15]

However, LaMountain's spectacular showmanship was not designed to impress the cautious senior officers of the Union army. Ironically, it might have gone down better with inspirational and inventive Confederate generals like Stonewall Jackson and Robert E. Lee. Unlike the well-organised and well-connected Lowe, he had difficulty in obtaining funds for further equipment. He did finally manage to lay his hands on another balloon, the *Saratoga*, but this was almost immediately lost through his own recklessness on 16 November 1861. He then tried to requisition one of Lowe's balloons from the Aeronautics Corps, but Lowe refused to cooperate, describing him bluntly as a man 'known to be unscrupulous and prompted by jealousy'. This antipathy was probably mutual. Each man found supporters in Washington, and the rivalry between the two grew. Finally, after further accusations and hostilities on both sides, in February 1862 General McClellan dismissed 'Professor' LaMountain from any further service to the Union military.[16]

3

For the rest of 1862 it was Lowe's Union Balloon Corps which operated exclusively during the Virginia phase of the Civil War. McClellan's basic strategy was to assault the rebel capital with a pincer movement. His plan was to transport the Army of the Potomac down to Fort Monroe, and then steadily roll back the Confederate forces up the ninety-mile

length of the Virginia Peninsula – through Hampton, Yorktown and
Williamsburg – until Richmond could be encircled from the south.
Meanwhile he would drive other scattered rebel elements back down the
west bank of the Potomac, and approach Richmond from the north.

The small but incredibly fierce battles which now took place between
the York and the James rivers came to be known as the Peninsula
Campaign. Lincoln's hope was that it would result in a short war. But
throughout 1862 McClellan's Army of the Potomac constantly threatened, but did not actually manage to take, the vital and symbolic rebel
stronghold. In the end there was no short war. Richmond did not finally
fall until April 1865, precipitating Lee's historic surrender to Ulysses S.
Grant at the Appomattox Court House (and the great culminating chapter of *Gone with the Wind*).

Lowe's balloons were present at the siege of Yorktown in May 1862; at
the Battle of Fair Oaks in May–June 1862; at the crucial Seven Days Battle
outside Richmond in June–July 1862; and at the bloody Battle of
Fredericksburg in December 1862. He also witnessed the famous rebel
victory by Robert E. Lee at Chancellorsville in May 1863. The last telegraph message Lowe sent from a balloon during the Civil War was from
Chancellorsville. It was time-dated 10.45 a.m. on 5 May 1863, and

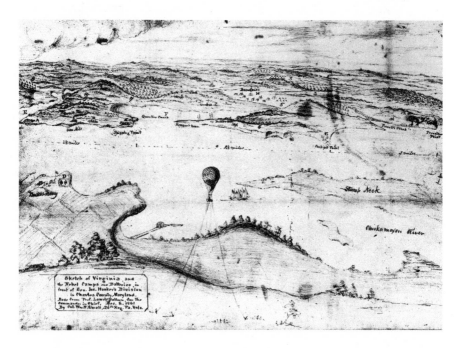

foretold the rebel victory: 'I am unable at this time to see a movement of the enemy except some wagons moving up and down the river. The enemy in force appears to hold all the ground they gained yesterday.'[17]

From his balloons, Lowe witnessed a new kind of American fighting. Rapid, violent, passionate and patriotic (on both sides), it was based on a swift exchange of attack and counter-attack. There were long days of immobility and siege, especially at Yorktown. But most characteristic was the constant manoeuvring of infantry and artillery across open countryside chequered with small townships, manufactories, farmsteads, mills, river bridges and railway junctions, any one of which could suddenly become a strategic key point, where thousands might die. Speed, and often dissimulation, were vital factors. Military intelligence – the knowledge of the enemy's dispositions, troop numbers, firepower, potential reinforcements, and above all its unexpected movements and hidden intentions – was paramount. Lowe always believed that balloons could supply this.

Lowe's observations during these early months also provided the first terrible evidence of the nature of this modern battlefield, and the new weaponry deployed upon it. He saw the long-distance impact of heavy artillery shells from the twelve- and sixty-four-pounder field guns; the effect of mines and the new-style hand grenades; the devastation of exploding canister shells on an infantry advance; and the terrible skittling effect of a volley of Springfield rifles fired at two hundred yards. The Springfield fired a big 0.58-calibre bullet capable of blowing off an entire limb at a thousand yards.[18]

He also saw the cruel deceptions increasingly practised with the new equipment. One of his telegraph officers, having bravely climbed a telegraph pole under fire to fix a broken line, climbed down and stepped directly onto a 'torpedo' or anti-personnel mine that had been buried at the foot of the pole by the retreating Confederates. It blew the man in half.

What Lowe observed was a new kind of infantry war, with great tides of men and metal constantly clashing. It produced casualty figures never before seen in American history. At Bull Run (July 1861) five thousand dead and wounded; at Fredericksburg (December 1862) seventeen thousand dead and wounded; at Chancellorsville (May 1863) thirty thousand dead and wounded; at Gettysburg (July 1863) nearly fifty thousand dead and wounded.

Yet Lowe rarely described the human details of what he saw.🎈
Instead, he confined himself to tactical reporting, like someone observing the moves in a vast, impersonal game of chess. But the unmistakable sounds of war came up to him – the boom of shells, the rattle of shots, the screaming of wounded. He wrote: 'It was one of the greatest strains upon my nerves that I ever have experienced, to observe for many hours a fierce battle.'[20]

Lowe was always active and adaptable. His balloons were brought into action from horse-drawn carts, from railroad wagons, and even from the decks of boats. At one stage he operated a tethered observation balloon from a coal barge, the *Rotary*, sailing up and down the Potomac River. He afterwards claimed it was the first 'aircraft carrier' – although his rival LaMountain had done the same thing on the James River. He was prepared to inflate his balloons from coal-gas mains, hydrogen field-generators, or cobbled-together barrels of sulphuric acid and metal shell-casings. On one emergency occasion he even used another balloon, 'transfusing' the gas from his small *Constitution* via a makeshift valve ('contrived from a convenient kettle') into the larger *Intrepid*.

Lowe himself had no doubts as to the impact of his Balloon Corps in the early months of the Peninsula Campaign. As he put it graphically: 'A hawk hovering above a chicken yard could not have caused more

---

🎈 Some things about the Civil War battlefields were peculiarly invisible from a balloon. For example, that more than half of all the ordinary, non-commissioned troopers were boys under seventeen; and that nearly two-thirds of all casualties of whatever rank were caused by disease – especially acute diarrhoea – not by battlefield wounds. Perhaps least visible of all was the fate of the black troops who fought so heroically for the Union. Of more than twenty-seven thousand deaths among these black soldiers, fewer than three thousand perished in combat – the rest died from disease, their conditions were so bad. Balloon observers also had little to report about the great question that had triggered the Secession: black slavery. Their occasionally flippant attitude to what they might have seen was caught in a facetious article that appeared at the home town of ballooning, in the *Cincinnati Gazette* for 22 October 1861: 'LaMountain has been sent up in his balloon, and went so high that he could see all the way to the Gulf of Mexico, and observe what [the Confederate troops] had for dinner at Fort Pickens, Florida ... A reporter asked him if he could see any Negro insurrections, and he said that he did see some black spots moving around near South Carolina, but found out afterwards that they were some ants which had got into his telescope.'[19] It is instructive that Cincinnati was a Union city, of apparently solid anti-slavery sentiment.

commotion than did my balloons when they appeared before Yorktown.'[21] Rebel sources seemed to agree: 'At Yorktown, when almost daily ascensions were made, our camp, batteries, field works and all defences were plain to the vision of the occupants of the balloons ... The balloon ascensions excited us more than all the outpost attacks ...'[22]

All Lowe's observation were made from tethered balloons. Except, that is, on one memorable occasion involving the unfortunate Lieutenant-General Fitzjohn Porter, whose balloon broke from its cable, 'and kept right on, over sharp shooters, rifle pits, and outworks, and finally passed, as if to deliver up its freight, directly over the rebel heights of Yorktown'. But miraculously it returned, having encountered a LaMountain-style air box, and crashed down onto some Union tents. Porter emerged from the heap of canvas, still brandishing his telescope, and was immediately serenaded by a nearby military band. It is not recorded what tune they played him.[23]

Lowe's field units typically consisted of two hydrogen gas generators and two balloon equipment carts (winches, cables, envelopes), pulled by a team of eight horses and manned by a detachment of fifty non-commissioned soldiers. There was also a field telegraph unit, and a team of runners.[24] He had two basic methods of observation, depending on wind

strength and direction. In fine weather he would fly directly above the
enemy positions, at an altitude between one and two thousand feet,
acting like a true 'spy in the sky'. In bad weather or contrary winds, he
would fly on double tethers at five hundred feet above his own positions,
where he functioned more like a traditional lookout on an aerial platform. In both cases he was regularly shot at, though remarkably none of
his balloons was ever brought down. But the lower platform position was
particularly unpopular with his detachment, as the moment the balloon
was seen rising above the trees, Confederate field-gunners would immediately try to shell its estimated ground position.[25]

Military observation with binoculars was a delicate art. A tethered
balloon was rarely stable – at five hundred feet in any kind of wind Lowe
found the balloon 'very unsteady, so much so that it was difficult to fix
my sight on any particular object'. At a thousand feet he could see twelve
miles and a whole battlefield, yet always 'indistinctly' because of the
dust, smoke and heat haze produced by masses of troops on the move.
Even more, the heavy silver-grey smoke produced by field guns in action
might temporarily block out the ground altogether.[26] ◊

Lowe communicated his observations by various means. Ideally, he
used a telegraph key operated from his balloon basket, transmitting
messages down a five-thousand-foot telegraph cable directly to
McClellan's headquarters. But bad conditions often made this impossible – the equipment could be too heavy to take aloft, or the cable could
break. In that case, Lowe would make notes and drop them in canisters
to a telegraph operator on the ground. He did the same with rapidly
drawn sketch maps of the enemy positions, which were then delivered to
McClellan by runner. On other occasions he used signal flags, coloured
flares or simply hand-gestures. If all else failed, he would have himself
winched down so he could deliver his observations in person, sometimes

◊ I was amazed to find Thaddeus Lowe's long, grey gunmetal observation binoculars,
much battered from use, at the National Air and Space Museum, Washington, DC.
They have been placed, with brilliant appropriateness, in a solitary plain glass cabinet
at the foot of an exhibition of modern observation satellites, dating from the 1960s.
The astonishing images taken from these orbiting space satellites, some military but
most civilian, suggest that they are the ultimate heirs to Lowe's observation balloons.
The dazzling photographs of the earth beneath, breathtaking in their detail and
formal patterning, are surely what Lowe must have dreamed of seeing one day from
his tiny, swaying, perilous platform.

galloping to the headquarters on his favourite grey mare, a procedure he apparently enjoyed.

Many of his most urgent observations consist of three or four short lines, obviously scrawled in haste, but carefully time-dated down to the nearest minute.[27] While observing the battle for Richmond from the *Intrepid*, he noted: 'I immediately took a high altitude observation as rapidly as possible, wrote my most important dispatch to the commanding general on my way down, and I dictated it to my expert telegraph officer. Then with the telegraph cable and instruments, I ascended to the height desired and remained there almost constantly during the battle, keeping the wires hot with information.'[28]

Lowe had promised to supply McClellan with strategic aerial photographs, which he said could be examined by giant magnifying glasses once delivered to the ground. He claimed optimistically that a three-inch-square glass negative would provide the equivalent of a '20 foot panoramic image' of a battlefield. He took aloft several professional photographers – '[Matthew] Brady the celebrated War photographer was also much interested in the work of the Corps, and spent much time with us'[29] – and the British balloonist Henry Coxwell also reported that 'some photographs were taken of the [Confederates'] position'.[30] Yet no such aerial photographs have survived. While there are thousands of Civil War photographs taken on the ground, there is not a single known

photograph of a Civil War battlefield taken from a balloon.[31] Probably cameras, glass negatives and chemical developing equipment proved too cumbersome for the tiny observation baskets. Or perhaps the whole process was simply too slow to be of any practical use. In the event, the mysteries of early balloon photography would be explored in Paris.

Timing was vital, because what Lowe had discovered was the highly *mobile* nature of battlefield observation. This transformed the traditional idea of the tranquil, all-seeing 'angel's-eye view' from a balloon. In warfare, the panoptic vision no longer provided the classic, unfolding 'map' of the world beneath (an image still used by LaMountain). Instead it revealed a constantly moving game-pattern, a shifting topography of hints and clues, secrets and disguises, threats and opportunities. An entire tactical situation could change within a matter of hours, or even minutes.

Visual clues had to be carefully sought out: smoke from campfires, rising road dust, sun glinting on armoury, newly dug patches of raw earth, the straight lines of fresh infantry trenches, the faint dimpled shadows of breastworks, the regular white circles of bell tents, the deep curving tracks left by heavy guns. Lowe writes on one occasion of taking a General 'to an altitude that enabled us to look *into the windows* of the

city of Richmond'. The battle landscape had to be *read* constantly, interpreted shrewdly, and summarised with the utmost speed.[32]

One of Lowe's most brilliant observational coups was the discovery
of the Confederates' secret evacuation of Yorktown, under the cover of
darkness, on the night of 4–5 May 1862. This gave the Union army one of
its most crucial tactical advantages in the whole Peninsula Campaign.
At the time it was thought that Yorktown was being resupplied, and stiff-
ening its defences against the Union's long siege. Lowe's account is vivid,
and turns on a single, precisely observed detail. First of all he sets the
scene: 'The entire great fortress was ablaze with bonfires, and the greatest activity prevailed, which was not visible except from the balloon. At
first the General [Heintzelman] was puzzled on seeing more wagons
entering the forts than were going out.'

This was apparently clear evidence of resupplying. Lowe, however,
observed and interpreted more carefully: 'But when I called his attention
to the fact that the ingoing wagons were light and moved rapidly (the
wheels being visible as they passed each campfire), while the outgoing
wagons were heavily loaded and moved slowly, there was no longer any
doubt as to the object of the Confederates. *They were withdrawing.*'

According to Lowe, his observations of this secret evacuation carried
instant conviction to the highest command level: 'General Heintzelman
then accompanied me to General McClellan's headquarters for a consultation, while I with the orderlies, aroused other quietly sleeping corps
commanders in time to put our whole army in motion in the very early
hours of the morning, so that we were enabled to overtake the Confederate
army at Williamsburg.'[33] The result was one of the few decisive victories
of the Union Army of the Potomac, which was otherwise becoming char-
acterised by its lack of decision and initiative. By the end of May
McClellan was within five miles of Richmond.[34]

4

One of Lowe's innovations was to take journalists up with him in the
basket, knowing what brilliant newspaper copy he could provide. Frank
Leslie, a pioneer of American illustrated magazines, paid for one of his
best war artists, Arthur Lumley, to go up in Lowe's balloon and draw
battlefield pictures.[35] George Townsend of the *Philadelphia Inquirer*, an

old friend from the City of New York days, was taken on an unexpected night ascent. His article begins with a romantic touch: 'Said the Professor: "Will you make an ascension with me tonight?" "Where to?" I answered, greatly astonished as to the meaning of the Professor's enquiry, "To the moon?"'

But soon Townsend was noting the bitterly cold air, the disturbing clarity of the stars, and the soldiers' campfires below like 'a handful of glowing embers'. Most disquieting was the silence, interrupted only by the unnerving 'grate' of the netting against the sides of the balloon, and the 'collapsing and expansion of the silk', like some enormous creature breathing, and 'highly suggestive of a break' somewhere unseen in the canopy. Though not a single shot was fired all the time they were airborne, it was an extraordinarily tense and disturbing experience.[36]

By contrast, an unnamed British reporter for the St James's Magazine in London, largely ignoring tactical matters, was delighted to turn in a sensational, all-action scoop, which he entitled 'Three Months with the balloons in America': 'The Confederates fire on the balloon and the first shell passes a little to the left, exploding in a ploughed field. The next, to the right, burst in mid-air. The third explosion is so close that the pieces of shell seem driven across my face, and my ears quiver with the sound ...'[37]

A military correspondent for the London Times, Lieutenant George Grover, reported more soberly from the later battlefield of Chickahominy: 'During the whole of the engagement, Professor's Lowe's balloon hovered over the Federal lines at an altitude of two thousand feet, and maintained successful telegraphic communication with General McClellan's headquarters. It is asserted that every movement of the Confederate armies was distinctly visible, and instantaneously reported.'[38]

However, by the end of May 1862, with the Union forces nearing the rebel capital, Richmond, the demand for recreational balloon trips became so great that General McClellan banned all further ascents by newspapermen, and even required Union officers to have his specific written permission to go aloft. Foreign journalists who could boast military backgrounds had better luck, and a notable series of reports were turned in by a British engineering officer, Captain Frederick Beaumont, RE.

Beaumont's articles appeared in Professional Papers of the Royal Corps of Engineers in 1863, and gave a realistic view of modern warfare. He was also deeply impressed by Lowe, 'a man celebrated in America as a very daring aeronaut ... from the earnest way in which he spoke, I felt convinced that he still intended to carry out his [Atlantic] scheme ... but the distracted state of his country obliged him to put it off for a while'.[39]

Lowe made many attempts to get senior military officers up in his balloons, but most of them declined his offers, especially after Lieutenant-General Fitzjohn Porter's experience. An exception was a dashing young cavalry lieutenant, a willowy, wild-eyed youth from the plains of Ohio. His long blond hair, silky moustaches and extravagant manner already marked him out. His name was George Armstrong Custer.

This was the man who would become the famous Indian fighter, and would die fourteen years later at the chaotic Battle of the Little Bighorn in 1876. At the time of his balloon ascent with Lowe in April 1862, Custer was aged twenty-two, and had been temporarily promoted captain by McClellan. He was already wearing his trademark red bandana, and had been noted for his reckless, showy courage in several local engagements on the west bank of the Potomac. But even Custer found the balloon experience oddly disconcerting. He was invited up by one of Lowe's assistants, James Allen, as he recorded in an unexpectedly witty and self-deprecating memoir.

My desire, if frankly expressed, would not have been to go up at all. But if I was to go, company was certainly desirable. With an attempt at indifference, I intimated that I might go along … The basket was about two feet high, four feet long … to me it seemed fragile indeed … the gaps in the wicker work in the sides and the bottom seemed immense and the further we receded from the earth, the larger they seemed to become … I was urged to stand up. My confidence in balloons at that time was not sufficient, however, to justify such a course, so I remained sitting in the bottom of the basket … [Mr Allen] began jumping up and down to prove its strength. My fears were redoubled, I expected to see the bottom of the basket giving way, and one or both of us dashed to the earth …

Custer was much struck by the dramatic panorama as he looked west up the entire length of the Virginia Peninsula towards Richmond:

To the right could be seen the York river, following which the eye could rest on Chesapeake Bay. On the left, and at about the same distance, flowed the James river … Between the two extended a most beautiful landscape, and no less interesting than beautiful, for being made the theatre of operations of armies larger and more formidable than had ever confronted each other on this continent before.

But he also noted how difficult it was to make precise observations:

With all the assistance of a good field glass, and when the balloon was not rendered unsteady by the different currents of air, I was enabled to catch glimpses of canvas [tents] through the openings of the forest … the dim outline of an earthwork half concealed by trees which had been purposely left standing on their front. Guns could be seen mounted and peering sullenly through the embrasures … while men [enemy soldiers] in considerable numbers were standing around entrenchments … intently observing [our] balloon, curious, no doubt, to know the character or value of the information its occupants could derive from [our] elevated post of observation.[40]

9

The rebel Confederate army was too poor to put any serious balloon opposition into the sky. But it did something almost as effective, by creating one of the most famous romantic legends of the South, the celebrated 'Silk Dress Balloon'. This remarkable balloon was said to be extraordinarily beautiful, and piloted with fantastic and cavalier daring during the defence of Richmond by the Confederates. It was composed of a shimmering mass of multicoloured silks, supposedly sewn up in homely patchwork fashion from dozens of gorgeous silk ballroom dresses. These dresses, so the story went, had been gallantly donated by the Southern belles of Richmond and the surrounding towns of Virginia, happy to sacrifice the last remnants of their antebellum finery to the cause of Dixie. They included not only the beautiful wives and daughters of the great, porticoed houses of the very wealthy, but also the poor but no less beautiful patriotic ladies of easy virtue. It was, in other words, a balloon made of gleaming Southern dreams.

It has been disputed whether this balloon ever actually existed. Certainly it sounds more like something dreamed up by Margaret Mitchell for Rhett Butler and Scarlett O'Hara in *Gone with the Wind*. Yet

it emerges that just such a small Confederate balloon, romantically named the *Gazelle*, mysteriously appeared overhead during the desperate battles to defend the rebel capital Richmond between May and July 1862; and its presence became symbolically associated with the final, decisive repulse of McClellan's Union Army of the Potomac.

Confederate official records and private correspondence have recently revealed that the *Gazelle* did in fact exist, and was secretly constructed at top speed in the workshop of a bankrupt armoury in Savannah, Georgia, in May 1862, at the request of General Thomas Drayton. It was built by two Southern patriots: Langdon Cheeves, a merchant from Charleston, Virginia; and Charles Cevor, an itinerant balloonist from Savannah who had once trained under John Wise. A letter from General Drayton, dated 9 May 1862, asks urgently: 'How soon will the balloon be finished? Put night and day work on it at your discretion.'[41]

The *Gazelle* was small and pretty, practically petite, standing about thirty feet high with a 7,500-cubic-foot capacity – roughly half the size of Lowe's regular balloons. It was composed of long, bright, multicoloured silk strips – yellow, green, white and dark red – some plain, and some carrying decorative patterns. It was sealed with a clear, vulcanised rubber coating removed from old wagon springs, which gave the silk an unusual glisten in full sunlight. It also leaked.

At a time when Union ships were successfully blockading all Southern ports, silk of any kind had been virtually impossible to find. But the merchant Langdon Cheeves had excellent business contacts in Charleston (incidentally, the home of Rhett Butler), and managed to locate a job-lot of twenty mixed bolts of unwrapped dress silk lying in the warehouse of Kerrison & Leiding, wholesale fabric merchants. For these he paid $514. He was still not sure that this would be enough silk for the balloon, and as he left Charleston for the Savannah workshop, he made a gentle, teasing joke to his daughters: 'I'm buying up all the handsome silk dresses in Savannah, but not for you girls.'[42] By the time the *Gazelle* arrived in Richmond, this wry Southern joke had started the gallant Southern legend.

Robert E. Lee, who would command the defence of Richmond all that summer, placed the *Gazelle* in the hands of a young Confederate artillery officer, Lieutenant-Colonel Edward Alexander. Alexander knew nothing about balloons, but was a brilliant tactical commander, who would later direct Lee's artillery at Gettysburg. With him as pilot was Charles Cevor.

Alexander decided not only to observe the enemy, but also to raise morale in Richmond.[43]

The *Gazelle* was kept inflated, and was run out daily on a railway car along the Richmond and York railroad, right up to the enemy lines beyond Fair Oaks. Everyone in Richmond could see the little balloon going out brazenly to flaunt the rebel flag in the face of the enemy. For once rail and balloon shared a mutual cause. From here Alexander could report on the threatened advances of McClellan's Union army over the Chickahominy river.[44] Tension increased as McClellan's massive forces manoeuvred into position for the crucial assault. But there was also excitement, as news of rebel reinforcements led by Stonewall Jackson were rumoured to be marching out of the Shenandoah Valley to relieve Richmond.

Just opposite Alexander and the *Gazelle*, at a distance of less than five miles on the other side of the Chickahominy, was Thaddeus Lowe in his *Intrepid* balloon. So for the first time, a Union and a rebel balloonist confronted each other, out of rifle shot but well within binocular and telescope range. On 27 June 1862, Lowe telegraphed with astonishment from the *Intrepid*: 'About four miles to the west from here the enemy have a balloon about 300 feet in the air.' An hour later, at eleven o'clock, the rebel balloon had mysteriously disappeared.[45]

Lowe was stationed at a strategic position known as Gaines Mill.[46] The proprietor of Gaines Farm was a medical doctor, and in fact a Confederate supporter, whose cornfields had been forcibly requisitioned by the Union Balloon Corps. It was a sign of the times that Dr Gaines had enough Southern pride to complain about his trampled crop, and that Lowe had enough Yankee style to apologise gracefully for it. He also promised to provide the farm with a protective guard against Union looting. Dr Gaines had a spirited teenage daughter, Fanny, who was fascinated by the enemy balloon camp set up in the middle of her father's meadow. She showed no nervousness in approaching Lowe, though she took the prudent precaution of addressing him as 'General'.

> *Every day the Yankees sent up a balloon in front of our house to see what was going on in Richmond. General Lowe was the man who ascended in the balloon, and he told many wonderful things that he saw going on in Richmond – such as people going to church, the evacuation of Richmond, wagon trains crossing Mayo's bridge.*[47]

Lowe evidently boasted to Fanny that McClellan and the Army of the Potomac would soon be in Richmond. When Fanny told this to an ancient neighbour, old Mrs Woody, she replied with true Southern scorn, 'Yes, Moses also viewed the promised land, but he never entered in.'[48]

Both balloon observers were soon reporting on a bloody series of skirmishes at the river crossing between Gaines Mill and Fair Oaks. The fighting was bitter and confused. It continued through the end of June and the start of July, eventually becoming known as the Seven Days Battle. The rival balloonists were attempting to advise their commanders on the ground about the rapidly developing troop movements and reinforcements, sudden advances and retreats, charges and counter-charges. But it was not easy to understand the action, or even to see it. Alexander believed he was better qualified, as a 'trained staff officer', and dismissed his opposite number with Southern scorn as belonging to 'the ignorant class of ordinary balloonists'.[49] But Lowe was experienced, and had the immense advantage of a direct telegraph line from the *Intrepid*'s basket. 'Immediately I ascended to a height of a thousand feet,' he recorded, 'and there witnessed the Titanic struggle. The whole scene of action was plainly visible and reports of the progress of the battle were constantly sent till darkness fell upon the grand but terrifying spectacle.'[50]

Inspired by Robert E. Lee, and supported by Stonewall Jackson, the Confederates pushed out from Richmond, crossed the Chickahominy river, and surrounded McClellan's troops at Gaines Mill. The fighting was unbelievably fierce: on that small wooded hilltop, more than fifteen thousand men were killed or wounded in a matter of hours. Hurrying down the hill from her father's farm, Fanny Gaines never forgot what she glimpsed: 'The dead were strewed on every side. I had to keep my eyes shut all the way to keep from seeing the horrible sights.'[51]

The position was eventually taken by the rebels, in a series of 'sublime' uphill charges led first by Jackson, and then consolidated by the flamboyant Texas Brigade from the deep South. The Union troops began to withdraw from Richmond, and Lowe hastily dismantled his balloon equipment. He always believed that his last reports saved hundreds of McClellan's troops from being cut off and massacred behind the Chickahominy, and that he had prevented retreat from becoming an outright disaster.[52]

But Lowe's balloon intelligence had not prevented Robert E. Lee from outmanoeuvring McClellan's much larger army, nor had it brought about the hoped-for Union victory. Richmond did not fall. The Seven Days Battle was to prove a crucial turning point for Lee. It forced McClellan to abandon his master plan to take the rebel capital, and instead to organise a full-scale Union retreat down the James River. It would also mark the end of McClellan's own military command, and the eventual disbandment of the Balloon Corps which he had championed.

Lowe was shattered. Exhausted by the strain of what he had witnessed, he succumbed to marsh fever. He did not fully recover his health for many months, and meanwhile his Balloon Corps languished in Washington. By contrast, Alexander and the *Gazelle* were immediately despatched by Lee on a new adventure. The reputation of the 'silk dress balloon' had evidently gained currency among troops on both sides, and the new assignment appears to have had a propaganda as much as a military motive. This time the *Gazelle* was teamed up with a boat, rather than a train.

On the evening of 3 July 1862 the *Gazelle* was secretly taken down to a wharf on the James River, and tethered aboard a small rebel tug ship, appropriately named the *Teaser*, which then steamed all night slowly towards the Union lines, with orders to make 'stealthy reconnaissance' of the retreat, and no doubt to add to the Yankees' shame. The next day, just before dawn, Alexander made a triumphant ascent, rising into the early-morning sunlight until the gleaming multicoloured silks must have been seen for miles. Despite losing gas, he was briefly able to observe the Union retreat, and flaunt his presence. But then things went wrong. On a falling tide, the *Teaser* ran aground on a mudbank, becoming stuck fast perilously close to the Union lines.

It is possible that Alexander had a chance to fly the *Gazelle* off from the deck of the trapped *Teaser*. It would have been a gallant, but probably a doomed attempt. Instead he decided to deflate the balloon completely, and sit tight until the tide turned, hoping to escape detection. It must have been a long wait. After eight hours, in mid-afternoon, just as the tide was flooding back, a Union gunboat, the *Maratanza*, steamed briskly into view. She immediately loosed off two rounds from her hundred-pound gun, both of which hit home. Having tried to scuttle their ship, the *Teaser*'s crew, including Alexander, leaped overboard and swam for the shore. There was no hope of saving the *Gazelle*. Alexander stood watching from the trees, and recalled regretfully: 'The *Maratanza* pulled [our ship] out of the mud and carried her off, balloon and all ... So I left the sailors, and struck out for the army, which I soon found, and General Lee also; and made my final balloon report.'[53]

But in a way, this was not the end, but the true beginning of the silk dress balloon. What both Alexander and General Lee soon realised was that the *Gazelle*, by being physically destroyed, had become mythically indestructible. As such, it was infinitely more valuable to the South, safely afloat as a provoking and permanent piece of rebel propaganda among the enemy. Indeed, it was the Union commander of the *Maratanza* who made the first known reference to a 'silk dress balloon' in print, thus unwittingly making its existence official. He reported to his superior officers on 16 July 1862: 'The Confederate officers and crew ... left everything behind. We got the officers' uniforms, swords, belts, pistols, muskets, silver chains, bedding, clothes, letters ... *We also found a balloon made of silk dresses.*'[54]

Lowe was clearly impressed by the truth of the story as well. He added a circumstantial, and curiously wistful, note to his memoirs, *My Balloons in Peace and War*: 'The fashions in silks at that period were ornate, large flowery patterns, squares and plaids in blues, greens, crimsons etc, and rich heavy watered silks; and the silk dress balloon was a

very brilliant and handsome object – a veritable Joseph's coat of many colours. It was taken to Washington and cut up, many pieces being given to Congressmen and others as souvenirs.'[55] Thus the myth was soon dispersed and spread, a sacred relic of the Old South: beautiful, gallant, self-sacrificing, doomed. ℘

Thirty years after the war, the Confederate General James Longstreet, who had fought alongside Jackson and Lee in the defence of Richmond, delightfully retold and embroidered the now well-established silk dress story, in cadences that catch the authentic voice of the Old South:

> The Federals had been using balloons in examining our positions, and we watched with envious eyes their beautiful observations as they floated high up in the air, and well out of the range of our guns. We longed for the balloons that poverty denied us. A genius arose for the occasion and suggested that we send out and gather together all the silk dresses in the Confederacy and make a balloon ... Soon we had a great patchwork ship of many and varied hues ... The balloon was ready for the Seven Days Campaign. We had no gas save in Richmond, and the custom was to inflate it there, tie it securely to an engine, and run it down the York River railroad to any point at which we desired to send it up. One day it was on a steamer down the James when the tide went out and left the vessel and the balloon high and dry on a bar. The Federals gathered it in, and with it

---

℘ Lowe added a further note towards the end of his life, while completing his memoirs as an old man in California: 'As I write, a good piece of it lies before me on the table, now frail and discoloured with age.' It is easy to imagine him turning this last fragment of the silk dress balloon carefully with his fingertips. I found it strangely moving to discover that very piece, a stiff section of dark-red material about the size of a playing card, folded into one of his letters, and still stored in the Library of Congress Archive, Washington, in 2010. It is not surprising that the story has inspired at least one modern romance, *The Last Silk Dress* (1988), by Ann Rinaldi, later retitled as *Girl in Blue*. Here the coloured silk dresses are even identified with their individual owners, as the balloon comes to life: 'The men on the dock were spreading the silk out, making it smooth. You could see the bright coloured patches, jumping and bubbling ... Lying there on the dock, the crazy patchwork mass of material was like something alive. The men had all they could do to hold it down as the gas hissed in ... We stood and watched as fold by fold of the balloon took life and it rippled into a mass of shimmering, bouncing silk ... Connie screamed in delight and pointed out Francine's green, Lulie's striped pink, and all the other colours we recognized from the girls at Miss Ballard's ... The balloon was taking on a life of its own.'[56]

the last silk dress in the Confederacy. This capture was the meanest trick of the War, and one I have never yet forgiven.[57]

For Longstreet, the capture of the *Gazelle* by the Union gunship on the James River was not an act of war, but an act of *ungallantry*, amounting almost to an insult, as if a fine Southern lady had been betrayed and abused by a Yankee carpetbagger.

The poetic possibilities of balloons also occurred to Walt Whitman, who was working during these months as a hospital orderly in Washington. Though he often saw Lowe's balloons on the horizon, he never had the chance to go up himself. But he imagined being in a balloon basket, calmly surveying the land and the rivers of the peninsula beneath. He sees no battlefields, but instead is given moments of detached, surreal vision, as described in *A Song of Myself*:

> *Where the pear-shaped balloon is floating aloft,*
> *(Floating in it myself and looking composedly*
>   *down,) ...*
> *Where the steam-ship trails hind-ways its long*
>   *pennant of smoke,*
> *Where the fin of the shark cuts like a black chip out of*
>   *the water.*[58]

By contrast, Stephen Crane, who wrote extensively about the Civil War, presents a balloon as viewed from below. It is seen by the ranks of an advancing infantry brigade, just about to go into battle. Though it is one of their own balloons, its appearance is unnerving, and almost repulsive. It seems an alien presence, presiding over a human sacrifice. It is the signal for the ritual of killing to begin.

> *The military balloon, a fat, wavering yellow thing, was leading the advance like some new conception of a war-god. Its bloated mass shone above the trees, and served incidentally to indicate to the men at the rear that comrades were in advance ... The first ominous order of battle came down the line. 'Use the cut-off. Don't use the magazine until you're ordered ...' A sound of clicking locks rattled along the column. All men knew that the time had come.*[59]

This story in fact draws on Crane's experience with troops in Cuba during the 1890s. No balloon appears in *The Red Badge of Courage*, which is a historical fiction, as Crane was born well after the end of the Civil War, in 1871. Nevertheless, the superb opening paragraph of the novel, a brooding panorama of an army encamped by a river like the Potomac, seems to draw on a balloonist's aerial view. The whole army is seen awakening from winter slumber (with a repeated imagery of opening 'eyes'); yet no individual soldier, horse or wagon appears. As from a balloon, what is seen is the pattern of mobilising forces. Both man and nature appear as anonymous combatants. It is just as Lowe might have seen the Union army in the spring of 1862.

> *The cold passed reluctantly from the earth, and the retiring fogs revealed an army stretched out on the hills, resting. As the landscape changed from brown to green, the army awakened, and began to tremble with eagerness at the noise of rumors. It cast its eyes upon the roads, which were growing from long troughs of liquid mud to proper thoroughfares. A river, amber-tinted in the shadow of its banks, purled at the army's feet; and at night, when the stream had become of a sorrowful blackness, one could see across it the red, eyelike gleam of hostile camp-fires set in the low brows of distant hills.*[60]

The Union Balloon Corps did have one significant effect on the future of warfare. Among the military observers sent to the Union army was a young Prussian officer, who took some part in the fighting around Fredericksburg in 1863, and saw the last of Lowe's balloons in action. Later he ascended in one of John Steiner's civilian balloons at St Paul, Minnesota, above the Mississippi. The experience was a great success. He admired the picturesque effect of the great river, but was even more impressed by the opportunities for 'military reconnaissance'. He noted: 'From the high position of the balloon the distribution [of the defenders' troops] could be completely surveyed ... No method is better suited to viewing quickly the terrain of an unknown enemy in occupied territory.' The only problem was the inability to navigate a spy balloon. This started a chain of ideas that would have a considerable effect on the use of balloons in the First World War. The young Prussian's name was Captain Ferdinand von Zeppelin.

10

Almost unnoticed, the Civil War brought to an end a golden era in American ballooning. For a brief period in the 1840s and 1850s, the balloon seemed to hold the magic key to the unlocking of the whole vast American continent, 'from sea to shining sea'. But the logistical demands of warfare had revealed the imperative necessity of modern communication networks: the telegraph, the railroad, the steamship. The business potential, as well as the visionary moment, of the balloon was gone.

The aeronauts responded to this each in their own way. John LaMountain, brave but cantankerous, struggled on briefly as a balloon showman, increasingly underequipped and underfinanced. After one of his worn-out balloons burst above Bay City, Michigan, in 1869, nearly killing him, he returned to New York State, and after various minor commercial ventures died, a forgotten figure, aged only forty-eight, in 1878.[61]

John Wise, nearly sixty by the end of the war, largely confined himself to giving lectures on the history of ballooning. In 1873 he published his admirable book, part history and part autobiography, *Through the Air: A Narrative of Forty Years' Experience as an Aeronaut*. His actual attempts to return to the air were less happy. In September 1879, aged seventy-one, he took off in another balloon from St Louis, in an attempt to commemorate the famous flight of the original *Atlantic* twenty years before. No doubt he hoped to make the full thousand miles to New York or Boston, but this time he strayed even further from his projected easterly course. Once again he headed northwards towards the Great Lakes, and once again he was caught in a gale. He was last observed near Chicago, sailing in high winds out across the enormous stormy expanse of Lake Michigan, and trusting in the Lord. Neither he, nor his balloon *Pathfinder*, was ever seen again. Perhaps it was the heroic conclusion that he desired. 'Astra Castra, Numen lumen', as he had written.

Despite the long after-effects of his fever, Thaddeus Lowe made the most successful career adjustment. Part scientist, part entrepreneur, part adventurer, his navigation skills stood him in good stead. First, he shrewdly turned down a tempting, but surely doomed, Brazilian government invitation to start a military balloon corps in São Paulo. Though

personally invited by Don Pedro II, Emperor of Brazil, with many blandishments, he passed the invitation on to his fellow balloonists the Allen
brothers, remarking with a wry smile, 'I think I am rather beyond the age
of adventure.'[62]

Next, he briefly promoted a spectacular 'Aeronautic Amphitheatre' in
New York City. Here his last ever balloon flight was made in 1866 for a
wealthy honeymoon couple, to take them, as he put it, 'nearer to heaven
than most clergymen ever get'. The flight lasted most of a summer's day,
and finished on a perfect idyllic note: 'We sailed high over cities, hills,
valleys and rivers, and came to earth with the sunset.'[63]

But Lowe the entrepreneur was far from finished. Still determined to
explore the vastness of America, he went west to Los Angeles, seeking
business opportunities. Here he used his knowledge of the chemistry of
hydrogen to invent a brilliantly successful ice-making machine. An early
version was used to refrigerate cargo ships steaming from San Francisco
around Cape Horn to New York. In its own way this fulfilled his original
transcontinental dream. Instead of mail, he was transporting food across
the whole continent. His perfected 'Lowe's Compressed Ice Machine' was
sold to all the newly opening stores and hotels of California, and made
him a dollar millionaire. It also won him the Franklin Institute's Grand
Medal of Honour in 1887.[64]

The same year, at the age of fifty-five, a successful businessman, he settled with his faithful wife Léontine and their extensive family in Pasadena, at the foot of the San Gabriel Mountains. Here he built a luxury *estancia*, and subsequently invested his fortune in a different form of elevation – a funicular, or scenic mountain railway, climbing into the hills above Altadena. It was to service an 'Alpine Tavern' and a forty-room luxury hotel, with panoramic terraces. This mountaintop complex, with its dazzling aerial views, was opened in July 1893 and publicised as 'The White City in the Clouds'. Perhaps, in a way, Thaddeus Lowe considered it his final version of an enormous tethered balloon.[65]

Lowe kept up his scientific contacts with the Smithsonian, continuing to correspond with Professor Joseph Henry, and wrote a long, racy aeronautical memoir entitled *My Balloons in Peace and War*. Typically, he never bothered to publish it. But the manuscript, treasured by his family and transcribed by one of his granddaughters, Augustine Lowe Brownbeck, eventually reached the Library of Congress in 1931.

Lowe lived on comfortably in Pasadena, saw the coming of the Zeppelin and the aeroplane, and died peacefully in 1913, at the age of eighty. But he never quite stopped dreaming. Almost his last act was to form a new business company: the 'Lowe Airship Construction Corporation'.[66]

The crossing of the American continent, so long dreamed of by Lowe, Wise and LaMountain, was never actually accomplished in the nineteenth century. ⸙ It was, however, achieved in fiction, as so often in ballooning. A novel by Jules Verne, *The Mysterious Island* (1875), opens

---

⸙ For the record, the nearest thing to an actual non-stop trans-American balloon flight did not take place until May 1980. The helium-filled balloon travelled west to east, just as Professor Henry and John Wise had always prophesied. The *Kitty Hawk* was flown by Maxie Anderson and his son Kristian, for four days and nights, from Fort Baker in California to Sainte-Félicité in Quebec, a distance of 3,313 miles. Anderson had also been the first to fulfil the transatlantic dream. In 1978 he flew with Ben Abruzzo and Larry Newman in *Double Eagle II*, from Maine to Picardy in northern France, a distance of 3,107 miles. Their insulated gondola is displayed in the Udvar-Hazy Center, Dulles International Airport, Washington, DC. His and Ben Abruzzo's names live on in the new Anderson-Abruzzo Albuquerque International Balloon Museum in Albuquerque, New Mexico, where the greatest of all American balloon fiestas is based. Anderson also founded the Anderson Valley Vineyards, which marketed a famous aeronautical rosé wine, christened 'Balloon Blush'. It is surprisingly dry.

with five men (and a dog) escaping by balloon from the besieged
Confederate city of Richmond in the spring of 1865. The balloon, at fifty
thousand cubic feet, is considerably bigger than the *Gazelle*, and the men
are escaping Union prisoners. To lend verisimilitude, one of them is a war
reporter for the *New York Herald*.

Stealing a rebel balloon at night, they launch in a terrible storm,
which is described by Verne with vivid details drawn from the apocalyptic
journey of Wise and LaMountain across the Great Lakes. But this storm
is blowing from east to west, and it sends the balloon at ninety miles per
hour right across America the other way: the Great Plains, the deserts, the
Rocky Mountains, and out over the Pacific Ocean, where eventually it
touches down on an unknown island. But ironically, because of the
stormclouds, its passengers spy absolutely nothing of that entire great
American land passing beneath them.

7

Gigantic Voyages

———

1

The lack of any aerial photographs from the American Civil War is a curious anomaly. For as early as 1858 the first aerial photograph of Paris had been taken from a balloon by a strange, daredevil figure who called himself Félix Nadar. Nadar (the name was one of his many inventions) eventually persuaded his friend, the distinguished artist Honoré Daumier, to draw a cartoon in celebration of this achievement. It appeared as a lithograph in the popular magazine *Le Boulevard* in 1862, and became one of Nadar's most treasured mementoes.

NADAR élevant la Photographie à la hauteur de l'Art

In it, Nadar's unmistakable gawky figure is shown perched in a tiny balloon basket above the rooftops of Paris. Top hat flying, he is clinging onto a spindly photographic apparatus mounted on a tripod. The balloon basket – emblazoned with his name – swings perilously above an imagined Parisian cityscape, which seems largely composed of photographic studios. The caption reads wittily: 'Nadar elevating Photography to the height of Art.'

Nadar had a genius for elevating things. He was one of the earliest masters of the new nineteenth-century art of visual publicity. He turned unknown people into celebrities, and gave new ideas public prominence, largely by fixing them with memorable imagery. Even his own name was created as a visual logo to publicise his work. The one-word signature 'Nadar' was carefully designed with a long, forward-racing letter 'N', to express his particular energy and enthusiasm. He actually copyrighted this logo signature, once fought a court case to retain control of it, used it on the covers of his books, and had it incorporated into an early form of red neon sign above his Parisian studio by Antoine Lumière.[1]

Born in April 1820 into a prosperous family of Parisian printers, recently established in the fashionable rue Saint-Honoré, Nadar was christened Gaspard-Félix Tournachon, and schooled in the latest techniques of printmaking and print advertising. Self-confident, sociable and highly original from the start, he broke away into the world of literary bohemia, and soon became known for his wild ideas, his immense height and his memorable shock of carrot-coloured hair. He was already a kind of walking logo. He plunged into radical politics and satirical journalism, and took part in the revolutionary street disturbances of 1848 in Paris, always remaining a republican. But his visual and business talents soon emerged, and by his mid-thirties he had established himself as a brilliantly inventive cartoonist, caricaturist and commercial artist.

From 1851 the Paris of the Second Empire saw an explosion of newspapers, illustrated journals and satirical magazines, such as *La Silhouette*, *Le Charivari*, *La Revue comique* and *Le Journal pour rire* (the equivalent of the modern *Le Canard enchaîné* and *Charlie hebdo*). Nadar exploited these with astonishing energy and flair. Commissioned by leading Parisian editors such as Charles Philipon, he drew cartoons of all the leading celebrities of the day. Already thinking in 'panoptic' terms, he chose figures from every artistic field of endeavour for which Paris was famous: painting, music, theatre, opera and literature. These early *portraits-chargés*

were signed by an early version of his logo, with his name altered from
Tournachon to 'Tournadard'. This might be translated as 'Jab-the-barb',
although his caricatures were for the most part gentle and humorous,
even flattering. His subjects faithfully collected the originals.

Nadar then had the genius to combine them all into a single, enormous cortège: Le Panthéon Nadar, a true panoptic vision. He published
this independently, as an expensive and hugely popular poster, in 1854.
It made his name, and established his logo. He was now almost universally known, especially among the writers of the day. Baudelaire, Hugo,
Gautier, Nerval, Dumas, the Goncourts, George Sand and – notably –
Jules Verne, were all to become personal friends. And they were all also to
be photographed.

Alive to all new technical developments, Nadar saw the possibilities
of photography – 'writing with light' – as a novel means of portraiture,
and also of publicity. With the arrival of the wet-plate collodion method
in the early 1850s, it was possible to take a photograph in less than thirty
seconds, rather than the several minutes that daguerreotypes had previously required. Individual photographic portraiture – not merely a super-
ficial 'likeness', but a study of personality in depth – became practicable.
Nadar saw both the immense artistic and commercial potential of this.
He swiftly mastered the techniques, and 'exchanged the pencil for the
camera'. In January 1855 he opened a fashionable photographic studio at
35, boulevard des Capucines. At the same time, he also fought a successful legal battle against his younger brother Adrien to retain exclusive
rights to his logo 'Nadar', and installed his first illuminated 'Nadar' sign
above his premises.[2]

It was a shrewd move. Nadar quickly became the most famous
portrait photographer in France. Between 1854 and 1860 he photographed
nearly all his celebrated friends, compiling a matchless Album Nadar. It
was said that while André Disderi was the official 'establishment' photographer (politicians, generals, aristocrats, wives), Nadar was the official
'opposition' photographer (writers, actors, painters, mistresses). Victor
Hugo – himself always a republican – sent letters to him from Belgium or
the Channel Islands addressed simply to 'Nadar, Paris'.[3]

Determined to master the whole photographic field, Nadar had been
secretly struggling with aerial photography, on and off, since 1855. He
worked from a small hydrogen balloon, provided by Louis Godard, tethered above an apple orchard in the village of Petit-Bicêtre, outside Paris.[4]

Since the *clichés*, or glass photographic plates, had to be prepared and painted with the wet collodion gum on the spot, that is to say while actually in the air, Nadar had immense difficulties. It was not a question of breakages, but of contamination. For a long time the escaping gas from the balloon contaminated his developing chemicals, and rendered all his plates black after exposure. After endless experiments, he hit upon the solution of closing the balloon's escape valve and fitting a special kind of thick cotton insulating tent to the basket.[5] He described the intense excitement of his first success in a memoir, *When I Was a Photographer*:

> *It's just a simple positive on a glass plate, very feeble in this misty atmos-*
> *phere, all blotchy after so many false starts. But what does that matter?*
> *The evidence cannot be dismissed. There beneath me, the only three houses*
> *in the little hamlet: the farm, the inn and the police station. It's the unmis-*
> *takable image of Petit-Bicêtre. You can clearly make out on the road a*
> *furniture cart whose driver has come to a halt directly below the balloon,*
> *and on the tiles of the rooftops two white pigeons which have just alighted*
> *there. Yes, I was right!*[6]

Nadar was much concerned to establish commercial priority for his invention, and made wildly extravagant claims about the exact date on which he achieved this technical breakthrough, for example as early as autumn 1855. But the only reliable documentary proof is his French patent, which was first registered on 23 October 1858. The patent referred to a 'special combination of equipment' taken up in a balloon, which would allow him 'to employ photography for the production of topographical plans and cadastral surveys, and also for the direction of strategic military operations, for the erection of fortifications, and the disposition of armies on the march etc.' The camera would be mounted 'perpendicularly in the balloon car', either fixed 'laterally on the outside', or else positioned 'at the bottom of the car using a perforated section'.

Nadar also patented a 'sliding lens shutter' to allow the camera to be operated automatically; a black cotton tent for preparing the plate, and also a yellow one for developing it afterwards. The whole operation could thus be performed within the balloon basket, and the aeronaut could descend to earth with the finished glass negative. The design was of course very far from the flapping, comic tripod wittily imagined by Daumier.[7]

The following spring, 1859, Nadar claimed he was able to take a further historic series of Paris photographs from several hundred feet above the Right Bank. He recorded these simply as 'several views of the Bois de Boulogne, Arc de Triomphe, and perspectives on the place des Ternes'. They are not of fine quality, but have a surprisingly long perspective to the north of the city, and have been proudly annotated to show the place des Ternes, the Parc Monceau and the heights of Montmartre in the far distance.[8]

But again, the date Nadar gave is not certain. If he really did take his first aerial photographs in 1858, it is curious that the Daumier cartoon is dated 1862. Nadar did not return to aerial photography until 1868, when a second, much-better-quality set, clearly showing the Arc de Triomphe, is well documented. So there remains some dispute over who actually took the first successful aerial photograph from a balloon. On 13 October 1860 James Black and Samuel King flew in their balloon *Queen of the Air* over Boston harbour, and took a series of high-quality photographs of the roofscapes and ships immediately beneath. The best of these was subsequently widely distributed in an oval frame as 'Boston, As the Eagle and

the Wild Goose See It', and soon gained the reputation of being 'the first
ever aerial photograph, 1860'. Yet no balloon photographs of Civil War
battlefields, over the next five years, are known.

As for Nadar, in a typical *volte-face*, he left the air and went underground to photograph the Paris sewers and catacombs, taking out a first
patent on 'photography by artificial lights' on 4 February 1861. He sold a
set of eight photographs entitled 'Paris Overhead and Underground' to a
minor Brussels publisher on 25 October 1866, which suggests his aerial
photos were well known by that date. A cartoon by Cham shows two
Parisians on a boulevard, one peering down into a manhole, the other
gaping upwards into the sky. The first turns to the second: 'Are you looking for Monsieur Nadar? He ain't up there! *He's down here!*'[9]

Nadar was rarely content to pursue one scheme at a time. His quick
mind had already moved on to a new publicity challenge. He began to
think about ballooning itself, and by extension the whole question of
sustained and navigable flight. Was it possible? He had always been fascinated by balloons, having first seen one in Paris at the age of eight or nine
during the Fête du Roi of 1829. It flew very low down the Champs-Elysées,

grazing the treetops and followed by a shouting, ecstatic crowd. He was much struck by the people's reaction, and thought afterwards that the publicity possibilities of such an event were huge.[10]

With undimmed boyish enthusiasm, Nadar threw himself at this age-old question in aeronautics. Could a balloon, a 'lighter than air' machine, ever truly be navigated? And if not, what kind of 'heavier than air' machine must be invented to replace it? To address this debate in his own unique way, Nadar undertook three quite different forms of publicity campaign. The first was fairly modest: he set up a discussion group, founding the Society for the Promotion of Heavier than Air Locomotion, which began meeting at his studio in July 1863; Jules Verne was a member of the steering committee. Then, the following month, Nadar launched a beautifully illustrated review, *L'Aéronaute*, dedicated to the challenge of flight in all its aspects. Its first issue had a cover designed by Gustave Doré. Finally he began construction of a truly enormous gas balloon, *Le Géant* ('The Giant'), to demonstrate both the capabilities and the limitations of aerostation.[11]

As part of his campaign, Nadar circulated photo-portraits of himself as a gentleman balloonist. These were not shot perilously *en plein air*, but safely in his studio at the boulevard des Capucines. He appears

incongruously in top hat and evening dress, with a warm tartan plaid casually over one shoulder, and an expensive pair of opera glasses to hand. His large frame is elegantly posed in a wicker observation basket against a background of delicately painted clouds. The effect is unexpectedly comic, but this may have been exactly Nadar's intention. It is in its own way a photographic caricature: the balloonist as aerial *flâneur*, a gentleman of the upper air, a *voyageur extraordinaire*.

In the same mode, Nadar published a spectacular 'Manifesto of Aerial Autolocomotion' in a winter 1862 edition of *La Presse*, announcing his *Géant* scheme. It is presented as a disinterested scientific project, undertaken purely at his own expense, but with obvious commercial potential internationally (for anyone who might care to invest). At the same time Nadar makes witty use of hyperbole, so it can also be read as a brilliant and playful piece of advertising copy:

I shall construct a balloon – the last word in balloons – in proportions extraordinarily gigantic, twenty times larger than the largest hitherto known. It will realise what has only been a dream in the American journals; and it will attract, in France, England, and America, those vast crowds that are always ready to run to witness even the most insignificant balloon ascents.

In order to add further to the interest of the spectacle – which, I declare beforehand, without fear of contradiction, shall be the most beautiful spectacle which it has ever been given to mankind to contemplate! – I shall dispose under this monster balloon a small balloon (or balloonette) designed to receive and preserve the excess of gas produced by dilation. Instead of losing this excess, as has hitherto been the case, this will permit my balloon to undertake genuinely extensive voyages, instead of remaining in the air two or three hours only, like our predecessors.

I do not wish to ask anything of any private investor, nor of the State, to aid me in this proposal of such general, and also of such immense scientific interest. I shall endeavour to furnish entirely by myself the enormous sum of two hundred thousand francs necessary for the construction of my balloon. Once my balloon is completed, I am confident that a series of well-publicised ascents and successive exhibitions at Paris, London, Brussels, Vienna, Baden, Berlin, New York (and anywhere else I can think of) will generate ten times the funds necessary for the construction of my next scheme: our first true [heavier-than-air] aerolocomotive. [12]

2

Le Géant was designed and piloted by the leading French aeronauts Louis and Jules Godard.[13] Nadar was careful to publicise the precise financial and technical details of its gigantesque construction, knowing that such lavish extravagance had a particular appeal in the Second Empire. The balloon was made of twenty-two thousand yards of silk, costing the equivalent of five shillings and fourpence a yard, making its price alone almost £6,000. This was cut into 118 gores, which were entirely hand-sewn with a double seam. Two hundred women were employed for a month in the sewing of the gores. For the sake of greater strength, the silk was doubled. In other words, there were two balloons of the same size, one within the other. The vast envelope contained 212,000 cubic feet of gas and stood approximately 196 feet in height when fully inflated. This was well over twelve storeys high (nearly the height of the first platform of the Eiffel Tower when it was built in 1887). It was a truly fantastic sight, visible for miles above most of the surrounding buildings of Paris. It was the biggest logo that Nadar had ever imagined.[14]

Nadar's deployment of all these hard facts and figures was carefully set against a personal narrative which appealed directly to the sympathies of his readers. He turned the whole construction process into a gripping drama.

> *I have set myself to work immediately, constantly and with great difficulties, suffering sleepless nights and daily vexations. These I have kept to myself up till now, but one day this winter when the most urgent part of my task is completed, I shall reveal all to my readers. I have succeeded in establishing my balloon, and in simultaneously founding this journal –* L'Aéronaute. *It will become the indispensable guide to aerial autolocomotion. And I shall have laid the basis of that which shall be, perhaps, the greatest financial operation of the age. Those who see and appreciate these labours, will I hope pardon their Nadar, for wiping the sweat from his brow with a little gesture of pride! In one month's time, a mere one month! – I shall announce: 'C'est fait! It is accomplished!'*[15]

Nadar commissioned a spectacular wickerwork gondola to house paying passengers in the greatest possible comfort. The gondola expressed something of Nadar's childlike enthusiasm and fantasy. It looked less like a conventional balloon basket than a fairy-tale cottage out of a children's illustrated book. (Admittedly, some critics said it looked like a small garden shed.) It was thirteen feet long, eight feet wide, and ten feet tall – but somehow looked bigger and more mysterious than these prosaic dimensions. It was constructed on two levels, with an open-top sundeck like a kind of aerial balcony, and an enclosed lower deck like a ship's cabin.

The cabin had a central entrance door, and several little windows or portholes. It could be divided by partitions into a maximum of six separate compartments, and the extensive fixtures were altered according to the requirements of different voyages. Besides the captain's compartment with a navigation desk, it contained at various times a set of guest bunk beds, a small printing press, a photographic studio, a galley kitchen and wine store, and – most important last touch of luxury – a portable lavatory. The upper deck was reached by an internal ladder. Nadar claimed that Le Géant could carry up to twenty people, and had a lifting capacity of four and a half tons. The original is still proudly kept on display at the Musée de l'Air at Le Bourget, Paris.[16]

The maiden voyage of Le Géant started from the Champ de Mars at 5 p.m. on 4 October 1863, with fifteen passengers and many crates of

champagne aboard. The flight lasted five hours, and came down near Meaux (the mustard capital) before midnight, instead of flying on till dawn as planned. But the occasion was a masterpiece of Nadar's commercial and publicity skills. Each passenger agreed to pay a thousand francs. Despite a 'no women or children' rule, Nadar shrewdly accepted at the last moment a fashionable and glamorous young aristocrat, la Princesse de la Tour d'Auvergne. He also distributed 'a hundred thousand' copies of a special number of his review L'Aéronaute to a huge crowd of paying spectators. His receipts amounted to thirty-seven thousand francs, although, as expected, these did not cover half his costs. But the press coverage was global, reaching even the Scientific American.[17]

Taking advantage of the immediate publicity, Nadar launched the second flight of Le Géant a fortnight later, on 18 October. This time he announced a sustained long-distance voyage eastwards: Germany, Austria, Poland, even Russia were all possible destinations. The balloon supplies were fully restocked, but this time there were only six people on board. It was something like a professional crew: the balloon's two designers, the brothers Louis and Jules Godard, a member of the Montgolfier family, a reporter, Théobald Saint-Félix, Nadar himself and – rather surprisingly – Nadar's young wife. Nadar's public relations had been masterly: the spectacular sunset lift-off was witnessed by no less than the Emperor Napoleon III himself, and the King of Greece, both VIP guests in a special enclosure.[18]

Le Géant sailed rapidly north-eastwards across Paris and towards the Belgian border. Nadar records that they took supper on the sundeck and opened no fewer than six different cases of vintage wine supplied gratis by the Paris wine merchant Courmeaux (just one of many examples of his skilful marketing). After nightfall, like Green in the Nassau, they watched with fascination as they passed over the fiery Belgian ironworks.[19]

Just before dawn they were over Holland, and mildly worried about floating out over the North Sea. But the wind veered to a more south-easterly direction, carrying them inland into Germany in the direction of Hanover. It also stiffened considerably, though this was not evident at four thousand feet, and when the sun came up the passengers took a tranquil breakfast of coffee and croissants on the sundeck. After some discussion, the Godards decided to attempt a landing in the open countryside before they reached Hanover. Nadar later claimed that he had serious reservations about this, and wanted to continue much further east, according to the original plan; waiting till the wind dropped, and at least crossing the Rhine.

As they valved gas and descended, it became clear how alarmingly fast *Le Géant* was travelling over the ground. It was a situation very similar to that experienced by Wise on the foreshore of Lake Ontario, though with a quite different outcome. Unused to the aerodynamic properties of the enormous balloon, the Godards had released too much gas, and left themselves too little ballast to recover height. They found themselves committed to a very rough landing near the village of Nimbourg. The passengers were instructed to come up from the cabin, and brace themselves around the sides of the sundeck, holding on tightly to the special leather hand-grips. They let down both the enormous grapnel irons, and hoped for the best.

The huge balloon brushed the ground and rebounded. It instantly ripped off both its grapnel anchors, and began careering over the open farmland at an estimated speed of thirty miles an hour. This might not sound very fast, but for Nadar and his passengers it was like being attached to a wild animal that had gone completely berserk.

It is worth remembering at this point that the balloon was nearly two hundred feet tall. As it was still partially inflated – the Godards had not supplied a rip-panel – it continually bounced fifty feet into the air and came crashing back down in a series of giant leapfrogs. When trees got in its path it simply tore off huge branches and lunged on. Everyone on board was paralysed with shock. From below them in the cabin came the sound of smashing crockery, furniture and equipment. One side of the sundeck was ripped off, so the passengers tried clinging to the balloon rigging above their heads. In the confusion, Nadar lost hold of the valving line. Jules Godard made three attempts to climb into the hoop to retrieve it. Meanwhile the nightmare journey continued.

In a narrative development that Verne would have appreciated, the balloon was now blown into the path of an approaching express train. They were travelling at right angles to the track; the alert train driver spotted the monster bounding towards him across the fields from the west, and correctly calculated that to avoid a fatal collision, he must slow down rather than accelerate. He applied the emergency brakes, and halted the train in time for the balloon to bounce across the tracks directly in front of the engine. They were so close to the driver's cab that through the cloud of steam they could hear him yell a warning in German.

Nadar worked out that he was shouting 'Mind the wires! Mind the wires!'[20]

Indeed, the balloon basket was now hurtling towards a line of telegraph wires running parallel to the railway track. Nadar glimpsed them approaching at head height, and a single grotesque thought flashed through his mind: 'Four electric wires – *four guillotines!*' He grabbed Madame Nadar.

> We lower our heads, crouching down ... By fantastic good luck at that precise moment we skim back towards the ground. The wires strike just above us, at the level of the balloon hoop and the small gabion fastenings. Only one or two of the leading balloon cables are cut by these slashing razors, which are immediately torn up and dragged along behind us – like the flying tail of a crazed comet ...[21]

To Nadar's amazement, the force of the balloon uprooted the two telegraph poles on either side of them, and the whole hissing tangle of wires and poles was dragged along for several seconds. Having shed them, *Le Géant* continued its terrible, headlong course for a further ten miles, continually leaping into the air and then smashing the gondola onto the ground. One by one the passengers jumped or were thrown out, until only Nadar and his wife remained, clinging to each other and curled up in a foetal position in a corner of the sundeck, their hands locked onto the remaining pieces of wickerwork.

Le Géant, now reduced to a long flailing tangle of ropes and silken canopy, finally tore into a dense thicket of woodland, and was shredded between its branches. The wreckage came to rest in a clump of trees, the battered gondola lying on its side, bodies and pieces of wickerwork strewn in a long trail behind it. It had travelled four hundred miles in fourteen hours, at speeds ranging between twenty and sixty miles per hour. Every one of its passengers was injured, although inexplicably none was dead.

News reports were telegraphed to journals all over the world, including *La Vie Parisienne*, the *Illustrated London News* and the newly founded *New York Times*. The *Scientific American* carried a gripping (but slightly inaccurate) stop-press item: 'Second voyage of the Géant. Seventeen hours and two hundred and fifty leagues. Landing near Nieubourg in Hanover. Balloon dragged for several hours. Nadar suffered fractures to both legs, his wife a deep wound to the thorax.'[22]

Paradoxically, this was just the kind of story that Nadar could turn into brilliant publicity. The moment he and his wife got back to Paris (by train) he began working on his *Mémoires du Géant* (1864). His detailed description of the crash-landing occupied no less than thirty-two pages, and ends with the dramatic admission that at first he thought he had been responsible for his wife's death. In fact she was slightly concussed, and cut under the chin, but not otherwise hurt.[23] He wrote the book while recovering from his own leg injuries, which were more severe, and took several months to heal. For good measure, he also brought a lawsuit against the pilot, Jules Godard, for incompetent balloon management. Surprisingly, none of the other passengers brought a lawsuit against Nadar.

Nadar saw that in narrative terms, the climactic moment was the encounter with the express train. Accordingly, he commissioned his brother Adrien to compose a semi-documentary drawing of this, carefully assembling all the elements into a single image of imminent peril. The viewpoint is dramatically from above, as if from another balloon. The dark shape of the *Géant*, bent over sideways and grotesquely distorted by the howling wind, almost fills the picture. It looks like some blind, maddened creature rushing across the fields, casting a huge shadow, and terrifying a flock of sheep and a pair of horses. The gondola drags along helplessly in its wake, leaving behind a trail of shattered trees. In front of it the steam engine, tiny by comparison with the huge balloon, thunders

down the track under a flattened plume of steam. The near-fatal telegraph wires are just visible in the balloon's path.

Nadar carefully copyrighted this image, which appeared in journals
all over Europe during the winter of 1863. He also used it as the frontispiece to his book when it was published eight months later, in the spring
of 1864. It had become the symbol – the logo – of the flight. He had the
balloon and basket repaired and refitted, and over the next four years *Le
Géant* made further demonstration ascents from Brussels, The Hague,
Hanover, Meaux and Lyons. When it flew from Amsterdam, the publicity
posters were distributed as far afield as Geneva and Marseille.[24] Nadar
had broken with the Godard brothers, and instead employed Camille
d'Artois as his pilot. There were no further aerial dramas. Despite the
understandable objections of Madame Nadar, he went up again himself
from the Hippodrome at Lyons on 2 July 1865.

3

What Nadar now decided to do was ingenious. He would use the story of
the *Géant* to demonstrate that ballooning was superannuated as a
concept. A lighter-than-air machine would publicise the heavier-than-air
cause. Indefatigable, Nadar set out to write a different kind of

campaigning tract, *Le Droit au vol*, 'The Right to Flight'. This made the case for the 'aero-locomotive' as against the balloon. The title also neatly implied that flight had somehow become one of the Rights of Man. He coopted George Sand to write the Preface. She did so with a flourish, but later wrote to him privately, on 28 September 1865: 'Dear Nadar, I must beg you to renounce these terrible balloon-*antics* that worry your friends far more than you realise. I beg you to go back to photographic portraiture! *Mine* for example ...'

Le Droit au vol promoted Nadar's new conviction that flight would only reach its full potential when a true 'aircraft', powered by an engine, was invented. Such an aircraft or airship, powered by a steam, electric or even gas engine, would necessarily be heavier than air, and would at long last provide the fully navigable vehicle that balloonists had been seeking for over a hundred years. Just like an ocean-going ship, an airborne ship could be steered by rudders working against the airflow. Thus it would become a '*dirigible*' – a new word that conveniently worked in both French and English. In fact, with rudders and elevators, it could be steered in three dimensions. Exactly what this machine would look like, no one yet knew. But it would not look like *Le Géant*.

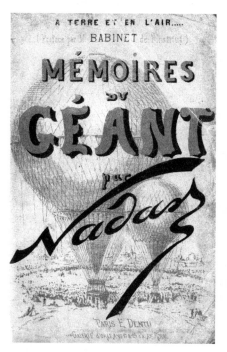
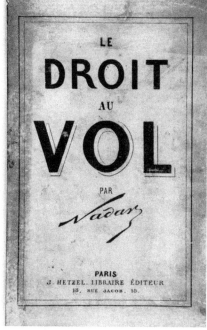

Nadar turned to the most controversial and celebrated writer in France to back his cause. Victor Hugo was a loyal friend from the days of the *Panthéon Nadar*, and Nadar wrote to him requesting some form of public statement in support of *Le Droit au vol*. Would Hugo underwrite his vision of the future of flight? Hugo had after all had such aerial visions himself, describing the astonishing view of medieval Paris 'as seen from the air' in the second chapter of his great novel *Notre-Dame de Paris*, entitled '*Paris à vol d'oiseau*' ('A bird's-eye view of Paris').[25] He had also recently featured in a flattering cartoon, published by the *Journal amusant*, showing him ascending heavenwards for earnest discussions with God, in a balloon marked 'Mankind'.

Hugo was controversial because he was a declared republican enemy of the Second Empire and all its works. To prove his 'eternal' opposition he was living in self-imposed exile from France in his famous clifftop residence, Hauteville House, on what he called the 'foam-lashed rock' of Guernsey. This exile had only increased his huge public following and readership back in France. Hugo was himself a past-master of publicity. His was exactly the name that Nadar needed.

Hugo wrote back one of those letters addressed simply to 'Nadar, Paris'.[26] He congratulated Nadar on his personal bravery during his balloon adventures – 'What courage, intrepidity, audacity!' *Le Géant* was, certainly, a typically monstrous product of imperial flamboyance and exaggeration, but 'the risk you took was magnificent! And *the risk* is the true example!' He did not mention the risk run by Madame Nadar and the other passengers.

Hugo announced that he would willingly put on his 'prophetic wings' for his reckless old friend. The concept of Flight was democratic, it was progressive, it was 'universal'. The long open *Letter on Flight* which followed was, in effect, the text of a brilliant popular tract specifically designed for Nadar to print and distribute under the auspices of *L'Aéronaute*. It was addressed, in a modest gesture typical of Hugo, 'To the Whole World'.

Its theme was radical, and its grandstand manner was crafted to appeal to the broadest possible readership. Like a modern tabloid editor, Hugo skilfully invented catchy slogans, coined neat catchphrases, and spun sensational headlines. His letter became a publicity brochure, a masterly piece of advertising copy for the art and science of flying. As he put it: 'Let us deliver mankind from the ancient, universal tyranny! What ancient, universal tyranny, you cry. Why, the ancient, universal tyranny of *gravity!*'[27]

Hugo began on a patriotic note, recalling the self-sacrifice of all the previous French aeronauts, from Pilâtre de Rozier onwards. Ballooning was indeed a specifically French gift to the world, and no foreign aeronauts – not even Charles Green – were mentioned. It was the French who had opened a new world, a new direction of travel. Rousing phrases were piled one upon the other. 'The iron bolt has been drawn back from the blue abysm.' The 'vertical journey' had become possible. Mankind would take possession of the 'fourth of the ancient elements', and be 'master of the upper air'. Man would be a bird, an eagle, a 'thinking Eagle with a Soul'.[28]

It was the heroic Nadar, wrote Hugo, who had conclusively demonstrated that the aerostat, the 'lighter than air' machine, could never fulfil the immense promise of flight. The terrible crash of the *Géant* in Hanover had proved once and for all that the aerostat was fundamentally flawed as a concept.

> *Today the balloon has been judged, and found wanting ... To be torn from the ground like a dead leaf, to be swept along helplessly in a whirlwind, this is not true flying. And how do we achieve true flight? With wings! ... For the* dream *of flight to become the* fact *of aviation, we have only to accomplish a small and relatively simple technical break-through: to construct the first true ship of the air* [le premier navire] *...*
>
> *Whoever you are, reading this declaration, lift up your heads! What do you see above you? You see clouds and you see birds. Well then, these are the two fundamental systems of aviation in operation. The choice is right in front of your eyes. The cloud is the balloon. The bird is – the helicopter!*[29]

The idea of the helicopter or '*helice*', as proposed by Hugo, was taken from both Nadar and Verne writing in *L'Aéronaute*. It was an idea gaining wide acceptance in the 1860s. Hugo popularised the fundamental distinction between the floating balloon (the cloud) and the driven airship (the bird). But how was the airship to be driven? How would its wings actually work? One solution was some form of spinning 'airscrew' or 'propeller' powered by an aerial engine. Like wings, but far more efficiently, the angled blades or paddles of such a device would have purchase on the air, and would drive or drag the craft through it, exactly like a marine

propeller driving a ship through water. One possible design was Ponton's 'helicopter', photographed by Nadar in 1863.

The two airscrews were apparently designed to produce both horizontal movement and vertical lift, though the parasol-parachute suggests some uncertainty about their efficacy. There is no indication of what engine might power this machine, but it clearly abandons the age-old chimera of 'flapping' wings, and is a step towards the propeller-powered 'airship'.🜨

Navigable flight, continued Hugo in his *Letter on Flight*, would be of huge scientific and social importance once it was achieved. In praise of this hypothetical future, Hugo let out all his rhetorical sails:

🜨 In fact the first true airship or dirigible had already been invented in France by the engineering genius Henri Giffard (1825–82). In September 1852, Giffard successfully flew his baguette-shaped hydrogen balloon with a three-horsepower steam engine slung beneath it, driving a '*helice*', or propeller. It was steered with an upright rudder, and flew twenty-seven kilometres southwards from the Paris racecourse to Trappes. However, it was slow and clumsy, and would not fly against contrary winds. Hampered by lack of money to develop his ideas, and despite receiving the Légion d'Honneur in 1863, Giffard spent the rest of his career doggedly building bigger and bigger captive hydrogen balloons. With equal irony, his wonderful steam engines were used to power their cable winches on the ground. His final balloon was a true captive monster, a King Kong of the balloon world. In his late fifties Giffard's eyesight began to fail, and, unable to contemplate a world without a visual horizon, he committed suicide. But his ideas influenced Charles Renard's airship *La France*, which made its maiden flight in 1884; and later still, as we shall see, Salomon Andrée in the Arctic.

It will bring the immediate, absolute, instantaneous, universal and perpet-
ual abolition of all frontiers, everywhere ... The old Gordian knot of gravity
will finally be untied ... Armies will vanish, and with them the horrors of
war, the exploitation of nations, the subjugations of populations. It will
bring an immense and totally peaceful revolution. It will bring a sudden
golden dawn, a brisk flinging open of the ancient cage door of history, a
flooding in of light. It will mean the liberation of all mankind.[30]

There was much more in this wild, heroic vein. The absurd error, or
perhaps the glorious naïveté, of Hugo's prophecies is striking. His vision
of the 'universal peace and freedom' arising directly from the conquest of
the air has a kind of innocence about it, which goes back historically to
the '*ballomania*' of the 1780s, and the declarations of poets like Erasmus
Darwin and Percy Shelley. Yet perhaps we are now too quick to view these
dreams of 'liberating all mankind' as entirely misplaced. The fact remains
that air travel and transportation, as well as satellites, are the *sine qua non*
of our global civilisation; and space flight may yet become the final means
of its salvation. It would be interesting to read Hugo on the Apollo space
missions of the 1960s, and the current Mars-landing programmes.

At any rate, Hugo now had the bit between his teeth. He unblush-
ingly compared Nadar's pioneering balloon flight to Hanover with
Christopher Columbus's sea voyage to discover America. Nadar's public
bravery – rather less obviously – was likened to the moral courage of
Voltaire and the religious iconoclasm of Luther. If Nadar could be accused
of self-promotion, of seeking publicity, of 'making a noise' with his
balloon adventures, then so could those other master spirits. Yet they
were each fighting for a worthy cause. With this thought, Hugo became
fully airborne:

People accuse you, Nadar, of 'just seeking to make a noise'. That is the
age-old sneer! The sneer of silence against speech, dumbness against
expression, castration against fecundity, nihilism against creativity, envy
against the masterpiece, egoism against the generous act, the tin-whistle
against the sounding horn, the abortion against the new born child ... But
I say the noise you make with Le Géant is A GOOD NOISE.[31]

Hugo closed his *Letter on Flight* on a more personal note, recalling a memorable exchange he had once had with the great French scientist François Arago (1786–1853). The outstanding astronomer and physicist of his generation, a staunch republican and a supporter of the earliest French scientific balloon ascents, Arago had died ten years previously, and was now widely regarded in France as something of a scientific visionary and secular saint. He had already been canonised by having craters on both the moon and Mars named after him, and was looked upon as a man who had seen the future.

According to Hugo's story, the poet and the scientist were walking one evening in the Luxembourg Gardens, along the path known, symbolically enough, as l'allée de l'Observatoire. It had been an official festival day and public holiday, and they were deep in speculative discussion. A large balloon passed unexpectedly overhead, having just taken off from the Champ de Mars. Its full, round, pregnant shape, touched with gold by the setting sun, was 'truly majestic', and filled them both with a moment of silent awe.

I said to Arago: 'There is the floating egg which is destined to become a bird! That bird is still inside the egg. But it will soon emerge!' Arago seized me by both hands, and fixed me intently with his large, luminous eyes. 'On that day,' he murmured, 'Geo will become Demos – the earth will belong to the People.'[32]

Hugo signed off with a final farewell to Nadar: 'I no longer address you personally, brave Nadar the Aeronaut. There is nothing further I can tell you, that you do not already know. Instead, I fling this open letter upon the four winds. I write on its envelope: *A tout le Monde!*'[33] As it turned out, Hugo was far from finished with balloons in 1865. The four winds would soon bring them back to him, as unexpected symbols of liberty, flying above the rooftops of Paris.

Over the next three years Nadar continued to send the repaired *Géant* around the cities of northern Europe, as much for tethered publicity displays as for actual flights. They were always accompanied by pamphlet

copies of Hugo's *Letter on Flight* and his own *Droit au vol*. Nadar even took the balloon to London, where it was exhibited at the Crystal Palace. The celebrated two-storey gondola was much photographed, and the cause of 'heavier-than-air' flight much discussed.

Balloons and flying became populist symbols again; the propaganda for aviation and republicanism were twinned. During these tours Nadar met many writers and opposition politicians, in exile like Hugo from the Second Empire. Among them were Charles Baudelaire and Armand Barbès, both in Brussels and both old friends from the *Panthéon Nadar* days. Nadar's aeronautical fame gave him a certain freedom of manoeuvre. When he was presented to Leopold II, the King of the Belgians, the following exchange took place. His Majesty: 'You are a republican I suppose, Monsieur Nadar?' 'Yes indeed, sire. And you?' 'Ah! Monsieur Nadar, my profession absolutely forbids it.'[34] Even so, the king could not be persuaded to ascend in the republican balloon.

The end of the *Géant* itself was strangely muted. Perhaps this was appropriate for a balloon dedicated to advertising its own superannuation. The last three ascents of the now ancient and creaking campaigner, damaged as much by overland as by aerial travel, took place in Paris during the Exposition Universelle of 1867. Captained by Nadar, it was launched from Les Invalides, amid great celebrations and thousands of spectators, but got little further than the Paris suburbs.

A balloonist of the new generation, Wilfrid de Fonvielle (1824–1914), who was on board, left a wry account of 'the *three last gasps* of the late, great *Géant*'.[35] The old balloon appeared 'striped like a zebra', encircled by long bandages of white silk crudely sewn on with black thread to cover some of its 'many wounds'. Fonvielle dreamed of going as far as the Danube, but finished up at Choisy-le Roi instead, and the balloon leaked alarmingly all the way. White smoke oozed from its top seams, 'like the steam that issues from the funnel of a locomotive'. Someone joked that it was just the old *Géant* 'smoking his pipe', but it was no laughing matter to Fonvielle: 'This pipe was being smoked over a barrel of gunpowder.' It was a relief when the balloon was finally deflated, folded and bagged up. The whole extraordinary legend of *Le Géant*, so carefully built up by Nadar, seemed finally reduced to a large dirty sack of 'so much matchwood'. Perhaps, Fonvielle concluded, it had rendered 'some slight service to the art of aerostation'.[36]

3

Yet the imaginative influence of Nadar's gigantesque balloon, and the publicity surrounding it, was far more subtle and widespread than this. It elevated and transformed the very idea of travel itself. Jules Verne had been following Nadar's adventures from the beginning, and in December 1863 published a long article in *Musée des familles*, 'Nadar the Aeronaut', praising his courage and vision. He said that Nadar was the man who had demonstrated that it was not enough merely to 'float in the upper air' passively. The true hero would actively 'fly through it' with purpose and energy, towards a definite destination. This destination could be a place, real or invented; or it could be 'adventure' itself.[37]

Jules Verne (1828–1905) had been an unknown freelance science journalist from Nantes when he first met Nadar in 1862. He was still struggling in his mid-thirties to establish his literary career in Paris, and had been supporting himself with miscellaneous legal work, writing plays, and publishing occasional short articles on travel and invention. For several years previously he had been looking for actual adventures, 'true stories', for his magazine articles. He slipped away 'day and night' from his job at the Paris *Bourse* to research at the Bibliothèque Nationale, which offered him 'endless resources'.[38]

Verne saw that travel in general, and ballooning in particular, potentially supplied fantastically rich material. For instance, in 1849 a young French aeronaut, François Ardan, had made the first daring crossing of the Maritime Alps by balloon, starting at Marseille and landing near Turin.[39] Verne did not forget Ardan's name when he later came to invent his scientific daredevil 'Michel Ardan', who flies to the moon.

Verne had also researched Julien Turgau's vivid study *Les Ballons: Histoire de la locomotion aérienne*, which had appeared in 1851. It had an inspired Introduction by the visionary poet Gérard de Nerval, and memorable steel engravings of airborne balloons. Equally, his friendship with François Arago's younger brother, the seasoned travel writer Jacques Arago, turned him towards the idea of fantastic expeditions. Early results were his articles 'A Balloon Trip' (1851) and 'Wintering on the Ice' (1853). By the end of the decade he was already discussing with Alexandre Dumas *fils* his new concept for what he called '*le roman de la science*'.[40]

When Verne met Nadar and the editor Pierre-Jules Hetzel – another member of the *Panthéon Nadar*, and the hugely successful publisher of Balzac, Hugo and George Sand – in the autumn of 1862, the crucial moment had arrived. He abandoned a negative, dystopian novella about Paris, and agreed to write instead a high-spirited balloon adventure.ᵠ Verne wrote excitedly to his friends at the *Bourse* bidding them adieu, and putting his future in appropriately commercial terms. He was writing a novel 'in a new style, truly my own – if it succeeds it will be a gold mine'.[42] The contract for *Cinq semaines en ballon* was signed with Hetzel in October 1862.

In fact, a crude first version of the story had already appeared as an article in a Swedish magazine under the title 'In a Hot-Air Balloon Over Africa'. It was never translated, and according to Verne's wife he had struggled to expand it in a manuscript version with the working title 'A Voyage Through the Air'. It caused him such despair that he had threatened to burn it.[43]

After many talks with Hetzel and Nadar, Verne rewrote and expanded the manuscript in his 'new style'. Inspired by Nadar's enthusiasm and Hetzel's editorial skills, he transformed his flat documentary 'magazine manner' into a brisk form of narrative. His chapters became brief and punchy: there are forty-three in the short novel. He added a mass of realistic scientific details and dramatic ballooning incidents: storms, condors, elephants, volcanoes, drought, hallucinations, mirages, wild tribesmen, and of course frequent near-misses and crash-landings.

ᵠ It is often forgotten that at this time Jules Verne still had profound doubts about scientific progress and technology. In this same year, 1862, he wrote the short dystopia entitled 'Paris in the Twentieth Century', about a young poet living in a world of skyscrapers, high-speed trains, gas-powered cars, chemical warfare and global telegraphy, where even the dead can be revived by electric shock treatment. Far from being happy, Michel Defrenoy is deeply depressed by the materialism of society; he can find no serious books or music, and he detests living off synthetic food made from coal. A new ice age engulfs the whole of Europe, and Michel dies in the snow outside Père Lachaise cemetery clutching an unpublished book of his poems entitled *Hopes*. When Verne presented the draft of this grim vision to Hetzel, he refused it: 'Wait twenty years to publish this book. No one today will believe your prophecies, and no one will care about them – It will damage your reputation.' Instead he persuaded Verne to turn to the balloon story. The manuscript was only recovered over a hundred years later, by Verne's great-grandson, in 1989. It was finally published in 1994, and paradoxically, its pessimism rather enhanced Verne's reputation.[41]

Above all he introduced a lively play of comic dialogue among his balloon crew. Verne saw that the balloon basket (like Ardan's later moon rocket, or Captain Nemo's submarine) was the ideal enclosed space in which to stage a drama, and draw out contrasting characters under pressure. As Nadar would experience, any balloon flight – especially a long one – was always essentially a piece of theatre. To exploit this Verne introduced three highly contrasted protagonists, and Hetzel soon appended a provocative subtitle: *Five Weeks in a Balloon; A Voyage of Discoveries in Africa by Three Englishmen*. The novel was swiftly completed and published in late January 1863, three months after Nadar's 'Manifesto'. It was the perfect moment: talk of ballooning and *Le Géant* was becoming all the rage.

4

Five Weeks in a Balloon is Jules Verne's first true science fiction novel, his first *roman de la science*. It proved to be the major breakthrough in his career as a popular author. With extraordinarily realistic details and statistics, it recounts a purely imaginary balloon expedition westwards across the whole of Africa, starting from the island of Zanzibar, in the Indian Ocean. Zanzibar, then a British colony, lies off the east coast of what is now Tanzania. The protagonists are three unflappable British types, an adventure formula that Verne would develop and repeat in many later books.

The first is the eccentric Dr Samuel Fergusson, dreamer and explorer, the commander of the expedition, ex-member of the Bengal Engineers, 'possessed by the demon of discovery'. Florid-faced, calm, stoic, wiry, immune to any disease or privation, Dr Fergusson has ranged restlessly through India, Australia and America, and has 'dreamed of fame like that of Mungo Park and Bruce, or even – I believe – like that of Selkirk and Robinson Crusoe'.[44] For Fergusson the balloon is the application of pure science to exploration. He is the prototype of all Verne's cerebral adventurers.

Next is his bluff and fearless friend Dick Kennedy, a big-game hunter and 'a Scotsman in the full significance of the word, open, resolute, and dogged'. They had met in India where Kennedy 'was hunting tigers and elephant and Fergusson was hunting plants and insects'. Kennedy is shrewd and practical, and thinks Fergusson's African balloon project is obviously mad. But when he finds he cannot prevent it, he decides to join Kennedy out of blind loyalty. '"Rely on me," said Fergusson, "and let my motto be yours: *Excelsior!*" "Very well, old man, *Excelsior!*" answered the sportsman, who didn't know a word of Latin.'[45] *Excelsior* means, of course, Ever Higher!

Finally, as the catalyst between the two, Verne conjures up their small Cockney manservant, Joe. In recognition of Victorian class distinctions, Joe has no surname, but he is far from being a cypher in the plot, or mere ballast in the balloon. (Indeed, he is the forerunner of Passepartout in *Around the World in Eighty Days*.) Smart, resourceful, quick-thinking and sharp-tongued, he is also a formidable gymnast, and so perfectly adapted to life in a balloon basket: 'Jumping, climbing, somersaulting, performing a thousand impossible acrobatic tricks, were child's play to him.' More than this, he is one of Nature's enthusiasts: fascinated by science, in love with balloons ('beautiful things'), and above all loyal to his master: 'If Fergusson was the brilliant brain of the expedition, Kennedy was the brawny arm, and Joe was the dexterous hand.'[46]

The novel opens with a remarkably convincing account of Dr Fergusson being summoned before the Royal Geographical Society in London, from whom he hopes to raise funds. However, satirical elements soon surface. Verne hints at his wide background reading by making disguised reference to Poe's balloon hoax story of nearly twenty years previously:

> *'Perhaps this incredible scheme is only intended to hoax us,' said an apoplectic old Commodore. 'What if Dr Fergusson didn't really exist?' cried another malicious voice. 'Then he'd have to be invented,' replied a waggish member of this learned Society. 'Show Dr Fergusson in,' said Sir Francis simply. And the doctor entered amid a thunder of applause, and without the least show of emotion ... His whole person exhaled calm gravity, and it would never have occurred to anyone that he could be the instrument of the most innocent hoax.[47]*

Verne then presents the mass of historical, scientific and statistical data that he was learning to deploy, sometimes quite mischievously, in order to provide the necessary frame of realism. Chapter 4 is a short history of African exploration, from Bruce to Speke; Chapter 7 a short treatise on balloon theory and practical navigation by east-to-west trade winds (which John Wise would have recognised). Some of it is straight-facedly pedagogical in tone: 'By giving the balloon a capacity of 44,847 cubic feet and inflating it not with air but with hydrogen which is fourteen and a half times lighter than air and weighs only 270lbs, a change of equilibrium is produced amounting to a difference of 3,780 lbs, which constitutes the lifting force of the balloon.'[48]

Other sections slip into pure romance, as when Dr Fergusson harangues Kennedy on the beauty and convenience of balloon flight. Here one can hear the voice of the authentic balloon geek. It is a voice partly inspired by Nadar, but also going back to the very roots of the aeronautical tradition and eighteenth-century *ballomania*.

> *I don't intend to stop until I reach the West coast of Africa! With my balloon there will be nothing to fear from either heat, torrents, storms, the simoon, unhealthy climate, wild animals or savage men. If I'm too hot, I*

*go up. If I'm too cold, I come down. If I meet an impassable mountain, I
fly above it; a precipice, I sail over it; a river, I float across it. If I encounter
a storm, I ascend above it; a torrent, I flit over it like a bird. I travel without fatigue and halt without need of rest. I soar over new cities. I fly with
the swiftness of a hurricane. Sometimes I rise to the very edges of the
breathable atmosphere. At others, I descend to skim over the earth at a few
hundred feet, so the great map of Africa unwinds beneath my eyes like the
mightiest atlas in the world.*[49]

Launched from Zanzibar, they do indeed follow 'the great map of Africa'
along the westward line of the upper Nile into the African interior. But
nothing goes as Fergusson had prognosticated. The picaresque adventures and mishaps of their balloon, the *Victoria*, unfold one after another,
helter-skelter, practically without pause for breath – or hydrogen. Verne,
brilliantly inventive and bold in his plotline, gives a new meaning to the
term *suspense*. The crew suffer from fever, are towed by a runaway
elephant, attacked by condors, worshipped by witchdoctors (who think
they are the moon come to earth). They rescue a missionary from canni-
bals, sail through a hail shower, survive a huge electrical storm, overfly an
active volcano, escape from clouds of locusts, beat off 'incendiary
pigeons', and find the source of the Nile, though not necessarily in that
order. In the end, with no more hydrogen left, they turn the *Victoria* into
a Montgolfier. By burning dry grass beneath the canopy, they just manage
to escape the final wave of 'infuriated natives', and skim across a river
into French-occupied Senegal and safety.

Infuriated natives in fact feature regularly throughout the novel.
Long before, Shelley had imagined that 'the shadow of the first balloon'
crossing Africa would be an instrument of liberation and enlightenment.
But for Verne the balloon is more a symbol of imperial command, and
scientific superiority.� *Cinque semaines* is paradoxically a work of colonial

---

♆ It is instructive, and poignant, to compare this with Percy Shelley's views some fifty
years earlier: 'The balloon promises prodigious faculties for locomotion, and will
allow us to traverse vast tracts with ease and rapidity, and to explore unknown countries without difficulty. Why are we so ignorant of the interior of Africa? – Why do we
not despatch intrepid aeronauts to cross it in every direction, and to survey the whole
peninsula in a few weeks? The shadow of the first balloon, which a vertical sun would
project precisely underneath it, as it glided over that hitherto unhappy country, would
virtually emancipate every slave, and would annihilate slavery forever.'[50]

exploitation, as much as exploration. It is blatantly – almost naïvely – racist throughout, and observations such as 'Africans are as imitative as monkeys' occur regularly throughout the story.[51]

Balloon height gives the crew not only safety, but also moral superiority. To them, tribesmen are savages, and the wildlife – the elephant, the blue antelope – is there largely for big-game hunting. *De haut en bas* gains a literal force. Africa, and Africans, can become just 'scenery'. It is the recognisable beginnings of a safari culture. Yet it is perhaps also relevant that this is a Frenchman writing about British imperialists, and their very balloon is named after their Queen.

> *The Victoria passed close to a village which the doctor recognised from the map as Faole. The whole population had turned out, and howled with rage and fear. Arrows were vainly shot at the monster of the air soaring majestically above all this impotent fury. The wind was blowing them south, but this did not worry the doctor, as it would enable him to follow the route taken by Captains Burton and Speke. Kennedy had now become as talkative as Joe. 'A bit better than travelling by coach,' he observed. 'Or by steamer,' replied Joe. 'I don't know I think much of railways, either,' continued Kennedy. 'I like to see where I'm going.' 'Balloons is priceless,' agreed Joe. 'You don't feel you're moving at all – the scenery just slides under you. Just begging to be gawped at!' 'Yes, a splendid view! Like dreaming in a hammock.' 'What about some lunch, sir?' said Joe.[52]*

Nevertheless, by using the balloon to open up a new, exotic world – with its special geography, anthropology, natural history, geology and climate – to a popular readership, Verne had a surprise best-seller on his hands. The book was quickly translated throughout Europe, and Verne was able to follow it up with astonishing rapidity. He had found his path, and from that point on he published two or more books a year for the next decade.

It was an amazing output. The most successful titles became equally celebrated in English as in French (and eventually as films). They included *Voyage au centre de la terre* (Journey to the Centre of the Earth, 1864); *De la terre à la lune* (From the Earth to the Moon, 1865); *Vingt mille lieues sous les mers* (Twenty Thousand Leagues Under the Sea, 1869); and *Le Tour du monde en quatre-vingts jours* (Around the World in Eighty Days, 1872). Hetzel quickly saw the enormous potential of the series and began to publish it collectively as *Voyages Extraordinaires* (Astonishing Voyages). From then on Verne could live on his writings, and his reputation was made.

Verne continued to work for Nadar's Society for the Promotion of Heavier than Air Locomotion, of which he eventually became Secretary. He also immediately repaid his fictional debt to Edgar Allan Poe, by publishing a short study, *Edgar Poe et ses oeuvres*, in 1864. He had, after all, trumped Poe by producing probably the most famous of all imaginary balloon flights of the nineteenth century, to which he later appended his Pacific balloon story *The Mysterious Island*, of 1875. Yet the novel is in a sense uncharacteristic of the rest of Verne's *romans de la science*. Like its successors, it is wonderfully inventive in its picaresque storyline, and it is obvious why Hetzel had such confidence in his author's power to produce incident-packed narrative.

Yet it contains very little scientific or technological prophecy. It is not a 'futuristic' novel. The journey of the *Victoria*, just like that of *Le Géant*, is presented as a contemporary marvel, almost as an extended news item, and is specifically described as 'the most noteworthy expedition of the year 1862'.[53] It is true that Dr Fergusson also has various invented devices for increasing the balloon's lifting power (such as a 'Bunsen battery' burner for raising the temperature of the hydrogen – a glimpse of the prototype propane balloons of the 1960s). But he also takes every opportunity to expatiate on the advantages of 'modern' flight over traditional land-based transport in the tropics. If, for example, they had attempted to travel overland like Mungo Park or Speke, they would have been overcome by disaster:

> 'Since leaving Zanzibar half our pack animals would have died of fatigue. We should be looking like ghosts and feeling desperate. We should have had constant struggles with our guides and porters, and no protection against their savagery. We would have suffered the humid, unbearable, disabling heat by day; and often intolerable cold by night. We would have been bitten by insects with mandibles capable of piercing the thickest canvas and driving men mad. Not to mention the wild animals and savage tribes ...' 'I'm in no hurry to try it,' remarked Joe.[54]

There is a premonition of future imperial romances, and the Lost World genre, especially of the British Boy's Own type – Rider Haggard's *King Solomon's Mines* (1885) and *She* (1887) are both set in Africa. But Verne's sense of authentic reportage remains strong. The novel closes on a deliberately dry, factual note: 'The chief result of Dr Fergusson's balloon expedition was to confirm in the most precise manner the geographical facts and surveys reported by Bath, Burton, Speke and others ... Thanks to the present expeditions undertaken by Speke and Grant, we may before long be able to check in their turn the discoveries made by Dr Fergusson in that vast area lying between the 14th and 33rd meridians.'[55] Verne cannot quite forbear to add another, more mischievous statistic: that a special edition of the London *Daily Telegraph* covering the balloon story 'sold out 977,000 copies' on the day of publication.

In consequence, one of the most striking things about the novel's enthusiastic reception was that many reviewers thought it might actually be a true story. With widespread knowledge of balloon adventures like those of Green, Godard and Nadar, an aerial safari over darkest Africa did not seem intrinsically unlikely. Verne might be writing non-fiction. The reviewer in the new Paris daily *Le Figaro* played elegantly with this idea: 'Is Dr Fergusson's journey a reality or is it not? All we can say is that it is as bewitching as a novel and as instructive as a book of science.

Never have the serious discoveries of celebrated travellers been summed up as well.'𝄞

5

The idea that an 'astonishing voyage' might be instructive, or frankly educational, was already in the air by the 1860s. Hetzel considered Verne's readership was primarily mass-market and adult, but this profile would soon alter. The later editions of *Les Voyages Extraordinaires* were more and more lavishly illustrated, which indicates that they were intended for an increasingly youthful audience. This was particularly true of *Cinque semaines en ballon*, which joined a flourishing new genre of educational children's books, using the balloon as a pedagogic device. These had begun appearing, both in France and England, in the 1850s. The balloon, with its bird's-eye or 'panoptic' view, its 360-degree *tour d'horizon*, and its ability to take its passengers across a huge variety of landscapes, became an instrument of, or even a metaphor for, encyclopaedic knowledge. In the right hands, Education itself could be presented as a kind of magic balloon journey.

One of the earliest attempts to do so was an American illustrated geography book, *The Balloon Travels of Robert Merry and his Young Friends*

𝄞 As so often with ballooning, the boundaries between fact and fiction remain curiously porous. In summer 1962, a real balloon voyage across Africa, starting from Zanzibar, was undertaken by three Englishmen to celebrate the centenary of Verne's *Five Weeks*. The story is told in Anthony Smith's delightful balloon classic *Throw Out Two Hands* (1963). (The title refers to the tiny amounts of sand ballast required to adjust the equilibrium of a hydrogen balloon.) The modern trip was presented as an African safari from the air, with great attention paid to wildlife and conservation. One of its notable feats was to overfly the Ngorongoro crater, though unlike Verne's volcano, this was not erupting at the time. As the expedition was partly sponsored by the *Sunday Telegraph*, Smith and his amiable camera crew, Douglas Botting and Alan Root, had considerable technical back-up, and were only too grateful to rely on the help of local tribesmen, rather than fighting them. The style of the book might be described as post-colonial jovial. The balloon itself is christened *Jambo*, and ballooning is presented as an eccentric European sport, whimsical rather than imperial in its manner. The next year Smith became the first Briton (following Nadar's avatar Michel Ardan) to cross the central Alps in a balloon.

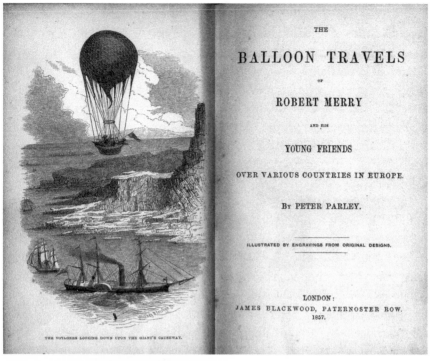

THE VOYAGERS LOOKING DOWN UPON THE GIANT'S CAUSEWAY.

THE

BALLOON TRAVELS

OF

ROBERT MERRY

AND HIS

YOUNG FRIENDS

OVER VARIOUS COUNTRIES IN EUROPE.

By PETER PARLEY.

ILLUSTRATED BY ENGRAVINGS FROM ORIGINAL DESIGNS.

LONDON:
JAMES BLACKWOOD, PATERNOSTER ROW.
1857.

(1857). This was written by the prolific children's author Peter Parley, as part of a series of educational adventures promoted by the Boston magazine *The Robert Merry Museum*. 'Peter Parley' was in fact the pseudonym of Samuel Griswold Goodrich (1793–1860), who liked to be known as 'the Pied Piper in Print'. He claimed to be the author or editor of 170 volumes, with total sales of seven million copies. His series, beginning as far back as 1827, embraced geography, biography, history, science and miscellaneous tales. In 1851 Goodrich – or Parley – became American consul in Paris, and adapted many of his books for French children, a development that did not escape Hetzel.

The Balloon Travels, written towards the end of Goodrich's career, was a kind of farewell to his young readers. It recounts a stately geographical sightseeing tour around the countries of Europe, each site illustrated by an aerial engraving. The frontispiece shows the balloon hanging dramatically over the famous Giant's Causeway, an extraordinary outcrop of basalt rock columns on the coast of County Antrim, in Northern Ireland. The young travellers learn that its legends connect it to Fingal's Cave in Scotland, and this kind of 'aerial perspective' gives them a new view of both history and geology.

In 1869 came a much racier French book, Jean Bruno's *Les Aventures de Paul enlevé par un ballon* – 'The Adventures of Paul Kidnapped by a Balloon'. It was also remarkable both for its twelve beautiful illustrations (the originals in watercolour) and its naïve imperial attitudes. Paul's epic

flight begins by chance one summer in southern France. He accidentally falls into the basket of a giant French balloon, the *Leviathan*, while its three adult aeronauts are on the ground, attempting to anchor it. In a dramatic picture, we see the balloon caught by a gust of wind, the aeronauts sent sprawling, the anchor ropes snapping free, and Paul sailing away over a hillside. Soon the balloon crosses the Mediterranean, and takes him over Algeria and North Africa.

Paul is subsequently swept helter-skelter right across the varied landscapes of Africa. He watches fascinated as its jungles and deserts and savannahs pass beneath him. He encounters storms, wild animals, and wilder tribesmen. Some greet him with friendly waves, others shout and threaten him with spears. One picture shows enraged natives hurling weapons at the balloon overhead, with the caption 'Paul could judge from this specimen of their behaviour what kind of greeting he could expect if he landed.'

Such imperial balloon adventures remained popular throughout the nineteenth century. When Mark Twain brought Huck Finn and Tom Sawyer over to visit Europe in 1894, a chaotic journey in an airship forms the main thread of *Tom Sawyer Abroad*, and appears on the striking scarlet cover design. In fact this seems like Twain's deliberate parody of Verne, as the balloon is piloted by a lunatic professor who insists on taking them to Africa: 'He said he would sail his balloon around the globe, just to show what he could do.' This vogue for imperial balloon voyages,

launched by Verne, only concluded seventy years later with Jean de Brunhoff's unforgettable *Le Voyage de Babar* (1932). In a gentle reproach to Verne, Brunhoff cleverly reverses the imperial perspective from the balloon basket. These native balloonists start from their home in Africa, and fly off to explore the wild boulevards of France. They are two charming and aristocratic jungle elephants (the young King Babar and his bride Celeste), who sail away in a bright-yellow balloon to spend their honeymoon shopping in Paris. This is a truly astonishing voyage. ⁜

⁜ The motif of the small boy carried away by a runaway balloon is so frequent in balloon history that it has almost reached the status of a myth to rival Icarus. Most recently a boy was carried off in a home-made balloon from a farm in Colorado in 2009; though significantly this later turned out to be a hoax perpetrated by his father. Ian McEwan makes memorable use of the runaway balloon in Chapter 1 of *Enduring Love* (1997), where it becomes a symbol of fatality, or fatal attraction, which brilliantly predicts the themes of his entire novel. The forces at work have a kind of 'geometrical' inevitability, in both physics and psychology. The fate of John Logan, lifted off his feet into the air when he gallantly tries to save the boy, and subsequently dropping to his death, is a bad case of falling upwards, which leaves the witnesses racked with guilt, while the small boy himself eventually floats back to earth quite unharmed. McEwan reflects further, and mordantly, on the fatal nature of balloons in his Introduction to *At the Mercy of the Winds*, by David Hempleman-Adams (2001). It is clear that, like Charles Dickens before him, McEwan profoundly distrusts them, even as instruments of the imagination, and possibly for similar reasons. The result is one of the most haunting opening chapters in contemporary fiction.

8

Vertical Explorations

———

1

In October 1864 the Scottish literary magazine *Blackwood's* published a satirical article on superfluous Victorian hobbies, especially extreme sports and the fashion for futile risk-taking. It was particularly fierce on the desire to rise above one's station. Having expended its ire on the contemporary cult of mountain-climbing, it concluded briskly: 'Next to the climbers of the Alpine Club, in order of utter uselessness are the people who go up in balloons, and who come down to tell us of the temperature, the air-currents, the shape of the clouds, and amount of atmospheric pressure in a region where nobody wants to go, nor has the slightest interest to hear about.'[1]

In fact this was an excellent summary of the remarkable revival of scientific ballooning in England during the 1860s. Since the Royal Society's sceptical investigation of the subject in the 1780s, and Gay-Lussac's two daring high-altitude ascents from Paris in 1804, there had been virtually no specifically scientific ballooning anywhere in Europe or America. The emphasis had been firmly on the long-distance journey, the horizontal exploration. The vertical voyage, with the specific aim of studying the upper atmosphere, or high weather formation, had languished. Neither Charles Green, nor John Wise, nor Nadar, for instance, had any such interests.

It was true that two French scientists had announced in June 1850 that they intended to fulfil the prescription for investigative ballooning, as set out by the great François Arago. Jean Barral was a professor of chemistry, and Jacques Bixio described himself as an 'agricultural journalist'. Amazingly, with no previous aeronautical experience whatsoever, they somehow obtained funding from the Académie des Sciences, a suite of barometrical instruments, and public support from Arago himself. In fact their real intention was less scientific than patriotic: to challenge the international altitude record, which was still casually held by the British, since Charles Green had nonchalantly ascended to twenty-one thousand feet above Lewes in the *Nassau* in 1838, while testing a new aneroid barometer for a friend.[2] Barral and Bixio proclaimed that their target was ten thousand metres (approximately thirty-one thousand feet), an altitude at which, though they did not know it, they would reliably die of asphyxiation.

With great ceremony they installed their balloon and a powerful hydrogen generator on the sacred front lawn of Arago's Paris Observatoire. Watched by a crowd of *savants*, they inflated the envelope, 'placed themselves in the car without testing the ascending power of the balloon, and darted off into the air like an arrow'. They promptly rose to twenty-three thousand feet, thereby taking the record from Green. But their balloon was over-inflated, and before they could make any further instrumental readings, it bulged down through its netting and pressed them flat on the floor of the basket, 'which unfortunately was suspended by cords much too short', effectively trapping them. They panicked, believing the balloon

would go on ascending till it exploded, and in attempting to open the valve line, tore a bottom panel from the swollen fabric. Overcome by the escaping gas, they both passed out, and awoke some time later lying peacefully in a vineyard on the edge of the champagne region in Lorraine. They were forced to admit that they had brought back absolutely no new scientific data for Arago, only some very good wine.[3]

In England the art of ballooning, without the example of Nadar or the Godard brothers, had also fallen into a certain disrepute. Since the successes of Charles Green, the glamorous firework ascents from Vauxhall and the Cremorne Gardens, and the giant captive balloons brought to London by Henri Giffard, English aerostation had gained a certain end-of-the-pier seediness, increasingly regarded as the province of showmen, 'artistes' and publicists. It was the subject of smoking-room jokes, witty cartoons in the newly founded *Punch*, and a number of rowdy music-hall songs:

> *Up in a balloon, up in a balloon,*
> *All among the little stars*
> *Sailing round the moon!*
> *It's something very jolly*
> *To be up in a balloon.*[4]

The idea that ballooning had dwindled to something 'jolly' was indeed a descent.

Green himself was firmly in retirement, having 'dropped his grapnel' at Aerial Villa. But he was now regarded, throughout the ballooning world of Europe, as something of an aeronautical prophet. So the French journalist Wilfrid de Fonvielle, after his experiences with Nadar's creaking *Géant*, made a pilgrimage to pay his respects to the man who described himself as 'an Ancient Mariner of the Upper Atmosphere'.

Fonvielle, now aged forty-four, was in search of inspiration. He had recently returned from a period of political exile in Algeria, where he had been banished for his extreme republican views. An anglophile, his political idealism was now directed into ballooning. He evidently expected the prophet's Aerial Villa to be a splendid residence, perched on some superb, wind-blown hilltop overlooking the famous Hampstead Heath. Instead he found a 'small, snug suburban cottage' pleasantly tucked in among a little grove of trees and houses on the gentle wooded slopes of Upper Holloway. Silver-haired and sprightly, the old aeronaut received him

ceremoniously, opened a bottle of fine French wine, and brought out a
bulging portfolio of balloon articles, letters, pictures and memorabilia.[5]

Green was obviously delighted by his French visitor, who spoke good
English and told him jokes about Nadar's ballooning. In response he
reminisced freely – not to say garrulously – till a late hour. His one major
regret was that he never succeeded in crossing the Atlantic as he had long
wished to. Otherwise he was proud of his achievements: 'I have made
more than six hundred aerial excursions. I have crossed the English
Channel three times; and I have had as many as seven hundred persons
in the car of my balloon at different times, among whom I could mention
some very distinguished names, and one hundred and twenty ladies.'

To Fonvielle's amusement, Green laid great and solemn emphasis on
the importance of women to the future of ballooning: 'The ladies have
always shown great courage in this respect. If you wish balloons to
become popular in France, – believe in the experience of an old man, an
Ancient Mariner of the Upper Atmosphere – begin by taking women in
your balloons. Men will be sure to follow.'[6]

Green then took Fonvielle to the end of a narrow court behind Aerial
Villa, and quietly opened the door of an outhouse. Lo and behold! There,
lovingly stacked and folded, were the treasured remains of the celebrated
Nassau – the canopy with its Latin Daedalus motto, the battered basket,
the mighty manila guide rope carefully slung from a wooden rafter. The
old aeronaut seemed 'quite overcome' when he stood before his splendid

aerostat. '"Here is my car," he said, touching it with a kind of solemn respect, "which like its old pilot, now reposes quietly after a long and active career ... And there is the tissue of the Nassau itself. – Poor old balloon, I love it like a child."'[7]

Then Green took another restoring glass of wine, and concluded with a wistful reflection that much struck Fonvielle. He said that what he really regretted in life was having had to make ballooning a commercial *business*, rather than a true scientific 'vocation'. He gripped Fonvielle by the arm: 'How happy you ought to be, to be able to carry on scientific and artistic researches in the air! I would like to have done so too, but I could not follow my own plans. I was an aeronaut by profession, and had to gain my bread by it. Lack of money prevented me from carrying out the numerous experiments that I had in mind ... Let me shake you by the hand and wish you every success. In the upper atmosphere ... there is so much still to be discovered!'

As Fonvielle left Aerial Villa, he fancied he saw 'a large round tear standing in the bright eye of the celebrated old man'. Green turned and walked slowly back towards the outhouse, which contained, as Fonvielle put it, 'all his daring adventures of forty years ago'.[8] Fonvielle went back to France, inspired by Green, and determined to continue with his aeronaut's 'vocation'. But was there any balloonist in Britain or France prepared to take up Green's challenge? Was there anyone capable of further, serious 'scientific and artistic researches' in the upper atmosphere?

2

In 1862 the British Association for the Advancement of Science elected a Scientific Committee to investigate 'Hygrometric and other Conditions of the Upper Air'. It decided to do this with a series of sponsored scientific balloon ascents. Among the committee's fourteen signatories were many leading names in British science, notably the Astronomer Royal, Professor Sir George Airy; Sir David Brewster, who had written Isaac Newton's biography; Sir John Herschel, the best-known public scientist in Britain; and John Tyndall, who had inherited Humphry Davy's position at the Royal Institution.[9] The man they chose to prosecute these researches, and put the science back into ballooning, was a fifty-three-year-old meteorologist named James Glaisher (1809–1903).

Glaisher did not exactly fit the profile of an aerial adventurer. A large, taciturn family man, solidly built and heavily bewhiskered, he was a Fellow of the Royal Society, an expert on the theory of magnetism, and for the last ten years had been Secretary to the Royal Meteorological Society. He lived comfortably with his extended household in a pleasant mansion at 22 Dartmouth Road on Blackheath, South London (now marked by a blue plaque). He strode daily over the Heath on his way to work at the nearby Royal Observatory at Greenwich. A proper Victorian patriarch, thoroughly regular and earthbound, he had ten children, the youngest of whom he was training to be a mathematician.[10]

But Glaisher had hidden qualities. For a start, he was a meticulous scientific investigator. Undertaking to draw up a fully documented balloon research programme for the BAAS committee, he compiled a thirteen-point 'Terms of Reference' document, setting out the main lines of proposed research in a manner reminiscent of Arago. These were detailed and complex, involving precise measurements of a number of phenomena: the earth's magnetic field; the solar spectrum; dew point; oxygenation of the atmosphere; atmospheric electricity; barometric pressure; cloud density; and air currents. The key to the whole project would be continuous, accurate measurement, using the latest scientific

instruments: that is, the production of *data*. His terms of reference also specified a working altitude limit of five miles, or about twenty-six thousand feet.[11]

Glaisher agreed with the BAAS to oversee a series of high-altitude ascents, and to organise the appropriate scientific equipment. But of course he had no intention of going up himself. Clearly he did not consider himself suitable for such exploits. Instead, after consultations with Charles Green, he decided to employ a professional aeronaut, Henry Tracy Coxwell (1819–1900), who had already made over four hundred ascents, many of them abroad. Coxwell was modest and unflappable, the kind of man who would joke that his main connection with gas was that he had once been a rather successful dentist.

Glaisher and Coxwell found that Charles Green's famous *Nassau* was now – like its owner – too old and fragile to be of use. So the BAAS commissioned Coxwell to construct a rugged new balloon specifically designed for high altitudes. The ninety-three-thousand-cubic-foot *Mammoth* was made of top-quality American fabric, and its basket fitted out to carry the most complex suite of laboratory instruments ever taken aloft. With the correct ballast, it was theoretically capable of reaching altitudes well in excess of thirty thousand feet. It was constructed over the next few months, while Coxwell trained up several young meteorologists to accompany him. Thus the first serious scientific balloon programme since Gay-Lussac was planned, financed and launched.[12] Glaisher had, perhaps unwittingly, taken up the baton from Green. ꙮ

ꙮ Charles Green lived to follow Glaisher's adventures, dying safely in his bed at Aerial Villa in March 1870. A 'Green' tomb, with a fine bas relief of a fully inflated balloon in Highgate Cemetery, East Division, is always considered his memorial. In fact it belongs to a different balloonist, Charles Green Spencer, who died twenty years later. But there is a touching connection. Charles Spencer was the eldest son of Green's early balloon assistant Edward Spencer, who named his boy after his beloved aeronautical captain. Charles Green Spencer himself was evidently proud of his father's connection with Green, and grew up to found a balloon-manufacturing business, Charles Green Spencer & Sons; on his death in 1890 he was described on this tomb as 'Aeronaut of Holloway'. As in most balloon baskets, there is room enough for two. Memorials took another form when Henry Coxwell acquired the remains of the actual *Nassau* after Green's death, lovingly restored it, and took it up to ten thousand feet above Holloway in Green's honour. Green never turned his memoirs into a book, but much of the contents of the balloon portfolio that Fonvielle saw are now held in the Cuthbert-Hodgson Collection at the National Aerospace Library, Farnborough, Hampshire.

At first Glaisher defined his view of the balloon, with deliberate dry British understatement, as 'an instrument of Vertical Exploration'.[13] Yet he later found himself expanding on this utilitarian view. The scientific hopes of the great French balloon enthusiasts, François Arago and Antoine Lavoisier, might yet be fulfilled. Balloon experiments, he became convinced, would begin to give us a completely new conception of the planetary envelope within which we all live. This atmospheric envelope was 'the great laboratory of *changes* which contain the germ of future discoveries'.

Such experiments, if boldly pursued, would throw light on 'the physical relation to animal life of different heights, and the form of death which at certain elevations waits to accompany its destruction'. In short, the balloon, properly placed in the hands of empirical British scientists, could become nothing less than 'a *philosophical* instrument' of enquiry.

In fact Glaisher began to see infinite perspectives opening up: 'Do not the waves of the aerial ocean contain, within their nameless shores, a thousand discoveries destined to be developed in the hands of chemists, meteorologists, and physicists? Have we not to study the manner in which the vital functions are accomplished at different heights, and the way in which death takes possession of the creatures whom we transport to these remote regions?'

This idea of an aerial ocean, with its 'nameless shores', its crucial boundaries of life and death 'at different heights', was effectively new and would have immense implications.[14]

3

How had James Glaisher come to this visionary viewpoint? Born in 1809, he was the son of a watchmaker from Rotherhithe, and had grown up with a keen interest in precision instruments and an almost religious respect for meticulous accuracy. As a young man he had worked for the British Trigonometrical Survey, and been sent to learn his craft in Ireland. Tramping over the mountains of Donegal in every kind of weather, he had become fascinated by that most imprecise and indefinable phenomenon – clouds. How, he wondered, could such infinitely changeable and complex things be precisely measured, and their subtle influence accurately and mathematically calculated?

On returning to England, he was appointed assistant at the Cambridge University observatory, where his skill and dedication were quickly noticed by Sir George Airy. At the age of twenty-nine he was recruited by Airy to head the newly opened Department of Magnetism and Meteorology at the Royal Observatory, Greenwich. He would hold this post for the next thirty-six years (1838–1874).[15]

Glaisher had an almost Platonic commitment to the power of mathematics. He was passionate about measurement, and believed that there was nothing in Nature that would not yield to it. He began to consider how the kind of precise mathematical and statistical observations essential to astronomy might be applied to the still-infant science of meteorology, concluding that the crucial requirement was accurate and systematic *data* – lots of it. Like Professor Joseph Henry at the Smithsonian, he soon realised that the key to gathering such data was the new telegraph system. For the first time the electric telegraph made *national* weather reporting a genuine possibility.

Within a decade Glaisher had established his own amateur network of some sixty local volunteers, using properly calibrated instruments, right across England. His dedicated team of weathermen were mostly doctors and clergymen, men tied to their particular parishes, who could be relied upon to take regular readings of temperature, barometric pressure, wind speed and atmospheric conditions (cloud, rain, sunshine) at precisely nine o'clock every morning.

The results were telegraphed to Glaisher at the Royal Observatory before noon. He could now command a daily picture of the evolution of weather systems right across the country. In August 1848 he began to contribute a national weather report to the *Daily News*. He was elected a Fellow of the Royal Society in June 1849, and became the Secretary of the Meteorological Society in 1850. His weather charts were shown at the Great Exhibition in 1851. He could track the weather, but still he did not attempt to *forecast* it.[16]

In the autumn of 1852, Glaisher followed four of Charles Green's 'last' ascents from Vauxhall, using a telescope, from the roof of the Greenwich observatory. Perhaps in response to Barral and Bixio, Green was deliberately going for height, and claimed to have reached nearly 22,930 feet. It struck Glaisher that high-altitude ballooning might be a possible way of radically extending meteorological research. But for the moment he was kept too busy by Airy to do anything about it.[17]

As he entered his fifties, Glaisher's meteorological career seemed distinguished and settled, but unlikely to go much further – or higher. He diverted himself with some exquisite studies of the formation of snowflakes, each one a perfect demonstration of mathematical symmetries in Nature. Yet, once asked to organise the BAAS balloon project with Coxwell, for the first time he admitted 'the desire which I had always felt for observations at high altitudes'.[18] Perhaps the spirit of his youth, or the spirit of the Donegal hills, breathed back into the whiskery face of the paterfamilias.

Glaisher's moment came when one of the younger meteorologists from Greenwich, chosen to accompany Coxwell on a training flight, 'declined' to ascend. There was nothing else for it – Glaisher would simply have to go up himself. He put it to the BAAS, and perhaps to Mrs Glaisher, as a matter of reluctant necessity: 'I found that in spite of myself I was pledged both in the eyes of the public and the British Association to produce some results in return for the money expended. I therefore offered to make the observations myself.'[19] But there can be little doubt that James Glaisher FRS, in his own quiet way, was truly delighted at the heady prospect.

Glaisher came to regard the clouds and the upper atmosphere – 'the great laboratory of changes' – as the natural extension of all his previous, ground-based work. Deciding that the balloon basket must become a miniature laboratory in the air, he assembled a remarkable array of twenty-four instruments designed for high-altitude research. Mounted on a large, tilted wooden control board fixed across the centre of the basket, they included five different types of aneroid and mercurial barometers (to indicate altitude), four types of thermometer, and two types of hygrometer. He also had a compass, a chronometer, a magnet, a pair of opera glasses, and 'scissors for cutting string'.[20]

He set himself the primary task of recording and comparing the readings from all his instruments, continuously for the entire duration of the flight, and logging them accurately against a chronometer. This required extraordinary speed and self-discipline. Indeed, once Glaisher teamed up with Henry Coxwell in the air, he would prove an exceptionally dauntless and phlegmatic observer.

At one hectic moment during a flight in July 1863, Glaisher logged in the space of sixty seconds seven readings from his aneroid barometers, accurate to the hundredth of an inch, and twelve readings of the thermometer, accurate to a tenth of a degree: an average of one reading every three seconds. In another flight, over an extended period of approximately ninety minutes, he logged the flight time on 165 separate occasions, and recorded on average one instrument reading every nine

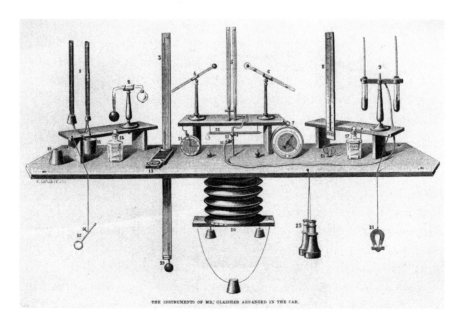

THE INSTRUMENTS OF MR. GLAISHER ARRANGED IN THE CAR.

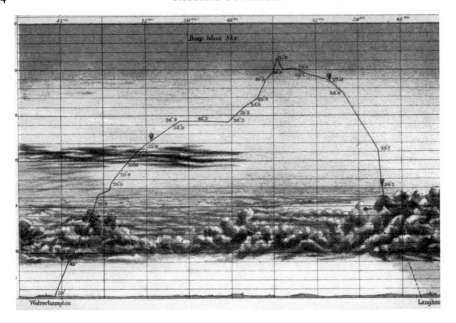

seconds.[21] From this mass of meticulous data he would compile detailed reports for the BAAS committee, which subsequently became the basis for his contribution to the most celebrated study of Victorian ballooning, *Travels in the Air*.[22] Not the least remarkable part of this data was individual flight 'profiles', in the form of illustrated graphs of hitherto unrivalled accuracy.

Speed and concentration were not the only requirements. Safeguarding his instruments under all conditions in the air was also paramount. As they were mostly glass, the great problem was to prevent them breaking during the impact of landing. Glaisher devised an ingenious method of securing each instrument to the board by a system of string laces. These provided a simple but highly effective quick-release mechanism, as they could be cut by scissors. If a landing promised to be rough, Glaisher could cut and dismount his entire instrumentation set in less than a minute, sliding each instrument into one of a series of flat, cushioned drawers enclosed in a heavily padded crash-box securely fixed in the centre of the basket. (In a sense this was a forerunner of the aircraft 'black box'.)

Glaisher and Coxwell would make twenty-eight experimental ascents between 1862 and 1866, usually from the small industrial town of Wolverhampton, but later over London.[23] Wolverhampton particularly suited their purposes, as it was close to the geographic centre of England, and thus prudently furthest from a sea coast in any direction. It also had a highly efficient and economical municipal gas company, whose chief engineer, Mr Proud, became one of Glaisher's warmest supporters.[24]

Their first ascent was made from Wolverhampton on 17 July 1862, and reached a little over twenty-two thousand feet in the space of two hours. Immediately on their descent, Glaisher wrote up a remarkable description of the physical conditions experienced in a balloon at high altitude. Although this was his first time aloft, the coolness and precision of his account is already characteristic. It is striking that he records instrumental and bodily measurements with equal objectivity.

At the height of 18,844 feet, eighteen vibrations of a horizontal magnet occupied 26.8 seconds, and my pulse beat at the rate of 100 pulsations a minute. At 19,415 feet palpitation of the heart became perceptible, the beating of the chronometer seemed very loud, and my breathing became affected. At 19,435 feet ... the hands and lips assumed a dark bluish colour, but not the face. At 21,792 feet I experienced a feeling analogous to seasickness, though there was neither pitching nor rolling of the balloon, and

through this illness I was unable to watch the instruments long enough to get a dew point [reading]. The sky at this elevation was a very deep blue colour, and the clouds were far below us. At 22,357 feet I endeavoured to make the magnet vibrate, but could not ... Our descent began a little after 11 a.m., Mr Coxwell experiencing considerable uneasiness at our close vicinity to the Wash.[25]

4

Information about the nature of the upper atmosphere was astonishingly vague until Glaisher's ascents of the 1860s. There was no clear idea of how high the respirable atmosphere reached, or whether the air would grow hotter or colder as an aeronaut got 'nearer' to the sun. In the legend of Icarus, of course, it got hotter. No one knew therefore how high a man (or indeed a bird) could safely fly before he was either asphyxiated, frozen, burnt or even electrocuted by static electricity in high clouds. Nor was it known how much warning you would get about the imminent arrival of any of these lethal conditions.

Certainly ballooning anywhere above four miles, or twenty-one thousand feet, was regarded as perilous, and entering upon unknown territory, the *terra incognita* of the upper air. Gay-Lussac, Barral and Bixio, and Charles Green had each flown somewhere in the 22–23,000-feet zone, though their barometric altimeters were nowhere near as accurate as Glaisher's, and their claims probably exaggerated. Most crucially of all, before Glaisher and Coxwell there was no clear identification of what we would now call the stratosphere, starting about six miles up, or thirty-two thousand feet. This is the critical point where the air can no longer be breathed, and human life cannot be sustained. Almost the only certain thing that aeronauts knew about the upper atmosphere was that the barometric air pressure dropped steadily as you got higher.

Information about the life of weather systems was equally sketchy. Up till now, the approach to weather had been strictly earthbound and empirical. Weather records had been built up over centuries by local observers, notably in England by country clergymen and squires such as Gilbert White in Hampshire or James Woodforde in Norfolk. But such 'meteorology' amounted to little more than the simplest, daily records of temperature, air pressure, wind speed and rainfall. The instruments to measure these had been widely available in England since the seventeenth century.[26]

The first mercury barometer, using a thin column of mercury sealed into a calibrated glass tube, was invented by Evangelista Torricelli in 1644. Robert Hooke at the Royal Society invented the aneroid or 'clock-faced' barometer, based on a coiled metal spring rather than on mercury, in the 1670s, together with an improved type of thermometer and hygrometer. Regular readings with such instruments could suggest broad seasonal trends and averages, over long periods of time. But this was purely empirical data. It implied no theory of weather formation, no concept of what forces or systems actually generated weather, and, crucially, no idea of prediction or 'forecasting'. So up to now the foretelling of weather had been essentially a local affair, the province of local expertise and memory, of folklore and tradition, of proverbial sayings and immemorial superstitions. ⓦ

Theoretical developments had been very slow. The philosopher and mathematician René Descartes had published an essay on La Météorologie in 1637. His word 'meteorology' was derived from Aristotle's Greek term, meteor, simply meaning anything 'high up'. So Descartes' meteorology was partly astronomy, including meteors, comets and shooting stars, as well as speculations on the nature of clouds, fog, sunshine, the formation

---

ⓦ Proverbial weather sayings and dates still persist. Among the most famous are 'Red sky at night, shepherd's delight' and 'If St Swithin's day be fine, it brings forty days of sunshine.' By contrast, rainfall on Napoleon's birthday indicates a wet autumn in France. Many of these weather proverbs do have a basis in scientific fact. For example a 'red sky' in the west at sunset indicates the presence of high-altitude cirrus of a stable, well established high-pressure or anti-cyclone system, which will continue fine weather for next twenty-four hours. The forty days of dry weather prophesied by a fine St Swithin's day (15 July) in England is roughly the time the summer anti-cyclone pattern needs to be established by the Gulf Stream coming over the Atlantic. This high-pressure system then tends to remain, bringing fine weather until early September. If, however, it ends early, by 15 August (Napoleon's birthday), then low pressure or cyclones tend to remain dominating the Gulf Stream for the rest of the summer, bringing cooler and less stable weather. These trends are evident on modern isobar maps, although increased melting of the North Polar ice cap in summer is predicted to alter this pattern in the future, and is already doing so around the globe. See internet 'Climate Change' sites of the Royal Society and the World Wildlife Fund.

of ice and the cause of storms. In 1686 the British astronomer and mathematician Edmund Halley drew on maritime records to construct an early chart of global weather systems, suggesting tropical and subtropical airflows over the major oceans and land masses.[28]

In the 1730s George Hadley, like Halley a Fellow of the Royal Society, developed the theory of trade winds, which took account of global rotation and thermal convection. Some hundred years later a Polish scientist, Heinrich Wilhelm Dove, a future director of the Prussian Meteorological Institute, suggested in a series of papers that storms were the result of warm-weather systems colliding with cold ones. Thus the idea of mobile weather 'fronts' (partly drawing their imagery from Napoleonic warfare) began to emerge. But all these observations still lacked any more general, unifying theory of meteorology.[29]

Even so, a growing fascination with the mystery and beauty of weather was becoming evident in the Romantic writers and painters of the time. John Ruskin, the future champion of the great atmospheric cloud paintings of J.M.W. Turner, was one unexpected enthusiast. At seventeen, while still an undergraduate at Oxford, the young Ruskin joined the newly formed London Meteorological Society. The following year, in 1838, he composed an elegant essay on the general problem of finding an empirical basis for true, predictive meteorology. He contributed it to the Society's *Journal*:

> A *Galileo* or a *Newton*, by the unassisted working of his solitary mind, may discover the secrets of the heavens and form a new system of astronomy. A Davy in his lonely meditations on the crags of Cornwall or in his solitary laboratory, might discover the most sublime mysteries of nature ... But the meteorologist is impotent if alone; his observations are useless, for they are made upon a point, while the speculations to be derived from them must be [extended] in space.[30]

Ruskin's central argument was that genuine 'forecasting' required a much broader understanding of weather 'systems', and multiple observers. It must depend on the collation of data *simultaneously* from many points over the land – and ultimately over the sea. He imagined whole teams of volunteer observers – in the European Alps, on the Atlantic Ocean, in the American prairies – recording weather information at agreed times. He would eventually be proved right by Glaisher, Fitzroy, Henry and others.

But until the 1860s, foretelling weather even one day ahead still depended on the supposed 'weather eye' of farmers, foresters, hunters, fishermen and sailors. Indeed, at a local level – where such people genuinely knew their own patch, a particular valley or harbour or line of hills – such short-term predictions could be impressively reliable (and still are, for example for modern yachtsmen). But there now appeared a professional race of Victorian weather prophets, who claimed to be able to do this scientifically, and who achieved national followings. Here the equilibrium between science and superstition – or plain mumbo-jumbo – was finely balanced.

On the north-east coast at Whitby, the curator of the Whitby Museum, Dr George Merryweather, became nationally famous throughout the 1850s for his Patent Tempest Prognosticator. It was shown at the Great Exhibition of 1851, at the same time as Glaisher's weather reports, and submitted to the Board of Trade for testing. It was not adopted, for reasons that soon became clear.

The Prognosticator was based on an idea current all along the stormy east coast of England, which was mentioned a generation earlier by the depressive poet (and temperamental balloonophile) William Cowper – himself extremely sensitive to atmosphere – while staying in East Anglia. It was based on the well-known response of leeches to sudden changes in barometric pressure. Their soft, gelatinous bodies were squeezed and made drowsy and inactive by normal air pressure, but low pressure refreshed and awoke them. As Cowper wrote: 'I have a leech in a bottle that foretells all these prodigies and convulsions of Nature. No change in the weather surprises him ... he is worth all the barometers in the world.'[31]

Dr Merryweather's Prognosticator was in fact an ingenious form of multiple leech barometer. It consisted of a circular display of twelve glass flasks, each containing a prize leech partially immersed in rainwater. The flasks were cunningly enclosed at the top with a system of whalebone springs, and these in turn were linked to a set of counterweights connected to metal hammers arranged to strike against an impressive brass bell mounted in the centre of the apparatus. It was another arrangement that would have delighted Heath Robinson.

When a sudden drop in barometric pressure occurred, indicating the arrival of a severe low-pressure system and hence the likelihood of a storm in the North Sea, all the leeches became animated and climbed rapidly to the tops of their individual glass prisons. Their combined

upward pressure triggered the whalebone springs, which in turn released the series of counterweights, which then drove the chime of small hammers against the bell. Thus a tempest was ringingly prognosticated. To add to its mystic appeal, Merryweather decorated the instrument with flutings, chains and curlicues, making it look like a cross between an Indian temple and a gypsy merry-go-round.

Ingenious and attractive as Dr Merryweather's device might appear, it did not greatly advance the theoretical understanding of pressure systems. In fact it rather obscured it, by introducing a false animating principle, and vaguely suggesting that the leeches consciously 'knew' about storms and were anxious to impart this valuable information to mankind.[32] Nevertheless, the Prognosticator reappeared a hundred years later at the Festival of Britain in 1951, and a replica can still be found at the Barometer World museum at Okehampton in Devon.

When the Royal Society was asked to look into the whole question of weather forecasting, it delivered a damning report in 1865: 'We can find no evidence that any competent meteorologist believes the science to be at present in such a state as to enable an observer to indicate day by day the weather to be experienced in the next 48 hours.'[33]

5

Glaisher and Coxwell's most celebrated high-altitude ascent took place from Wolverhampton on 5 September 1862.[34] They left the ground at 1 p.m. on a fine, cool autumn afternoon with an air temperature of 50 degrees. Forty minutes later they had risen through the twenty-thousand-foot barrier, where breathing begins to become harder, the temperature drops towards freezing, and the sky above is dark Prussian blue. At 1 p.m. and 54 minutes Glaisher meticulously recorded an adjusted barometric pressure of 9¾ inches, and a height of twenty-nine thousand feet, with the balloon still rapidly rising. So far all had gone smoothly, but now Coxwell saw that the continuous rotation of the basket had tangled the release-valve line overhead. Panting hard, he climbed into the balloon hoop to free it.

During the next three minutes both men began to suffer from the extreme cumulative effects of oxygen deprivation. Glaisher found his eyesight blurring, so he could no longer read the fine barometric scale on the instruments, even with the aid of a magnifying glass. This was followed by rapid loss of muscular power in his arms and feet, and then in his neck: 'In looking at the barometer my head fell on my left shoulder.' When he tried to straighten it, it fell doll-like onto his right shoulder. He tried to reach for a bottle of brandy, but his arm went slack and fell across his instrument table. His legs gave way, and he slumped back against the side of the basket. He found he could no longer move his head at all, and could only stare upwards at Coxwell struggling in the balloon hoop.

> I dimly saw Mr Coxwell and endeavoured to speak, but could not. In an instant intense darkness overcame me, so that the optic nerve lost power suddenly, but I was still conscious, with as active a brain as at the moment while writing this. I thought I had been seized by asphyxia, and believed I should experience nothing more, as death would come unless we speedily descended.

Meanwhile Henry Coxwell, suffering from similar loss of muscle power, was struggling in the balloon hoop trying to disentangle the trapped valve line. He kept losing his grip and nearly falling backwards out of the balloon. Pulling the valve line was the one way they could release gas and get the balloon to descend.

Coxwell eventually freed the line so it hung down into the basket, but he now found his hands were completely frozen and incapable of grasping the line or pulling it. He also noted that his hands had gone black. He was unable to grip any rope sufficiently to manoeuvre himself back down into the basket. But after some experiments, he found that by locking both elbows across the diameter of the hoop, he was just able to lower himself down inside it, and then drop safely into the basket. Here he found Glaisher lying completely unconscious, 'his face quite tranquil', and could not wake him. He glanced at the small aneroid barometer fitted on Glaisher's desk, and though his sight was so blurred that he could not read the scale, he noted that the brass pointer was exactly parallel to a particular string tether tied next to it.[35]

Seizing the swaying valve line in the crook of his arm, Coxwell made one last effort, and caught the rope between his teeth. He pulled hard, 'dipping his head two or three times'. High overhead he heard the gas begin to vent, and knew they might just be saved. He rested for some moments, panting against the side of the basket, then turned to the sprawled and immobile body of his friend Glaisher and began to shake him.

Coxwell recalled: 'Never shall I forget those painful moments of doubt and suspense as to Mr Glaisher's state when no response came to my questions. I began to fear that he would never take any more readings.'[36] But Glaisher was not dead. To Coxwell's immense relief he began to stir and mumble. Glaisher's first impression on returning to consciousness was of a tall figure bending over him, and gently enquiring about temperature readings and instrumental observations. Glaisher realised that Coxwell was 'endeavouring to rouse him' by asking strictly scientific questions. Even at such a desperate moment Victorian politesse and scientific etiquette were wonderfully sustained.

'I then heard him speak more emphatically,' remembered Glaisher, *'but could not see, speak or move ... I heard him say: "Do try to take temperature and barometer observations, do try" ... I heard him again say, "Do try, now do." Then the instruments became dimly visible, then Mr Coxwell, and very shortly I saw clearly. Next I arose from my seat and looked around as though waking from sleep, though not refreshed, and said to Mr Coxwell, "I have been insensible." He said, "You have, and I too, very nearly." I then took a pencil in my hand to begin observations. Mr Coxwell told me that he had lost the use of his hands, which were black, and I*

Félix Nadar, cartoonist, balloonist and photographer; a radical and inventive spirit throughout the Second Empire, the siege of Paris and the Third Republic. A characteristically penetrating self-portrait taken c. 1854.

'Mr Glaisher insensible at the height of seven miles', 1862. This dramatic engraving shows the meteorologist James Glaisher slumped unconscious in the basket of the *Mammoth* while his pilot Henry Coxwell clambers into the hoop to secure the line of the gas-release valve. On landing they walked seven miles to a country pub for a pint. Their altitude record stood for the rest of the century.

'Paul is swept away in the *Leviathan*'. Opening image from *Les Aventures de Paul*, a popular book for children by Jean Bruno, 1858. The theme of the small boy in the runaway balloon has become universal.

A poster for *Le Ballon-poste*, the first ever airmail newspaper, published in Paris by *Le Figaro*, 1870–71. It was flown out weekly by *ballon monté* (manned balloon) over the Prussian siege lines, and contained 'a complete Journal of the week's events, and two columns of Private Correspondence', price twenty centimes.

Le Ballon, celebrated siege painting by Pierre Puvis de Chavannes, 1870, showing the armed figure of Marianne standing defiantly on the Paris ramparts while a balloon disappears westwards over the Fort Mont-Valérien towards the Prussian lines.

The dashing Major-General George Custer, US Army, 1865. One of the
few Union officers brave enough to go up in a balloon. He recalled that
he preferred to remain 'sitting in the bottom of the basket'.

Thaddeus Lowe with his famous military balloon the *Intrepid*, which made
him the leading Union balloon observer of the American Civil War, and
afterwards a legend. A modern drawing by Mort Künstler, 1991.

'In one bound we pass through the thick layer of cloud'. One of a series of sublime balloon engravings and weather studies produced by Albert Tissandier for *Travels in the Air*, edited by James Glaisher, 1871.

The famous pair of French aeronautical brothers, Gaston (right) and Albert Tissandier, with their balloons the *Zénith* (high-altitude), the *Jean Bart* (the siege balloon) and *La France* (a prototype airship powered by a German electrical engine).

poured brandy over them.' (Glaisher does not add what was surely the case, that they both swallowed large mouthfuls of the brandy too.)[37]

Afterwards, Glaisher minutely calculated what the final height attained in those lost moments might have been. His last accurate barometer reading of 9¾ inches indicated 'above 29,000 feet', 'implying a height of about 5¾ miles'. It had been made at precisely 1 p.m. and 54 minutes. This was just before the crisis with the valve line.[38]

At this point the barometer was still visibly dropping, so the balloon was still rapidly ascending. Glaisher calculated the climb rate as 'about one thousand feet a minute'. He became 'insensible' three minutes later, at 1 p.m. and 57 minutes, by which time the balloon would have risen another three thousand feet, reaching approximately thirty-two thousand feet. Glaisher remained 'totally insensible' for a further seven minutes, returning to consciousness at 2 p.m. and 4 minutes, having been roused by Coxwell. But, as he noted meticulously, he was only able to read his instruments accurately again at 2 p.m. and 7 minutes.

This gave a total elapsed time between the two accurate readings (1.54 p.m. and 2.07 p.m.) of exactly thirteen minutes. At that point his barometer read eleven inches, and was 'increasing quickly'. This indicated that the balloon was now at an altitude of approximately twenty-seven

thousand feet, and descending rapidly.[39] The crucial question was, what happened during those lost thirteen minutes between the two accurate barometer readings of twenty-nine thousand feet (ascending) and twenty-seven thousand feet (descending)?

Glaisher estimated as follows. The balloon probably continued to rise for at least half of the elapsed time, say six or seven minutes, before Coxwell succeeded in pulling the valve line, and starting their rapid descent. Therefore, at the climb-rate of one thousand feet a minute, this gave a possible increase in height of between six and seven thousand feet, or a maximum altitude of about thirty-five or thirty-six thousand feet. In fact one of their other instruments, a max-min thermometer which recorded height in terms of decreased temperature, suggested a maximum altitude of thirty-seven thousand feet, or just over seven miles. Glaisher also checked the position of the aneroid barometer scale against the tether that Coxwell had seen. This also suggested seven miles. But in a letter sent to *The Times*, Glaisher modestly claimed a maximum altitude of thirty-two thousand feet, or six miles. This record would stand for the rest of the century.[40]

The significance of the ascent was not immediately realised. On 6 September 1862 *The Times* published a brief paragraph simply headed 'A Recent Balloon Ascent'. It reported without comment that Mr Glaisher and Mr Coxwell had 'attained an altitude of six miles', and had used a

new Negretti & Zambra mercury barometer, which was more accurate than any previous instrument. Four days later it published the letter from Glaisher, briefly outlining the circumstances of the ascent, but drawing particular attention to Coxwell's climbing into the hoop and cool handling of the crisis with the valve line. Glaisher remarked that this was completely 'in character' for Coxwell, and that he always had complete confidence in his ability to handle any crisis. He also praised the excellent quality of the coal gas provided by Mr Proud, the chief engineer of the Wolverhampton Gasworks.

Of the maximum altitude attained, Glaisher confirmed the figure of thirty-two thousand feet, or just over six miles, though suggesting that it could well have been higher, if only he had been in a fit state to read his instruments. In the graph profile of the flight he later published, the last eight thousand feet are merely indicated with a dotted line, with the single descriptive note 'Intense blue Sky'.

His letter closed on a typically dry and cautionary note: 'It would seem from this ascent that five miles [26,400 feet] from the earth is very nearly the limit of human existence ... I think that prudence would say to all, whenever the barometer reading falls as low as 11 inches, open the valve at once; the increased information to be obtained is not commensurate with the increased risk.'[41] ꙮ

Glaisher later published a fuller version of this letter in his classic book, *Travels in the Air* (1871). His laconic account became one of the most famous documents in the history of ballooning. It was completed with the characteristically dry observation that 'no inconvenience' followed his and Coxwell's nearly fatal asphyxia and frostbite, though very

---

ꙮ To give a sense of what Glaisher's figures meant, the height of Mount Everest, first measured by British Trigonometrical Survey in 1849, was generally accepted as 29,002 feet. It was first climbed with oxygen in 1952. It is agreed, even among trained and acclimatised modern climbers, that a 'death zone' commences at twenty-six thousand, almost exactly the figure of five miles given by Glaisher. At twenty-nine thousand feet air pressure is approximately one-third that at sea level, and oxygen is absorbed into the blood at one-third the normal rate. This leads to an increase in breathing rate from a normal twenty breaths per minute to sixty or more per minute, a rapid panting or hyperventilation which is itself exhausting and painful. Oxygen deprivation also leads to muscular failure, blurred eyesight, and great difficulty in thinking quickly or clearly. Glaisher remarked perceptively that under these conditions it was particularly hard to 'take a decision'.[42]

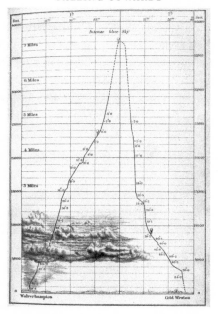

regrettably they lost five of their six carrier pigeons. On landing safely in the deep countryside near Ludlow in Shropshire, the two aeronauts could find 'no conveyance of any kind', so Glaisher stoically walked 'seven or eight miles' to the nearest village inn at Cold Weston for a pint of beer. Ever after he kept the small aneroid barometer which had recorded the unofficial seven-mile maximum altitude as a good-luck charm in his pocket, and would not fly without it: 'It was this instrument that Mr Coxwell read when we were seven miles high, and I at the time in a state of insensibility. It has been up with me in every high ascent since.'[43]

Glaisher's full report was published by the BAAS, and reproduced all over Europe and America. But the ascent was initially made famous by a magnificent third leader in *The Times* which appeared on 11 September 1862. On a page largely dominated by depressing news of the American Civil War, it turned its readers' attention to higher matters, opening in fine style: 'Poetry has described some famous *descents* into the subterranean world ... But we have just had an *ascent* such as the world has never heard of or dreamed of. Two men have been nearer by some miles to the moon and the stars than all the race of man before them.'

The leader gave details of Glaisher's scientific reasons for making such a perilous attempt, but then turned to reflect on Coxwell's peculiar brand of courage: 'For ten whole minutes Mr Coxwell ascended alone – or

rather worse than alone, with his companion insensible before his eyes
– in a region six miles distant from the earth. That is a very extraordinary
ten minutes, if we think of it, that solitary command, without a rival, of
the boundless regions of space ... It deserves its place among the un-
patrolled junctures and critical and striking moments of war, politics or
discovery.'

The thought of this terrible, lonely ascent into the outer planetary
darkness had evidently seized upon the journalist. He was soon imagin-
ing, for the benefit of his readers, the possible nightmare outcome:
'Another minute and Mr Coxwell would have been stretched by the side
of his companion, and a car containing two human bodies, would have
been mounting to worlds unknown, and encountering aerial storms and
shipwrecks so removed from all sublunary experience that we can hardly
form the faintest image of the reality.' For a moment, the fascinating
possibilities of science fiction overtook the *Times* journalist as he glee-
fully imagined the two aeronauts, rather in manner of an Edgar Allan Poe
story, sailing on upwards into outer space, their balloon eventually to be
'dashed upon the bleak shore of another planet', or like a second Noah's
ark, finding 'a resting place upon some Ararat of the moon'.

Little space was given to any of the actual scientific data obtained
by Glaisher; or the clear implication that the life-sustaining envelope of
breathable oxygen which surrounded the entire earth was alarmingly
thin. In fact it was less than the distance that could be walked along a
road in two hours (or from Glaisher's point of landing to the inn at Cold
Weston). But it was evident that the outer atmosphere was far less
inviting, and far more frightening, than the beauties once imagined by
poets.

What counted for *The Times* was the experience – particularly
Coxwell's experience – which was that of the intrepid explorer. It demonstrated moral courage, a courage quite unlike that of the soldier: 'The
courage of men of science deserves to have a chapter of history devoted
to it ... they are solitary, deliberate, calm and passive ... The man of
science has to fight alone and by himself against the faintness of nature,
without men shouting, or flags flying, or trumpets clanging around him.'

The leader concluded with reflections on the pioneering purposes of
science, the admirable urge to explore new frontiers and limits, and the
peculiar phlegmatic bravery of English scientists in a tight spot. It was
the moment for a little patriotic hyperbole: 'The cool feats of our

scientific men are known to us all – such as that of Sir Humphry Davy inhaling a particular gas with an accurate report every minute or two of its successive effects upon his brain and sense. The aerial voyage just performed by Mr Coxwell and Mr Glaisher deserves to rank with the greatest feats of our experimentalizers, discoverers and travellers ... They have shown what enthusiasm science can inspire and what courage it can give.'[44]

This theme was taken up in a popular lithograph, 'Mr Glaisher insensible at the height of seven miles'. It was reproduced in numerous journals in Britain, France and America. The image has a strange quality of nightmare suspension, and presents scientific research as pure Victorian melodrama. Its two protagonists grimly complement each other: the one active, the other passive, but both in imminent peril.

Coxwell is overhead, perilously astraddle the balloon hoop, struggling to untangle the valve line, and likely to pitch backwards into the fathomless abyss at any moment. Glaisher is slumped below, fallen back semi-conscious and apparently dying against the balloon basket, clutching at his throat with one hand, the other dangling limply over the side,

hanging into the void. Around them are strapped an orderly scientific array of barometers, compasses and other precision instruments, a mapping board, and a touchingly domestic pigeon cage. But all of them are now useless. Even the massive coil of manila guide rope attached to the side of the basket, with its purposeful iron grapnel designed to cling to the earth, looks fragile and impotent.

The picture seems to carry an allegorical message, skilfully developed from that of the *Times* editorial: the Scientist presses the extreme limits of Nature at his peril. The well-ordered world of the Victorian scientific laboratory is transported into the hostile chaos of the upper atmosphere. Order and purpose become disorder and terror. The message may even have owed something to Charles Darwin's new vision of Nature. But what is most fearful about this picture is the sense that death will come not by falling downwards, towards the familiar clouds and the earth seven miles beneath, but by *falling upwards* into the endless, empty, unknown blackness of space above. ᛃ

6

It is remarkable that Glaisher and Coxwell continued their scientific flights throughout the 1860s, and at least once, in June 1863, made another attempt to break the five-mile barrier. At just over twenty-three thousand feet, with a temperature of 14 degrees (Fahrenheit) below zero, Glaisher urged Coxwell to go on, despite nearby stormclouds: 'Mr Coxwell

---

ᛃ Just how lethal this ascent might have proved was demonstrated over a decade later by the fatal *Zénith* high-altitude attempt above Paris in 1875, when two of the three aeronauts were killed. Here is the sole survivor, Gaston Tissandier's, account of losing consciousness: 'I now come to the fateful moments when we were overcome by the terrible action of reduced pressure (lack of oxygen). At 22,900 feet torpor had seized me. I wrote nevertheless, though I have no clear recollection of writing. We are rising. Croce is panting. Sivel shuts his eyes. Croce also shuts his eyes. At 24,600 feet the condition of torpor that overcomes one is extraordinary. Body and mind become feebler. There is no suffering. On the contrary one feels an inward joy. There is no thought of the dangerous position; one rises and is glad to be rising. I soon felt myself so weak that I could not even turn my head to look at my companions. I wished to call out that we were now at 26,000 feet, but my tongue was paralysed. All at once I shut my eyes and fell down powerless and lost all further memory.'[45]

knew better, and I was met with a negative. "Too short of sand. I cannot go higher; we must stop even here."'

Glaisher looked around regretfully and observed that 'the blue was not the blue of four or five miles high, as I had always before seen it'; while the air below appeared murky and curiously uninviting. Yet at this height they could still hear a train. Glaisher then became entirely absorbed by a snowstorm as they descended: 'The snow was entirely composed of spiculae of ice, of cross spiculae at angles of 60 degrees, and an innumerable number of snow crystals, small in size but distinct and of well-known forms, easily recognisable as they fell and remained on the coat.' He barely noticed that Coxwell was struggling to control the speed of their descent. When he looked round again they were well under ten thousand feet, and flying down laterally through a veil of marsh mist over the mighty towers and lantern of Ely Cathedral.[46]

On all these ascents Glaisher took hundreds of temperature and pressure measurements, made pictures of different types of clouds, constructed and published vertical graph profiles of their flights, and several times tried photography, though apparently without success.[47] As more and more of their ascents were reported in the press, he and Coxwell became something like national heroes, and their launches now took place further south, from Windsor Great Park or the Crystal Palace. Here they drew increasingly large crowds. Coxwell, as a professional balloonist, valued this growing publicity, but Glaisher as a scientist eschewed it; and this occasionally led to a certain friction between the two friends, especially after one of Coxwell's balloons was mobbed by an over-enthusiastic crowd at Leicester in 1864.[48]

Glaisher still considered his first concern to be the BAAS and the Meteorological Society, and continued to publish only technical papers on the nature of the upper atmosphere. These covered the relationship between altitude and decrease in temperature; the appearance of the solar spectrum; barometric information about pressure systems and weather 'fronts'; changes in the magnetic field; changes in wind direction and velocity, and the possible causes of these; the propagation of sound in various atmospheric conditions; and above all 'physiological observations', or the reactions of the human body (and mind) to high altitudes.[49]

It was exactly this kind of data that Blackwood's had derided; and exactly this kind of data that was indispensable to the eventual under-standing of the nature of weather systems, upon which the future of genuine weather forecasting would depend. But Glaisher did more than this. He was a genuine explorer of the vertical. His meticulous records and measurements first mapped the general limits of the breathable atmosphere above the earth, the perils of the 'death zone', and the existence of what is now known as the stratosphere.[50] In particular, he showed that Gay-Lussac's observations that air temperature decreased steadily with altitude (at about one degree every three hundred feet) did not hold good above six miles. Here Glaisher had glimpsed the existence of a new atmospheric regime or 'layer', a true *terra incognita*.[51]

Supported by Coxwell, Glaisher's calm, heroic 'physiological observations' of his own body, even to the point of death, also opened up the subject of high-altitude asphyxia, with its accompanying decompression injuries and mental distortions. He had pioneered the subject of travel to the edge of space, and turned centuries of fiction into fact.[52] From this, the whole idea of the layered, vital, atmospheric skins of our planet, and the small, fragile, vital biosphere contained within them, would ultimately derive.

The concept of atmospheric zones, or layers ('stratos'), encircling and protecting the earth, like concentric shells or the skins of an onion, is one of the radical discoveries of nineteenth-century science. It begins with the ascents of Glaisher and Coxwell, but is not fully developed until the work of the French meteorologist and balloonist Léon Teisserenc de Bort (1855–1913) at Versailles. What is most striking about these protective skins or layers is their unexpected thinness, or fragility. In simplified form, modern meteorology divides these layers into four. First, the troposphere, capable of sustaining animal life, extends to about six miles up. Second, the stratosphere, which provides the protective ozone layer, continues up to thirty miles. Third, the mesosphere, which is the zone where meteors or 'shooting stars' are for the most part safely burnt, continues on to fifty miles high. Finally, the fourth layer, the ionosphere, where auroras and disruptive solar magnetic storms are largely held in check, stretches up to three hundred miles. Both birds and insects have been detected at the lower edge of the stratosphere, around seven miles up. But in general the sustaining troposphere (or 'biosphere') is not thick. Moreover, the upper limit of the ionosphere, where true planetary space begins, is only slightly 'higher' above the surface of the earth than the distance between London and Paris, or New York and Washington. The so-called 'low-earth' orbits of the Hubble Space Telescope and the International Space Station remain here, on the edge of the ionosphere, around three hundred miles up. Such man-made satellites (approximately three thousand of them in 2012) represent perhaps the ultimate metamorphosis, and historic destiny, of the balloon. See Douglas Palmer, *The Complete Earth: A Satellite Portrait of Our Planet* (Quercus, 2006).

7

Encouraged by Henry Coxwell, Glaisher did occasionally allow himself to savour the sheer delights of low-level ballooning. Flying one afternoon in July 1863, after a launch from the Crystal Palace, they drifted southwards until they were in sight of the South Downs and the sea near Arundel and Newhaven. Skimming along at eight hundred feet over the remote Sussex villages, with the shadows lengthening under the elm trees, Glaisher forgot his instruments long enough to record a kind of rural soundscape, or aural tapestry of country life as heard on an idyllic midsummer evening. Here the young, footloose surveyor of the Donegal hills could be glimpsed once more.

> The cheering cry of children was frequently heard above all other sounds. Geese cackled, and frightened scuttled off to their farms; pheasants crowed as they were going to roost ... and packs of dogs barked in the wildest state of excitement. Journeying in this way was most delightful; all motion seemed transferred to the landscape itself, which appeared when looking one way to be rising and coming towards us, and when looking the other as receding from us. It was charmingly varied with parks, mansions and white roads ...[53]

On this flight they had taken along the young Christopher Hatton Turnor, who would become one of the great historians of Victorian ballooning, celebrated in his huge anthology Astra Castra: Experiments and Adventures in the Atmosphere (1865). Turnor immediately sent an amusing account in a letter to The Times, dreamily entitled 'Coasting in a Balloon'. He noted that Glaisher took two photographs, one over Epsom and the other over Horsham, and seemed completely relaxed and at home in the basket. Similarly, Coxwell was in an adventurous mood, and kept mischievously proposing to extend their flight across the Channel.

Finally Coxwell settled on the more convenient landing site of Goodwood Park. They descended at 8 p.m., with dusk just falling: 'Mr Coxwell, after throwing a rope to a cricketer, landed us so gently that we could not have crushed a daisy. We were afterwards drawn by a rope to the front of the house for the benefit of a few gazers.' A substantial dinner followed in the pavilion.[54] Nevertheless, even on this idyllic occasion the indefatigable Glaisher did not omit to record a precise temperature variation of thirteen degrees between eight hundred and 2,800 feet, and also a ninety-degree shift in wind direction from east to north, which he found 'very informative'.[55]

In a series of later flights, in October 1865, Glaisher found a new delight in ascents over London, taken at various times of day and night: 'When one mile high the deep sound of London, like the roar of the sea, was heard distinctly ... but at four miles above London, all was hushed; no sound reached our ears.' Here the balloon provided a platform for human as well as meteorological observations. These were similar to those of Mayhew, though less radical, and more clearly touched with Victorian pride in the economic triumph of the city. Glaisher observed the great arterial movements of commerce: railways, river boats and barges, and horse-drawn traffic, spreading out to the suburbs of Kent and Essex.[56]

One sunset ascent over London on 9 October 1865 was especially memorable. At seven thousand feet directly above London Bridge, 'the scene around was one that probably cannot be equalled in the world ... with one glance the homes of three million people could be seen ... the Thames dotted over its winding course with innumerable ships and steamboats, like moving toys ... the coast around as far as Norfolk ... Smoke, thin and blue, was curling from [the city] ... I have often admired the splendour of the sky scenery, but never have I seen anything which surpassed this spectacle. The roar of the town heard at this elevation was a deep, rich, continuous sound – the voice of labour.'[57]

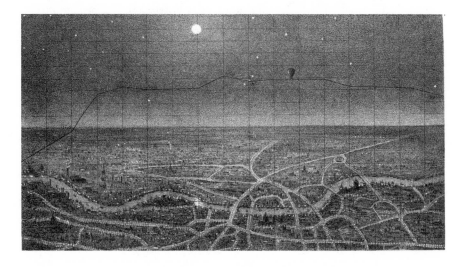

Glaisher had also made a notable night ascent over the city on 6 October 1865. Here he gave a much more impressionistic and surprisingly poetic account of his experiences. For the early part of the ascent he had, characteristically, been 'wholly occupied by the instruments', but at a thousand feet he suddenly looked up and saw stretching beneath the balloon basket a view that amazed and enchanted him, 'a starry spectacle of such brilliancy as far exceeds anything I ever saw':

> On leaving Charing Cross I looked back over London, the model of which could be seen and traced – its squares by their lights; the river, which looked dark and dull, by the double row of lights on every bridge spanning it ... It seemed to me to realize a wish I have felt when looking through a telescope at portions of the Milky Way; when the fields of view appeared covered with gold-dust, to be possessed of the power to see those minute spots of light as brilliant stars; for certainly the intense brilliancy of London this night must have rivalled such a view.[58]

Glaisher was elected President of the Royal Meteorological Society in 1867. A sign of the times, he had also become a member of the Aeronautical Society of Great Britain in 1866. Later in life he would give thrilling lectures on his balloon experiences. His 'Address to the Young Men's Christian Association' in 1875 reconstructed a typical flight with Coxwell, with the dry banter between the aeronaut and the scientist. But it also added some of those emotions he had so carefully and effectively suppressed, and for so long.

> Four Miles High [twenty-one thousand feet]. We are now far beyond the reach of all ordinary sounds from the earth ... Up to this time, little or no inconvenience is met with; but on passing above four miles, much personal discomfort is experienced; respiration becomes difficult; the beating of the heart at times is audible; the hands and lips become blue ... and it requires the exercise of a strong will to make and record observations ... Six Miles High [thirty-two thousand feet]. The balloon is now lingering as it were, under the deep blue vault of space ... We now hold consultations, and then look around, giving silent scope to those emotions of the soul which are naturally called forth by such a widespread range of creation ... Highest Point [thirty-five thousand feet]. Then in silence, for here we respire with difficulty and talk but little; in the centre of this immense space; in solitude, without a single object to interrupt the view for 200 miles or more

all-around; abstracted from the earth; upheld by an invisible medium; our
mouths so dry we cannot eat; a white sea below us ... I watch the instru-
ments, but forcibly impelled again, look round the centre of this immense
vacuity, whose bounding line is 1,500 miles ... I wave my hand and say,
'Pull!' A deep resonant sound is heard over head ... It is the working of the
valve which causes a loud booming noise, as from a sounding board, as the
springs force the shutter back ... a drum-like sound ... it is cheering, it is
re-assuring, it proves all to be right.[59]

In this final retrospective account, Glaisher emphasised the particular scientific virtues required by ballooning: meticulous care and accuracy, calmness and detachment, stoic self-discipline; and a kind of spiritual openness to the wonders of Creation. Ballooning, in other words, has moral value. The lecture ends with a short, uplifting sermon on the disinterestedness of Victorian science: 'We tell our story – how we have travelled in the realms of space, not for the purposes of pleasure, not from motives of curiosity, but for the advancement of science and the good of all mankind.'[60]

9

Mariners of the Upper Atmosphere

———

1

By the late 1860s, a younger generation of Parisian aeronauts had begun
to form a loose group of enthusiasts around the Godard brothers and the
Société Aérostatique et Météorologique de France, originally founded in
1852. They were a new intellectual breed, quite unlike the previous showmen and barnstormers. They regarded the Godards as the balloon professionals, Nadar as the balloon publicist, and Henri Giffard as the master
of the tethered balloon. But what they themselves dreamed of was free,
beautiful flight in the upper air.

They were amateurs in the true sense. Most of them had academic or scientific backgrounds, some had considerable private wealth, and several had strong republican sympathies. Among these younger aeronauts who regularly flew with each other were Camille Flammarion, the brothers Gaston and Albert Tissandier, and Jules Duruof, all in their twenties. Another was the republican journalist Wilfrid de Fonvielle, the man who came back with news of Charles Green in London, and who liked to ask, wherever his balloon landed, 'Are we still in France?'[1]

These men were strongly aware of the great balloon tradition, and the details of its history, about which many of them later wrote. They also contributed numerous balloon stories and articles to the French newspapers. Their later memoirs show how much they admired the feats of English aeronauts like Charles Green, James Glaisher and Henry Coxwell. Nevertheless, they regarded ballooning as almost exclusively French, and in this sense a 'patriotic' science. They had particular sympathy for Nadar, who despite his disasters, and without being a scientific aeronaut, had raised the profile of French ballooning throughout Europe, and even in America. They came to regard the skies overhead in a new way, as national territory, peculiarly and historically French, and probably republican too. These feelings were to prove highly significant when the French Second Empire met its sudden crisis in 1870.

A leader among the group was Camille Flammarion (1842–1925), brother to the great literary publisher Ernest Flammarion. Camille was both the poet and the scientist of the band, a charismatic and eccentric figure, with a wild bramble of hair, a romantic heart and strong republican sympathies. Unlike his brother, he had little time for French imperial ambitions, writing angrily: 'In France alone 250 times as much money is spent in the art of destroying the human species as is expended on education and science. That is why the projects and experiments of honest men remain so long in the state of dreams.'[2]

For Flammarion, ballooning was one of these neglected dreams. He said he first fell in love with ballooning at the age of sixteen, 'young and full of passion for discovery and adventures'. Like Nadar and many others, he recalled a balloon-conversion experience. He was out walking early one 'pure blue morning' in the Jardins du Luxembourg, when suddenly a dazzlingly beautiful balloon appeared over the treetops and flew low over his head. He could hear the two passengers – a man and a pretty woman – talking and laughing together. They leant over the basket and waved at

C. FLAMMARION.

him, then sailed silently away over Paris, and his heart went away with them. 'I would have given the world to be in the car of that balloon; and long afterwards I could think of nothing but a journey *into the atmosphere.*'[3]

As a young man, Flammarion trained first as a priest (which always left a certain mystical turn to his view of the world) then as an engraver, and finally as a mathematician and astronomer at the Paris Observatoire. All these *métiers* left their mark on his subsequent work. At the Observatoire he was regarded as unduly flamboyant, and had to leave when, aged twenty, 'enflamed with the fiery ardour of a teenager', as he put it, he published a controversial two-franc, fifty-four-page pamphlet, *On the Plurality of Inhabited Worlds.*[4] He joined the staff of a new scientific magazine called *Cosmos*, and like Jules Verne adopted the new career of science journalist. He produced expanded versions of his pamphlet, vigorously arguing the case for a universe teeming with extra-terrestrial life, and referring to beliefs held by Indian, Chinese, Arab and Greek philosophers, as well as to modern astronomy. It became a best-selling book which ran to thirty-five impressions.[5]

Flammarion soon made his name both as a prolific popular-science writer and as the founder and first president of the Société Astronomique de France. His wide-ranging interests took in everything from ballooning to speculative cosmology and science fiction. He wrote novels, short stories and scientific treatises, eventually publishing over fifty books (virtually launching his brother's firm single-handed), and from his royalties set up his own spectacular domed observatory at a château just south of Paris, at Juvisy-sur-Orge, to which he would invite students and fellow enthusiasts. Inscribed in letters of gold over the entrance gate were the words *Ad Veritatem per Scientiam* – 'To Truth through Science'.[6]

It explains a lot about Flammarion that he was an early proponent of the existence of an alien civilisation on the planet Mars, and that he was sure it would be 'much more intelligent' than that on earth. In his collection *Real and Imaginary Worlds* (1865), published when he was only twenty-three, he wrote brilliantly about extra-terrestrial life, reincarnation, psychical research, and even the end of the world. Although he was always fascinated by the wilder shores of scientific research, he would eventually subdue these heterodox interests to write two classics of largely conventional popular science, *L'Astronomie populaire* (1880) and *L'Atmosphère: Météorologie populaire* (1888). The English-language edition of the latter had a long, admiring Preface by none other than James Glaisher FRS, though Glaisher edited out many of what he called 'Flammarion's rhapsodies'.[7] Both books became Flammarion company best-sellers, and Flammarion was awarded the Légion d'Honneur for what was indulgently termed '*haute vulgarisation de l'astronomie*'.

Flammarion believed that astronomy was good for the soul. It might also bring universal peace and harmony. He promulgated such idealistic views with characteristic panache:

> What thoughtful spirit could look at brilliant Jupiter with its four atten-
> dant satellites, or splendid Saturn encircled by its mysterious ring, or a
> double star glowing scarlet and sapphire in the infinity of night, and not
> be filled with a sense of wonder? Yes, indeed, if humankind – from humble
> farmers in the fields and toiling workers in the cities to teachers, people of
> independent means, those who have reached the pinnacle of fame or
> fortune, even the most frivolous of society women – if they knew what
> profound inner pleasure awaits those who gaze at the heavens, then
> France, nay, the whole of Europe, would be covered with telescopes instead
> of bayonets, thereby promoting universal happiness and peace.[8]

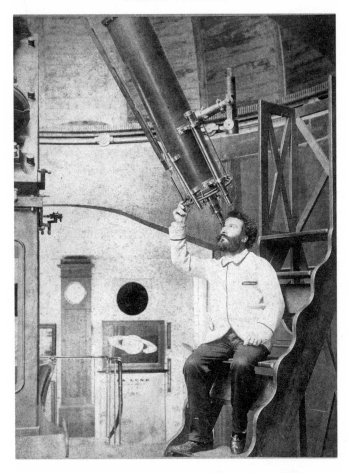

L'Atmosphère was powerfully inspired by Flammarion's balloon experiences. While studiously surveying all 'the great global phenomena of nature' and the latest meteorological research on such scientific topics as wind, rain and air pressure, the book also succeeded in projecting a certain otherworldliness. Flammarion would later tell a story of how the manuscript of the 'Wind' chapter was blown off his desk and out of his window one stormy night, only to land inexplicably at the printer's office the next day.

The book became celebrated for printing as its frontispiece a large mystic image showing the place 'where earth and heaven meet', Flammarion's poetic conception of Glaisher's rather more prosaic 'upper air'. This was a coloured engraving of striking beauty, made to look like a medieval woodcut, showing a pilgrim clambering from the warm, sunlit earth into the great icy-blue vault of the stars. Later this became known as 'The Flammarion Pilgrim', and was thought to symbolise man's eternal desire to explore ever upwards, into the upper air and beyond the stars. It had of course a particular appeal for aeronauts: *Excelsior!*

To enhance the mystery, Flammarion kept the name of the artist anonymous, although the image was quite possibly his own design. He accompanied it with a visionary commentary, which gives a good impression of his highly-coloured style and polyvalent approach, mixing science with history and mysticism. It was one of his better 'rhapsodies'.

> *Whether the sky be clear or cloudy, it always seems to us to have the shape of an elliptic arch; far from having the form of a circular arch, it always seems flattened and depressed above our heads, and gradually to become farther removed toward the horizon. Our ancestors imagined that this blue vault was really what the eye would lead them to believe it to be; but, as Voltaire remarks, this is about as reasonable as if a silkworm took his web for the limits of the universe. The Greek astronomers represented it as formed of a solid crystal substance; and so recently as Copernicus, a large number of astronomers thought it was as solid as plate-glass.*

Flammarion connected these early scientific conceptions with traditional literary and religious beliefs:

> The Latin poets placed the divinities of Olympus and the stately mythological court upon this vault, above the planets and the fixed stars. Previous to the knowledge that the earth was moving in space, and that space is everywhere, theologians had installed the Trinity in the empyrean, the glorified body of Jesus, that of the Virgin Mary, the angelic hierarchy, the saints, and all the heavenly host ... A naïve missionary of the Middle Ages even tells us that, in one of his voyages in search of the terrestrial paradise, he reached the horizon where the earth and the heavens met, and that he discovered a certain point where they were not joined together and where, by stooping his shoulders, he passed under the roof of the heavens.[9]

For Flammarion, ballooning was an idealistic and healing pursuit, which might indeed 'pass under the roof of heaven'. In what he, like Nadar and Fonvielle (not to mention Hugo), regarded as the corrupt and materialistic atmosphere of the French Second Empire, he felt a longing for the 'freshness and democracy' of the upper air. Typically, he celebrated his honeymoon in August 1874 by taking his young bride on an overnight balloon flight from Paris, landing at Spa, over the Belgian border, near Liège. 'What more natural,' he told the American magazine McClure's, 'than for an astronomer and his wife to fly away like a couple of lovebirds?'[10]

To him, although balloons were essentially a French discovery, and despite the historic support of Arago, their original promise had been shamefully neglected under the imperial regime. So ballooning had become almost a patriotic duty. It was vital to press on with the exploration of 'the vast atmospheric ocean at the bottom of which we live'.[11]

Flammarion saw this in the light of both science and of history:

> This splendid and marvellous means of locomotion was once hailed as an infallible method of obtaining a thorough knowledge of the earth's atmosphere ... The illustrious Benjamin Franklin foresaw the meteorological importance of a balloon. Whilst passing through Paris he spoke to several members of the Académie des Sciences on the scientific future for aerostation. This future was then supposed to be near at hand; but even now, in the seventieth year of this century, who can say we have remotely realised it?'[12]

Flammarion joined the Société Aérostatique in 1867, and undertook about thirty ascents, in various borrowed or hired balloons, between then and 1880. Some were made with Nadar, some with the Godard brothers, and some with his friends Fonvielle and Tissandier. They became more scientific and experimental as he progressed. His very first was launched, with a certain symbolism, from the Paris Hippodrome on Ascension Day, 1867. This was the Roman Catholic feast of the Ascent of the Risen Christ into Heaven, actually a moveable feast, on the fortieth day after Easter, but usually in late May.[13]

Flammarion was initially under the tuition of Eugène Godard, 'Aeronaut to the Empire' as he delicately put it, and used the balloon belonging to the Société Aérostatique. Ironically, this was originally constructed on the orders of the Emperor Napoleon III, as part of his imperial war adventures in north Italy. When these martial ambitions totally failed (a warning of things to come), the balloon was demobbed, so to speak, and sold off cheap to the Société. Flammarion made clear his anti-imperial sympathies by heading his account of the ascent with verses written by the exiled Victor Hugo:

> Où va-t-il ce navire? Il va du jour vêtu,
> A l'avenir divin et pur, à la vertu,
> A la science qu'on voit luire ...[14]

> Whither sails this ship? It sails with daylight
>   clothed,
> Towards the Future, pristine and divine; towards
>   the Good,
> Towards the shining light of Science seen afar ...

Flammarion based his experiments on the scientific programme origi-nally set out by François Arago, but also 'after perusing the results obtained by Gay-Lussac, Robertson and Glaisher'. While criticising balloon showmen and publicists, he recalled that Arago had prophesied that 'beautiful discoveries will reward those who make scientific excur-sions in balloons'. Flammarion himself regarded with particular awe the British ascents: 'For the finest and most productive series of scientific expeditions into the atmosphere we are indebted to James Glaisher,

Fellow of the Royal Society, the results of which are published in the volumes of the British Association.'[15] ⦿

In an early essay written in 1867, 'A Sketch of Scientific Ballooning', Flammarion set out his hopes for what he called the 'application of balloons' to meteorological investigations:

> This marvellous world of air, so mild and yet so strong, where tempests, whirlwinds, snow, and hail are elaborated, was henceforth opened to the inhabitants of the terrestrial soil. Its secrets would be disclosed, the movements of the atmospheric world would be counted, measured and determined as scrupulously as astronomers can determine those of celestial bodies; and man, once placed in possession of this terrestrial mechanism, would be able to predict rains and storms, drought and heat, luxuriant crops and famines, as surely as he can predict eclipses, and thus ensure an ever-smiling and fertile earth![17]

⦿ Glaisher's high-altitude ascents continued to be admired by French aeronauts and meteorologists for the rest of the century. They soon aimed to rival and supersede his discoveries, as with the ill-fated Zénith team of 1875. Eventually, the term 'stratosphere' ('sphere of layers') was introduced to redefine Glaisher's 'upper air' by Léon Teisserenc de Bort, head of the Central Meteorological Bureau in Paris, but not until 1899. Experimenting at similar heights to Glaisher, he sent up over two hundred balloons from his estate at Versailles, though all of these were unmanned, and automatically parachuted back their instruments. With these de Bort confirmed that above approximately six to eight miles the temperature, which drops steadily from sea level to that altitude, remained constant, or even began to increase. This surprising data indicated the existence of a new zone, or skin, of atmosphere (a vague term originally adapted from the Greek, 'the sphere of surrounding vapours'). In fact Glaisher had already submitted examples of these unexpected temperature gradients, though he had not specifically named the new layer. In 1902 de Bort suggested that the atmosphere was in effect divided into two shells or skins. He named the lower skin, containing breathable air and all active life, the 'troposphere' (the 'sphere of changes' – a phrase adapted from Glaisher). It was soon realised that all clouds, winds and pressure systems were largely confined to this troposphere. Thus the possibility of genuine long-range weather forecasting, on the basis of developing cyclones and anti-cyclones within this relatively narrow band, became a real possibility. But above all it was realised, as Glaisher and Coxwell had first demonstrated, how thin and fragile this vital band of planetary life really was.[16]

2

Flammarion's own declared scientific objectives were gloriously ambitious. Like Glaisher, he felt that only ballooning could supply the mass of data necessary to make genuine 'predictions of weather' possible, and thereby develop 'a true meteorological science worthy of comparison with her eldest sister Astronomy'. He planned to establish 'the various strata of the air', and the nature of atmospheric 'gradients' in terms of temperature, electricity and barometric pressures, and to gather all sorts of information and analytical measurements, though his list was a little vague and poetic compared to Glaisher's.

He intended to investigate the following: 'the moisture of the atmosphere, solar radiation, meteoric phenomena, the forms of clouds, the colour of the sky, the scintillations of the stars, the chemical composition of the air at various altitudes, the laws of sight and sound in these high regions etc. etc.' He also believed that it would be possible to compile complete maps of the consistent wind currents at various altitudes, depending on the geographical location, the time of day, and the seasons of the year, rather like three-dimensional maritime tide-tables of coastal waters.[18]

His instrumentation – basically various forms of barometer and thermometer – was amateurish in comparison to Glaisher's sophisticated aerial laboratory. At night, for example, to view his instruments, instead of a Davy lamp as used by Glaisher, Flammarion ingeniously employed a little glass jar which he had stocked with glow worms.[19] He obviously took pleasure in such an eccentric arrangement, poetry mixing with meteorology. Reading his elegant accounts, it is difficult to believe that the thrill of flying was not just as important for Flammarion as gathering data for the Société Aérostatique.

Flammarion had a wonderfully fresh eye, and he constantly picked out and delighted in unusual phenomena. As he put it, 'It seemed rational to "go and see" for myself what is being done in these higher regions.' Once, while aloft, he noted a strange cloud of dust over Paris, 'whitened by the rays of the sun', and thought at first that it was ordinary city pollution. Then he realised it was kicked up by the exceptional crowds visiting the National Exhibition, far below. As he put it, the democratic air 'bore witness to the excitement and pleasures of ordinary people': running feet, dancing horses' hooves, and flying carriage wheels over the *gravillon*. Another time, over green fields in the evening, with the sun low and behind him, he saw the balloon shadow 'completely surrounded by a yellowish white aureole, such as is seen painted round the heads of saints'. The air beatified the balloon with a halo.[20]

He delighted when he entered a thick cloud with a particularly high hygrometer reading, and suddenly found himself in the middle of a concert hall in which 'excellent orchestral music' was playing. It turned out that the dense, humid atmosphere was especially suitable for collecting the sounds from a village band playing in the central square of Boulainvilliers, a little country town lying invisible 3,280 feet below.[21]

He noticed the different colours of river waters, due to different soils, and how they did not always immediately intermingle on meeting: 'The water of the Marne, which is yellow now as it was in the time of Julius Caesar, does not mix with the green waters of the Seine, which flows to the left of its current; nor with the blue water of the canal which flows to the right.' The result became a single tricolour of river water which flowed for several kilometres, yellow in the centre and green and blue on either side. 'If travelling in balloons were commoner than it is at present,' he

remarked pointedly, 'what facilities it would confer on topography and surveying in general!'[22]

Flammarion carefully observed other creatures in the air, besides birds or his own carrier pigeons. He glimpsed moths, beetles, spiders, but especially butterflies: 'Butterflies hover round the car of the balloon. Until today I imagined that those little things passed their short existence among the flowers of the fields, and that they never rose to any great height in the air. But in fact they rise higher than any of the birds of our forests, and soar to many thousands of metres above the ground ... And another thing strikes me: they do not appear to be frightened by the balloon as birds are. Why is this?'[23] ◊

Sheep, horses and ducks, and sometimes the small children tending them, were also frightened by the balloon. The old superstitious cry of 'It's a Devil!' was still occasionally heard when Flammarion crossed remoter countryside, much to his embarrassment. But on the whole the balloon was favourably regarded almost everywhere in the provinces. He and his companions frequently heard church bells rung to greet them as they passed over villages, and saw local mayors putting on their official sashes and running out to salute them from the steps of the *mairie*.[25]

◊ The short answer is that they are too busy migrating. Modern studies of airborne insects have continued with tethered 'sampling' balloons, and most recently with a special type of 'vertical' radar. Occasional high-altitude bird and insect flights occur close to the stratosphere. But what has been discovered most recently are massive, seasonal 'airflow populations' of migrating insects up to about nine thousand feet. These include moths, ladybirds, lacewings, locusts, hoverflies and ground beetles, as well as Flammarion's fearless butterflies; and they may travel hundreds of miles. The numbers involved are astonishing, and give a wholly new idea of the richness (and hence vulnerability) of the troposphere. According to one study, a conservative estimate of the 'total bioflow' over a one-kilometre stretch of the southern English countryside is an astounding three billion insects per month. This is the equivalent to approximately 'one metric ton of insect biomass' regularly flying overhead, an idea at which Baron Munchausen would have rejoiced. The statistics are significant as they also throw light on other vital atmospheric phenomena. These include the study of bird migration patterns (especially those of swifts and swallows, which eat insects on the wing), the methods of insect navigation by magnetic field or even stars, and the impact of air pollution. As Flammarion also noticed, with fellow feeling, butterflies are more like balloonists than birds, because (above a certain height, say three hundred feet) they depend utterly on the wind currents for their heroic journeyings.[24]

Railway trains signalled to them 'by a joyous whistle from the locomotive' as they flew above the tracks, to which the crew replied with
merry – but faintly mocking – waving of flags. 'What dust and what an
infernal noise they make,' Flammarion reflected of steam engines. 'After
all, how slowly they go in comparison with the rapidity of our smooth and
silent course through the pure air!' Most satisfactory of all, as they passed
over large country estates, lordly invitations were frequently shouted up
to them: 'Ahoy, Monsieur! Do land here if you can, and come to dinner at
the château!'[26]

At night Flammarion revelled in the extraordinary brilliancy of the
stars. On one occasion Jupiter seemed far brighter than the moon – it was
'the sceptre of the night'. On another, the clarity of the craters and mountains on the moon's surface, even without a telescope, was hypnotic, and
reminded him of Tycho Brahe's 'naked eye' observations centuries ago in
Scandinavia. Indeed, he could see and worship the 'radiating mountain'
named after Tycho himself.[27]

At the same time, like John Wise in America, he could tell what kind
of ground they were flying over in the darkness, simply by listening carefully: 'The frogs indicated peat bog and morasses; the dogs were
evidence of villages; absolute silence told us we were passing over hills
or deep forests.' Smells and scents could give similar information: crops,
pine trees, cattle fields, duckponds, even rooftops (chimneys), all yielded
up their distinctive identifying 'perfumes'. These night flights, frankly
more impressionistic than scientific, cast the most sustained magic.
One lasted eleven and a half hours, and covered over three hundred
miles from Paris to Larochefoucauld in the Limousin. They landed in a
country lane just before dawn: 'We sank slowly down like a lazy bird,'
overwhelmed by the sweet smell of vines and cornfields all around
them.[28]

For all his mathematical training at the Observatoire, Flammarion
seems to have spent little time on data. He had a poetic and philosophical turn of mind, largely lacking from Glaisher's meticulous reports. At
six-thirty one perfect summer evening on 10 June 1867 he was floating at
exactly 10,827 feet above the river Loire, south of the forest of Orléans. He
was slightly higher, he noted characteristically, than Mount Olympus, the
home of the Greek gods. The air was perfectly clear, the sky perfectly blue.
All the central district of his beloved France was spread out beneath him,
like a magnificently painted geographical map of many colours. It was

'the most magical panorama which fantastic dreams could evoke'. All was rich, glowing, peaceful. He could even see back as far as the geometric alleys of the Luxembourg Gardens, where his love affair with ballooning had begun.[29]

In a sort of trance, he rose from his seat in front of the instruments, grasped the edge of the balloon basket with both hands, and leaning out as far as he could, gazed downwards into 'the immense abyss'. But the thoughts that came to Flammarion now were not what he expected, on that idyllic summer evening above France, 'in the midst of these blue heavens':

> Down below, at 10,000 and odd feet beneath me, exist the universal radiations of life and activity; plants, animals, and men are breathing in the lower strata of this vast aerial ocean, whilst here above animation is already on the decline. Here we may contemplate Nature, but we repose no longer on her bosom. Absolute silence reigns supreme in all its sad majesty. Our voices have no echo. We are surrounded by a vast desert. The silence that reigns in these high regions of the air is so oppressive that we cannot help asking ourselves if we are still alive. But death does not reign here; we are impressed only by absence of life. We appear to appertain no longer to the world below ... This absolute silence is truly impressive; it is the prelude to that which reigns in the interplanetary space in the midst of which other worlds revolve. The sky here has a tint which we never saw before ... Planetary space is absolutely black.[30]

3

Flammarion's fellow aeronaut, Gaston Tissandier (1843–99), appeared to be an altogether more conventional and earthbound character. His background was academic, and his manner deceptively restrained, sober, even pedantic. He would eventually become the greatest nineteenth-century French historian of ballooning, and his huge, meticulous collection of balloon pictures, letters, articles, books, documents and other memorabilia would form the major aeronautical archive in the Library of Congress, Washington, DC. But all this was deceptive. Beneath the calm professorial exterior, with his neat pedagogic beard, beat the heart of a wild, chaotic and dauntless balloon enthusiast.

Born in Paris in 1843, Tissandier studied chemistry at the Conservatoire des Arts et Métiers, and graduated from the Sorbonne. A brilliant young chemist, dedicated and serious-minded, he was appointed Director of the Laboratoire National d'Essai et d'Analyse in 1864, at the early age of twenty-one, and began lecturing to teenage students at the Association Polytechnique. Five years later, when already embarked on his ballooning adventures, he published a successful popular textbook for them, *Traité élémentaire de chimie* (1869).

The formative balloon experience came relatively late to the young professor, and was at first insidious rather than dramatic in its effects. He happened to witness one of Nadar's later Paris launches in autumn 1866: 'It was the *Géant* that drew me definitively into what I may term my *aerial vocation*. I shall never forget the ascent of that magnificent aerostat from the Champ de Mars, accompanied by the little *Imperial*. I have still before my eyes that mighty balloon awaiting the signal to rise into the air and soar through the clouds like an eagle ... I still see the *Géant* rising magnificently: a shower of sand falls from the wickerwork car, and the balloon is soon lost to sight in a thick cloud of vapour. Around me arms are uplifted on all sides, shouts of excitement fill the air, hearts beat fast, and

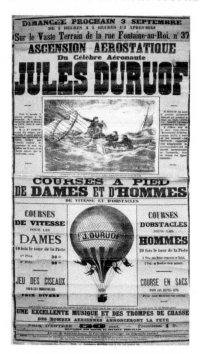

everyone returns home thinking of nothing but those aeronauts.'[31] Anyway, Gaston certainly did, although his teaching duties allowed him no immediate chance to pursue that unexpectedly tantalising vision into the 'thick clouds of vapour' above his head.

Two years later, in August 1868, while on a sedate seaside holiday with his parents and his elder brother Albert (1839–1906) in Calais, Tissandier spotted a 'great red placard' advertising a local balloon launch, scheduled for the next afternoon, in the place d'Armes. It instantly reawakened all his suppressed 'aerostatic tendencies'.[32] His account of what followed suggests something of the restless, impetuous spirit that secretly possessed the young chemistry professor at that time, as well as the continuing power of balloons to fascinate and intoxicate dreamers like him.

The balloon to be launched was the *Neptune*, piloted by the controversial amateur aeronaut Jules Duruof (1842–99). The ascent was to celebrate the annual Fête de l'Empereur, Napoleon's birthday on 15 August, the very day when the autumn weather was meant to declare its intentions. Its declared intentions over Calais were evidently stormy that year. This suited Duruof, still in his twenties, who had already made a reputation for madcap flights and disastrous landings, mostly on the north coast of France, or in the sea beyond it. He was even said to have once 'purposely' exploded his balloon, to create a sensation with a seaside crowd.[33] He was regarded by many as an irresponsible, bad-weather balloonist, a 'hooligan' of the clouds, who took too many risks, and accordingly had a faithful popular following.

Tissandier had never even heard of Duruof before, but quietly slipped round to his rooms at the Hôtel Dunkerque, and introduced himself. After 'a quarter of an hour's animated conversation', they were 'the best friends in the world', and Tissandier had been offered the third place in the Neptune's basket. 'I was transported with joy on leaving him.'

Tissandier's family were appalled, and spent the rest of the day trying to talk him out of it. 'This part of the world, they said, is particularly fatal to balloons and aeronauts. Pilâtre de Rozier lost his life not far from here, and Deschamps was nearly killed on the same coast; the wind is always violent and uncertain along the shore, and it is pure folly on the part of anyone to undertake such an adventure.'[34] But Tissandier held firm to his resolution, although he secretly purchased 'lifebelts and floaters' from the Calais Humane Society, which dealt with drowned holiday-makers and suicides.

All night Tissandier lay awake, tossing and turning with 'extraordinary dreams' of mocking crowds, bursting balloons, and falling into the sea. Pale-faced and exhausted, he staggered out like a condemned man next morning at 5 o'clock to the place d'Armes. Here he found a howling wind, a blinding rainstorm and a roaring sea, with Jules Duruof in high good humour: 'Don't worry, I had a disaster launching from this spot last time, but this time intend to *take my revenge* on the weather – we'll make our ascent *whatever happens.*'[35]

By midday the Neptune was inflated, but it was almost uncontrollable in the wind blowing across the square: 'The soldiers who lent a hand at the ropes were continually pulled off their feet and suspended like bunches of grapes in the air.' An incredulous crowd gathered to watch the proceedings, the best entertainment to be had in Calais on a wet holiday afternoon.[36]

Duruof sent up a small trial balloon, 'its course followed by a thousand eyes'. It fled horizontally across the square, gaining no height, and struck the bell tower of one of the town houses opposite. Then it bounced off the roof, and shot straight across the promenade and out over the North Sea, quickly disappearing into a line of black thunderclouds.

Tissandier turned to look at Duruof. He was still 'calm and resolute', with a faintly quizzical expression on his face. At 4 p.m. a municipal band in the shelter of the arcades struck up.[37]

Still wearing their rain-soaked clothes, the aeronauts – now only two of them – climbed into the basket, Tissandier shivering slightly, and launched. The cobbles of the square dropped away, and the bell tower seemed to tilt over and rush towards them. Some lines from Aristophanes' comedy *The Clouds* flitted into Tissandier's scholarly brain as he prepared to die: 'So let us scale the snow-capped mountains, keeping calm heads, losing no breath!'[38] Then suddenly the entire square disappeared and they were out over the sea.

From that moment, the flight took on a dreamlike quality. Duruof had adopted a typically maverick method of launching. Knowing that the soldiers could hold the *Neptune* down, he had offloaded a mass of ballast just before he ordered the release. The balloon was then so buoyant that it shot upwards almost vertically, 'like a cork from a champagne bottle', easily clearing the fatal bell tower, and rising in a few minutes to 5,900 feet, well clear of the immediate rainclouds and into a calm, peaceful, sunlit zone with a temperature of 59.6 degrees Fahrenheit. Tissandier could glimpse Calais rapidly receding, with 'a mass of microscopic spectators running along the jetties like a family of ants', among whom were his anxious parents.

The balloon was heading in a generally northern direction out over the Channel. In a kind of trance, Tissandier observed the sea 'like a vast field of emerald' below, and beautiful 'violet-coloured' cirrus clouds infinitely high overhead. Duruof was 'plunged in thought' as he watched the turning compass. 'We are making for the coast of England,' he first announced. Then a little later he corrected himself with a wry smile. They had turned north-east, and were heading straight out over the North Sea on a bearing which would take them, in 'undisturbed serenity', all the way to Scandinavia. Tissandier tried hard not to panic.[39]

After some time, Duruof asked Tissandier to take note of the wind direction at different heights. At their comparatively high altitude it was blowing steadily north-east; but several thousand feet below the troop of cumulus clouds were moving in exactly the opposite direction, south-west. Duruof had identified a classic 'box' (as so often used by the American balloonists), which offered a guaranteed return ticket to Calais, along what sailors called the reciprocal bearing. 'We can continue our excursion over the sea as long as we want,' he said, 'and return to shore whenever we like.'

Tissandier was astonished and relieved by this new promise of aero-nautical magic. They continued 'towards Scandinavia' for about an hour (in fact a modest distance of about twenty miles), then valved and dropped very low, to four hundred feet. By the end of the second hour they were skimming in over the breakers and sailing back directly over Calais. Here Tissandier had the satisfaction of spotting his brother Albert on the jetty, waving admiringly – and perhaps enviously. For him too it was a memorable flight.[40]

Despite the cheers of the holiday crowd, and pressing invitations to descend – or rather precisely because of these – Duruof coolly threw out more ballast and sailed on towards Boulogne. At sunset, using a guide rope and a grapnel, they managed to make a perilous but beautifully timed landing on the rocky beach just below the lighthouse at Cap Gris Nez. The lighthouse-keeper ran out to greet them in such a hurry that he forgot to put on his shoes, and cut his feet on the shingle. The next day they solemnly walked up to visit Pilâtre de Rozier's tomb, and pay their aeronautical respects. 'I shall never forget the humble stone that marks the spot,' wrote Tissandier. He telegraphed his brother Albert with news of their safe arrival. He had confirmed his 'aerial vocation'.[41]

Albert Tissandier soon came to share his younger brother's fascination with ballooning, but was characteristically more circumspect.
Trained as an architect, photographer and illustrator, he represented the
artistic side of the family. For him ballooning was essentially a source of
visual images. He would often accompany Gaston on his future ascents,
and they would make great play of their friendly rivalry: science competing with art. For the next two years the Tissandier brothers learned everything they could about balloons. Gaston wrote up scientific notes, while
Albert worked on his pictorial technique, combining drawings with
photography.

They met up with the journalist Wilfrid de Fonvielle, who had
returned from interviewing Charles Green in London to take up a
teaching post at the Conservatoire des Arts et Métiers, and was full of
outrageous aeronautical tales. Over dinner they 'spoke much about the
scientific use of balloons, and the numerous experiments which might be
made in them'. Fonvielle pointed out what brilliant newspaper stories
balloons could provide. Every flight was a potential drama: 'the launch,
the flight, the landing!' Moreover, they could be given a subtle anti-
imperial slant – 'the freedom of the skies, the irrelevance of borders, the
democracy of the air, and so on'. To prove it, he immediately dashed off
– 'whilst over dinner' – a vivid account of Gaston's Calais–Cap Gris Nez

adventure, and sold it the very next day to the radical paper *La Liberté*.[42] From then on Fonvielle and the Tissandiers became a band of ballooning brothers.

4

Between autumn 1868 and summer 1870 this three-man team – often accompanied by one of the Godards, or Jules Duruof, as their instructors – undertook a regular series of ascents from Paris. The more hazards they encountered, the better stories they came back with. They experienced violent snowstorms above Normandy, blinding fog over the North Sea, a freezing night in a captive balloon (with Glaisher) above London, a gale-force flight over Belgium, a long-distance trip into Germany, a burst balloon, and innumerable crash-landings.[43] They also made a logistical discovery that would soon turn out to have unsuspected significance: that because of the prevailing winds, the most efficient place from which to launch a balloon in Paris was the gasworks at La Villette, on the north-eastern outskirts of the city.

They broke several balloon records, including the most sustained 'platform' flight, which ironically turned out to be forty-eight hours spent almost stationary in the air between Paris and Compiègne. But in 1869 they also established the fastest average balloon speed: ninety miles per hour, achieved during a thirty-five-minute trip 'dragged along by the force of a furious gale' beyond Meaux into the flatlands of Flanders. They landed 'covered in bruises and more or less stunned'. But a quick calculation showed that no train had ever matched that 'astonishing celerity'.[44]

In spring 1870 Gaston began publishing a landmark series of monthly articles in the mass-circulation journal *Le Magasin pittoresque*, entitled 'Histoire d'un ballon'.[45] These soon attracted a broad popular readership, who identified with the spirit of adventure, and the celebration of the French countryside over which Gaston and his companions flew, as much as with the ballooning itself. Though Gaston provides a short history of ballooning, and various miscellaneous scientific observations in passing (on snowflakes, high-flying spiders, cloud structures and light diffraction), the central interest remains the aerial adventure and the unfolding vision of France, *la Patrie*.

Much space is given to one particularly refractory, but much-loved, balloon, called *L'Hirondelle* ('the Swallow'). She becomes a sort of mischievous character in the narrative, and is perhaps the spirit of Liberty herself. She takes them on various hair-raising flights over the remotest countryside, *la France profonde*, and out to the surrounding coastline. The unheard-of names of the tiny villages where they often land at dusk come to resemble a pastoral or patriotic litany. Once they even touch down in French Algeria. At the same time the high cloud-scapes above, the 'Alhambra palaces' of the upper air, become a sort of sublime extension of national dreams and longings.◊

All this was wonderfully illustrated by Albert Tissandier. Beginning with precise technical drawings (for example, of the exact workings of a balloon barometer or a sprung venting valve), he soon found his true subject in extraordinary panoramic pictures of the balloon in the clouds.

Albert captured the peaceful, visionary atmosphere of mid-nineteenth-century ballooning better than any other artist. Drawing on the great tradition of French landscape painting – which was just about to metamorphose into Impressionism – he invented something quite new: the extended aerial 'cloudscape'. These are not views *from* the balloon, but breathtaking panoramas from some imaginary viewpoint *outside* it.

Using a brilliant combination of fine engraving and photography, Albert Tissandier invented a new kind of sublime. Great oceanic stretches of iridescent clouds are dramatised by sunlight or moonlight. They are like enormous stage sets, upon which a single balloon – usually seen at a great distance – appears as the only actor, the only human point of reference and of visual scale. Varied meteorological effects – snow, fog, rain, sunset or sunrise beams – suggest a kind of transformed, celestial upper

---

◊ They planned eventually to fly right around the Mediterranean, an aerial version of the Grand Tour, starting in Morocco. It is possible that this trip was inspired by Jean Bruno's *Les Aventures de Paul enlevé par un ballon* (1858). However, *L'Hirondelle* was suddenly and shockingly incinerated when one of the launch crew casually lit his pipe next to the hydrogen generator. Though no one was hurt, this sudden conflagration in the summer of 1869 was an ill omen: 'Your beautiful balloon canopy, your genera-tor, your storage hangar, your basket, all gone up in flames! – nothing but a heap of cinders!' The image of a burning balloon over Paris – which goes right back to Sophie Blanchard – would soon become a political symbol of great power.[46]

Fig. 1. — Baromètre métallique compensé, pour les hauteurs de 6 000 mètres. — Extérieur et intérieur.

world. It is secular, even pagan; but shot through with feelings of infinite longing or melancholy or loneliness or hope. It is the dream world of the mariners of the upper atmosphere.

This lyrical, visionary age of French ballooning was to be transformed by the terrible catastrophe of the coming Franco-Prussian War of 1870–71. But just before its outbreak, these young and idealistic aeronauts came together to produce the greatest book of nineteenth-century ballooning ever published: *Travels in the Air*. It was, symbolically, a collaborative work, involving four authors. What's more, it was that rare thing, a Franco-British publishing project. Optimistic in tone, progressive in outlook, it was innocent of any reference to the recent conflict in America, and the coming conflict in France. Indeed, it could be said to have its collective head magnificently in the clouds.

The moving spirit was Camille Flammarion, who contacted James Glaisher in London and got him to agree to a collaboration. There is some suggestion that there was also an attempt to coopt Charles Green, but he was now too ill to write any kind of memoir. The first version, entirely in French and under Flammarion's editorship, appeared under the title *Voyages aériens* in 1870. Parts I and II consisted of long autobiographical pieces by Glaisher (translated) and Flammarion, including their nicely contrasted histories of 'scientific ballooning' in Europe, as seen respectively from an English and a French perspective. Part III added racier, miscellaneous accounts by Gaston Tissandier and Fonvielle, often co-signing their contributions.[47]

The revised and expanded English version, *Travels in the Air*, was published in London by Richard Bentley in 1871. This historic volume was given a distinctive appearance by a set of 118 magnificent aerial illustrations, largely by Albert Tissandier. It presents its overall editor as James Glaisher FRS, who has evidently exercised some editorial discretion.

The different styles of the aeronauts become particularly noticeable. Glaisher's dry 'English' description of his high-altitude flights contrasts strikingly with the light-hearted 'Gallic' touch of Tissandier's lively narratives, much emphasised by the nimble translation, while Fonvielle's witty, irreverent reminiscences (including his rides in the creaking *Le Géant* and his visit to the equally creaking Green) have an almost music-hall flavour in English. Each is attractive in its own way, yet none achieves quite the solemn poetry of Flammarion's observations of the upper air, equally effective in either French or English.

The tone of the whole collection was uplifting and visionary:

This book, we sincerely hope, will mark an epoch in the history of aerostatics, for it is the first time that a series of aerial scenes have been published as observed by the aeronauts themselves. It is also the first time that artists

have gone up in balloons ... reproducing these incomparable panoramas,
these magnificent scenes, before which the Alps themselves grow small,
earthly sunsets are eclipsed in splendour, and the oceans themselves are
drowned in an ocean of light still more vast ...[48]

The Franco-British entente was not without its tensions. While Glaisher
recognised Flammarion's distinction as a scientist, and regarded Gaston
Tissandier as 'agreeable, active and intelligent', this was not the case
with Wilfrid de Fonvielle. In a marginal note on his editorial copy,
Glaisher described Fonvielle as 'a Red republican', and 'over-excitable'.
His balloon writing was 'flippant and in bad taste', and Glaisher did not
approve of him contributing to leftist newspapers like *La Liberté*. In
return, Fonvielle clearly thought Glaisher was a snob and an imperialist
(quite unlike the amiable Green), and chastised him for slow editorial
work and not replying to his letters.

But these vague political irritations largely dissolved once they were
in print together, and airborne in history. It is clear that they were all
immensely proud of the book. In retrospect – after the earthly catastrophe of 1870 – it took on a sort of dreamy, utopian afterglow. The oceans
of the upper air would never again seem so free, so boundless, so
sublime.[49]

10

Paris Airborne

———

1

When Camille Flammarion had looked over his beloved France in June 1867, from 10,287 feet above the river Loire, he had seen a clear, sunlit horizon, with no remote stormclouds gathering to the east. Yet this was precisely the year in which the fifty-two-year-old Otto von Bismarck was appointed Prussian Chancellor, and became the driving wind of European politics. Far below, political events began moving with sinister inevitability. Beyond the river Moselle, in the borderlands of Strasbourg and Metz – borders invisible to balloonists – the newly founded North German Confederation was flexing its demographic muscles. Arcane diplomatic disputes, like the Hohenzollern candidature for the Spanish throne in 1868, produced a ping-pong of diplomatic insults between France and Prussia. They culminated in the deliberately provocative Ems telegram of July 1870, described by Bismarck as 'my red rag to the Gallic bull'.

In August 1870 the Emperor Napoleon III, confident of the strength of his three enormous armies, had marched eastwards across the Moselle and the Meurthe with the intention of claiming disputed territories beyond Alsace-Lorraine. Arrogantly, almost casually, he declared war on the fledgling Prussian state, ignoring the fact that this was precisely what Bismarck desired in order to unite a new German empire. Napoleon was also ill-informed about the small but highly efficient Prussian armies, equipped with superb Krupp weapons, well practised at mobilising by train, and commanded by the brilliant strategist General Helmuth von Moltke.

The shock military defeat of the French First Army took place on 2 September 1870 at Sedan, in the Ardennes, where – almost unbelievably

– the Emperor was captured. The Second Army retreated to Metz, where it was effectively blockaded. The remaining Third Army gallantly attempted to fight a rearguard action, retreating grimly westwards down the Loire. Some took their stand in Paris, others melted away westwards to fight again another day. These swiftly unfolding events were accompanied by the ignominious abdication of the Emperor Napoleon, whom Bismarck despatched politely to exile in England. On 4 September the Third Republic was declared under the republican politician Jules Favre, who had opposed the war. While some patriotic leaders, notably Léon Gambetta, remained in the capital, the official seat of government fled south-westwards, first settling in Tours on the Loire, and eventually in Bordeaux. Bismarck attempted to negotiate an armistice, but was refused, and proceeded to invade France at high speed, the first example of a true German *blitzkrieg*.

Accordingly, the Prussian 2nd and 3rd Armies under von Moltke contemptuously left the remaining French forces besieged in Metz, under the vacillating Marshal McMahon, and advanced steadily westwards towards Paris, along the lines of the Meuse and the Marne. The Prussian invaders split into a classic pincer movement, and by 10 September were pressing on the northern and southern suburbs of the city, halting only at the ring of isolated forts surrounding it. For the Parisians the air was full of the sound of approaching guns and the smell of burning.

Gaston Tissandier described the desperate rush of people from the surrounding countryside into Paris, with piled handcarts, aged relatives in wheelchairs, and scampering animals led on lengths of rope. He watched them hurrying through the city's gates and ramparts 'like a Biblical scene from the Flight into Egypt'.[1] With terrifying speed and efficiency, the Prussians had virtually surrounded Paris in less than a fortnight, by 15 September 1870. Bismarck, a master of both strategy and symbolism, ordered the occupation of Versailles, and installed heavy artillery in preparation for shelling the city into submission.

European, and notably British, opinion was initially sympathetic to the Prussians, since it was the French who had declared war, first invaded at Saarbrücken, and then refused Bismarck's offer of an armistice. The celebrated British war correspondent, William Russell of *The Times*, who had famously covered the Crimean War, significantly chose to report from the Prussian headquarters. This, for him, was the story his British readership wanted.

It was strategically important to Bismarck to retain this sympathy. The French under Napoleon III were regarded as foolish aggressors, and Paris under the Empire was widely viewed as the seat of fashion, frivolity and sexual licence, quite incapable of offering any resistance to an invader. It was a lax city of 'luxury and pleasure', as *The Times* put it with a certain relish. The newspaper's Special Correspondent sent a dramatic despatch datelined Palace of Versailles, 30 September 1870, which opened: 'Here at 7.30 p.m. the German Crown Prince and his Staff are comfortably quartered ... close at hand they have an angry Paris agitated by a hundred passions ... impotent in her rage and fierce vindictiveness.'[2]

In fact many pro-republican French people themselves felt ashamed of what France had become. Félix Nadar and Fonvielle both expressed such views. Victor Hugo, who had remained in exile since 1851, returned to Paris on 5 September, immediately the Third Republic was declared. He brought with him his new book of poems, symbolically entitled *Les Châtiments*. It meant, literally, 'The Punishments'; or perhaps, 'The Reckoning'.

Bismarck's strategy was to take advantage of these international views, and avoid any direct military attack on the civilian population of Paris, who, as he put it bluntly, could 'stew in their own juice'.[3] Moltke contented himself with shelling the ring of huge, isolated forts – like Mont-Valérien and the Fort d'Issy – that surrounded the city, a psychological as much as a military tactic. The master plan was to close down all communication between Paris and the outside world. The Prussians intended to silence, humiliate and punish Paris – but to do it largely in secret. Without too much fuss, they would quietly starve her citizens into submission within a matter of weeks. It was therefore vitally important that the minimum information about actual conditions in Paris should be allowed to filter out, and that the French government should remain paralysed. The blockade was a military necessity, but even more a diplomatic one.

On 17 September a *Times* editorial noted grimly: 'In a few days we shall know nothing of what is passing in Paris ... The Prussian army is large enough to destroy regular and effective communication between the invested city and the rest of France ... Will the whole body politic be paralysed?' Nonetheless, the commentator remarked, it was possible that 'Paris will show fight'.[4]

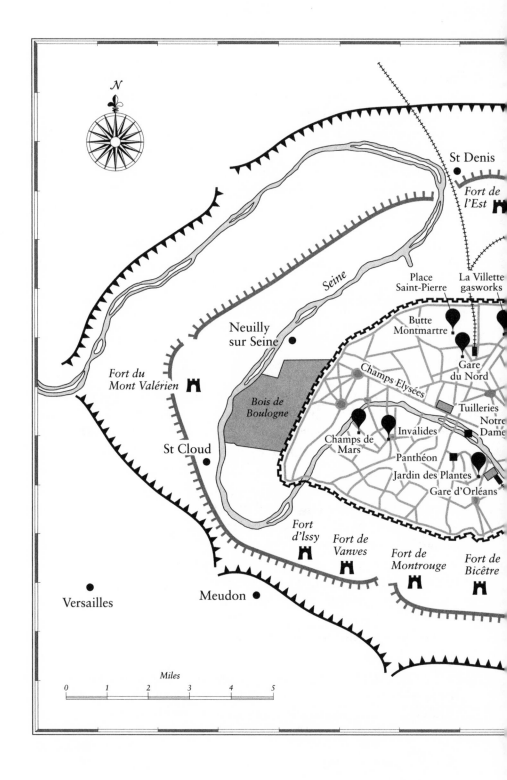

N

St Denis

Fort de
l'Est

Seine

Place
Saint-Pierre

La Villette
gasworks

Butte
Montmartre

Gare
du Nord

Neuilly
sur Seine

Champs Elysées

Tuilleries

Notre
Dame

Fort du
Mont Valérien

Bois de
Boulogne

Invalides

Champs de
Mars

Panthéon

St Cloud

Jardin des Plantes

Gare d'Orléans

Fort
d'Issy

Fort de
Vanves

Fort de
Montrouge

Fort de
Bicêtre

Versailles

Meudon

Miles

0   1   2   3   4   5

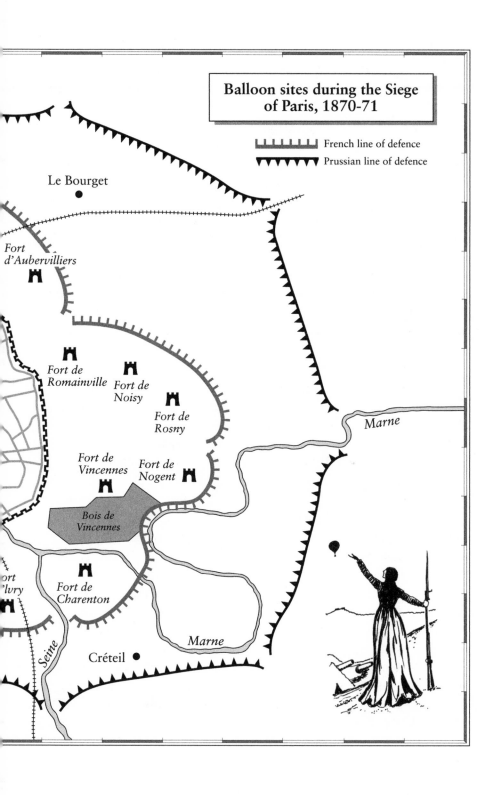

Balloon sites during the Siege of Paris, 1870-71

L⊔⊔⊔⊔⊔⊔⊔ French line of defence
▼▼▼▼▼▼▼ Prussian line of defence

Le Bourget

Fort
d'Aubervilliers

Fort de
Romainville   Fort de
              Noisy
              Fort de
              Rosny

              Marne

Fort de   Fort de
Vincennes   Nogent

        Bois de
        Vincennes

'ort
'lvry

         Fort de
         Charenton

                    Marne
         Créteil

It is difficult to imagine how complete such a blockade could be, in an age before electronic communications were universally established (let alone before radio, mobile phones or the internet). All the French telegraph lines were above-ground, and were easily identified and severed. In a matter of days, all roads were closed, all bridges were blown up or guarded, all railway lines cut, and all river traffic on the Seine strictly controlled. Even the smallest skiffs were caught in Prussian nets. There was a desperate last-minute attempt to lay a submerged telegraph line along the bed of the Seine from Paris to Rouen. This was hastily imported from England at vast expense in the last days of August, but within a week of its arrival the Prussians had discovered and destroyed it, its position betrayed by a French collaborator.[5]

One young British journalist daringly slipped into the city at the last moment, the bilingual Henry Labouchère of the *Daily News*. His last despatch noted that he had no idea how he could send his next one, and that he would keep a *Diary of a Besieged Resident* instead. For the moment Labouchère observed only a phoney war: 'The cafés are crowded ... In the Champs-Elysées the nursery maids are flirting with the soldiers ... there is universal drilling by the militia ... at the Rond-Point an exceptionally tall woman was mobbed because she was thought to be an Uhlan [Prussian dragoon] in disguise ... but no one took it seriously.'[6]

But things soon became very serious. On 19 September the last telegraph line was cut, and the official Paris mail coach was turned back with a contemptuous volley of shots by the Prussian pickets. Twenty-eight postal runners were sent out the following day, but all except one were captured or shot.[7] Every village in the Ile de France surrounding Paris, within a distance of some thirty miles, was either permanently occupied by troops or regularly swept by the Prussian cavalry, the dreaded Uhlans. The Prussian encirclement was complete, and the siege was now absolute.

It was expected to last a matter of weeks, and would certainly be – like all wars – 'over by Christmas'. In fact it lasted for five increasingly agonised months. Two and a half million people were crowded and sequestered within the walls of Paris. As Gaston Tissandier wrote in the *Magasin pittoresque*: 'Our mighty capital city was surrounded, cut off not only from France but from the entire outside world ... Two million human beings were shut away, silenced, and fenced in by a bristling ring of bayonets.'[8]

The siege tightened its grip in early October. The Prussians began a ceaseless shelling of the twelve Paris forts from the surrounding heights, the regular booming explosions wearing away at nerves, and making nights sleepless. On 13 October, when the French gunners of Mont-Valérien tried to fire a salvo at a Prussian encampment, they inadvertently scored a direct hit on the beautiful palace at Saint-Cloud and flattened it. Paris seemed to be collaborating in its own destruction.

City life became constrained, pinched and, above all, cold. One by one the theatres were closed, the street gaslights were dimmed or extinguished after 10 p.m., horse-drawn cabs virtually disappeared from the boulevards. Well before dawn, all the bakeries were thronged by endless queues. Makeshift ration books were issued, covering milk, coffee, bread and sugar. Meat was rationed to thirty-five grams – slightly over an ounce – per head per day. All luxury goods, including candles, became rare, or very expensive. Most of the trees in the Bois de Boulogne were cut down for firewood. The bitter comment went round that Paris was still a beautiful woman, but she had shaved her head in penitence.

Most of the cafés remained open, and wine alone remained surprisingly cheap and plentiful: many of the poor, especially the soldiers, were drunk for much of the time. As food supplies became seriously short, most domestic animals in the city were eventually eaten, except for a few prudent cats. There was a cull of ducks, carp and goldfish from the municipal ponds. Over forty thousand horses, and most of the rare animals from the zoo at the Jardin des Plantes, including two elephants and two zebras, were butchered.[9] A surviving restaurant siege menu proposes elephant soup, kangaroo stew, roast camel, antelope terrine, and baked cat with rat garnish.

By December, the poet Théophile Gautier found himself writing a tragic appeal to the municipal authorities on behalf of his horse. This horse, a family friend and a faithful old servant, 'a poor and perfectly innocent being', was due to be forcibly dragged off to the city abattoir in the next twenty-four hours. His appeal, beautifully phrased and genuinely touching, did not succeed. Finally, all inhibitions breaking down, there was a brisk trade in rats. In a later letter, Gautier remarked on the curious flavour of rat pâté.[10]

Hunger and humiliation were the chief Prussian weapons against the Parisians. There was a dangerous collapse of morale, and a mood of hopelessness and cynicism. The emergency government of Jules Favre seemed paralysed, and a Council of National Defence was formed. Metz surrendered. The French army and the National Guard, under General Trochu,

did little more than hold parades along the empty boulevards. Despite the presence of thousands of National Guards, there was no immediate attempt at a breakout. When this did come, on 27 November, under the vainglorious General Ducrot ('I will only return dead or victorious'), it was a catastrophic failure, costing more than twelve thousand lives (but not Ducrot's).[11] There was talk of setting up a revolutionary Commune, and on several occasions the Hôtel de Ville was surrounded by a hostile crowd, and General Trochu was threatened and barracked.

Above all, there was a growing sense of utter isolation. All communications with the outside world had been severed. No post, no despatches, no newspapers; no Reuters cables, no weather reports, no London stock exchange figures; no Italian magazines, no American scientific journals. Most demoralising of all was the crushing of ordinary private life. From beyond the steely Prussian lines there came no family letters, no news of grandchildren or aged parents, no cheering get-well postcards, and no lovers' *billets-doux*. Paris, the centre of European civilisation and enlightenment, was psychologically shrunk and physically silenced, just as Bismarck had planned.

Worse than that, it was mocked. *The Times* summarised the situation: 'The Germans have on their side all the organized apparatus of modern warfare, strong discipline, a unanimous will; while on the side of France there is wild fury, alternate fits of overweening confidence and blank despondency, no mutual faith, no truth, and a suicidal tendency to universal social dissolution.' In the circumstances, *The Times* concluded icily, the immediate, peaceful and silent surrender of Paris to the Germans would be 'a triumph of civilization'.[12]

2

It was precisely at this low point of collapse and chaos that the fantastic story of the Paris siege balloons began. In the space of four months, between 23 September 1870 and 28 January 1871, no fewer than sixty-seven manned balloons were successfully launched from the encircled city, finding a new method of breaking a modern siege.[13] In many ways these balloons represented the apotheosis of aerostation in the nineteenth century. They achieved what had never been done, or even fully imagined, before. Without being military balloons or carrying weapons, they changed the conditions of human warfare. They were the first successful civilian airlift in history.

Even as the Prussians advanced, the small group of aeronauts saw that there was one route out of the city that no one had really considered, and that remained unguarded: the air. The practical, and especially the propaganda, possibilities of balloons suddenly inspired them with a vision. Gaston Tissandier wrote a moving declaration on behalf of his band of brothers:

> The silencing of Paris would be the death of France. Our besieged city would be lost irrevocably if she cannot find some way of making her voice heard abroad. Whatever the cost, we have to find a means of avoiding the slow torture of psychological encirclement [l'investement moral], as well as establishing communications with the army of the Loire. All ground routes being blocked, all river routes being barred, there remains but one other dimension open to the besieged – THE AIR! Paris will be reminded that balloons are one of the chief glories of the scientific genius of France. The mighty invention of Montgolfier is destined to come to the aid of la Patrie in this hour of mortal danger.[14]

This was all very fine. But initially there was no official response from the new Council of National Defence, or the harassed ministers of the Third Republic. Neither General Trochu's army nor the Paris militia owned a single military balloon. The Minister for War, General le Flô, had his eyes fixed bleakly on the ground, desperately concentrating on supplying the Paris forts, and building up the city's ramparts and artillery defences. Even the young and vigorous Minister for the Interior, Léon Gambetta (1838–82), the man of vision and energy, the hope of France, had no strategic policy on the air. There was plenty of coal gas in the enormous gasometers of La Villette and Vaugirard, but nothing to put it in.

Besides, ballooning out of Paris was regarded as a high-risk, and probably suicidal strategy. No one knew if a balloon could escape the inevitable Prussian fusillade, and the legendary power of the new Krupp field guns. A balloon would pass with agonising slowness over the armed siege lines, more than sixty miles deep in many directions, and would doubtless be pursued by Prussian cavalry, the notorious Uhlans, in the spirit of a murderous fox hunt. Moreover, the Prussians had made it clear that anyone caught crossing the siege lines would be shot out of hand as a spy.

Equally, no one was sure which was the safest wind direction to try. The Prussians were known to be bivouacked in most of the towns and villages to the north and east of Paris, and out along the Marne as far as the newly occupied Alsace-Lorraine. To the south and west, where von Moltke had set up his headquarters at Versailles, approximately twelve miles from Paris, there were huge field garrisons and artillery supply dumps. Prussian cavalry units were constantly foraging and burning, and shifted quarters without warning. Officers had also forcibly requisitioned and, as it later emerged, looted and despoiled, many of the châteaux throughout the region, as far as Normandy. Moreover, the sympathies of the local farmers and peasantry could never be entirely relied upon. ⍦

Finally, even if balloons were available, could actually fly out of Paris, and land safely, what exactly could they achieve? Clearly there was no question of using them to move troops *en masse*, as Benjamin Franklin and later Napoleon had once imagined. But might it be possible to establish communications with the Army of the Loire? Or to make contact with members of the government in exile at Tours, nearly a hundred miles to the west? Beyond these purely military considerations were other tantalising possibilities: could they deliver news despatches, a civilian postal service, or even some form of propaganda campaign?

Just four balloons were ready in Paris in September 1870. All were in private ownership, and most were rather the worse for wear. The first was the *Neptune*, the same balloon that had flown out of Calais in the storm, still manned and owned by Jules Duruof, now twenty-nine. The second was *Les Etats-Unis*, owned by the Godard family, and piloted by the comparative veteran Louis Godard, though he was still aged only forty-one. The third was *Le Céleste*, the scientific balloon now owned by Gaston

---

⍦ The ambiguous aspects of the Prussian occupation, and the issue of collaboration, were savagely portrayed by Guy de Maupassant, who fought against the advancing Prussians as a twenty-year-old volunteer, and later wrote a number of short stories based on his experiences. 'Mademoiselle Fifi' (1882) has a typically ironic title, which is the nickname of a particularly sadistic Prussian officer. While entertaining willing French girls from the local village, he destroys the beautiful interior of his château billet, out of a mixture of contempt and brutal drunkenness. In 'Boule de Suife' ('The Suet Dumpling', 1880), a fat, kindly French prostitute is forced to sleep with a Prussian officer for the convenience of her fellow coach passengers. This happens in Normandy, near Rouen, over sixty miles from Paris. Maupassant's most moving reflection on the times appears in 'Two Friends' (1882). Two Parisians, driven by the hunger and boredom of the siege, get a little tipsy together on cheap wine, and set off for an innocent afternoon's angling on the lower reaches of the Seine, near the Fort Mont-Valérien. They inadvertently cross into no-man's land (hauntingly described), and after a few minutes' idyllic fishing, are captured by a Prussian patrol. They are cross-questioned, skilfully tempted to betray each other by revealing a password, and when this fails, casually shot as spies. Their bodies are thrown unceremoniously into the Seine, while their little catch of fish, still leaping and gleaming in the late sunshine, is symbolically emptied from the Parisians' net into the Prussian stockpot.

Tissandier, aged twenty-seven. The fourth was *Le National*, lent by the Godards but piloted by Albert Tissandier, aged thirty-one. But the man who largely organised the whole initiative was none other than Félix Nadar.

3

It was done with all Nadar's characteristic panache and flair for publicity. Within a week of the outbreak of war, he had created, virtually out of nothing except a piece of headed notepaper, the 'No. 1 Compagnie des Aérostiers' – The Number One Company of Balloonists. Starting with Jules Duruof and his balloon the *Neptune*, he began enthusiastically recruiting among his friends, and anyone who had known *Le Géant*. At the same time he wrote directly to the Council announcing that he was setting up, on his own initiative, an observation-balloon service on the heights at Montmartre. From here his No. 1 Aérostiers (still basically himself and Duruof) would guard the northern approaches to the city by day, and by night mount high-powered electric searchlights, adapted from arc lights from his photographic studio.

By 8 September 1870 Nadar had established a camp on the place Saint-Pierre, in the north Paris district of Montmartre. Besides Duruof, nine other balloonists had rallied to his call, including the Parisian balloon-maker Camille Dartois, the mechanical engineer Eugène Farcot, the doctor Emile Lacaze, and the militant socialist Jean-Pierre Nadal, who would go on to become Directeur Aéronautique for the Paris Commune during the tragic uprising of 1871. Apart from three bell tents and the tethered *Neptune*, Nadar's main equipment consisted of a supply of headed notepaper, and a fine metal stamp with the company's logo on it: 'République Français; No. 1. Compagnie des Aérostiers. Nadar-Dartois-Duruof'.[15]

The place Saint-Pierre was an ideal balloon site, both strategically and psychologically. It was a flat platform of open wasteground, three hundred feet below the Butte Montmartre. A pagoda-like monument known as the Solferino Tower dominated the northern ridge of the Butte, and a steep bank of tussocky grass on which goats grazed and ragamuffin Parisian children played led down to Nadar's encampment below on the Place. From the top of the Butte, the highest point in the whole of Paris,

one could look north as far as the little villages of Saint-Denis and Le Bourget, now occupied by the Prussians.

In the opposite direction there was an unobstructed view from the Butte southwards, over the rooftops of Paris towards Montparnasse. All the heart-stirring monuments of the city were laid out for contemplation. The Aérostiers could see, in a single great panoramic sweep from left to right, the Vendôme column, the towers of Notre Dame, the dome of the Panthéon, the spires of the Institut, the green copper roof of La Madeleine, and the glittering golden cupola of Les Invalides. It was a place from which a patriot could see what he was fighting for. ◊

◊ This is the spectacular view the modern visitor still sees from the panoramic terrace of the Sacré Coeur. In those days neither the Basilica nor the terrace existed, and the Butte, or Heights, were only marked by the Solferino Tower, commemorating one of Napoleon III's victories. There was also a popular restaurant, or guinguette, which charged extra for the panoramic view. Both were demolished in the winter of 1871, when it became clear that Prussian gunners were using them as sighting points for their bombardment. The Sacré Coeur, commenced in 1873, was originally intended to commemorate the Franco-Prussian conflict and 'to expiate the crimes of the Commune'. Its completion was endlessly and symbolically delayed until 1919. Today its stark white cupolas, uneasily tethered on the skyline of Montmartre, still seem to retain some memory of its earliest aeronautical associations. On the place Saint-Pierre below, there is usually a children's carousel which marks almost exactly the spot of Nadar's balloon launches. The garden at the side has been renamed the square Louise Michel, in honour of the famous female communard. But Nadar and his heroic Aérostiers have no monument or memorial. Perhaps he would prefer the children's carousel anyway, as it inspires so many happy photographs.

Nadar began his observational ascents in the tethered *Neptune* on 16 September, going up six times by day and three times by night. He had small cards printed off for his observation notes. These showed an outline map of Paris upon which he marked in all the military dispositions he could identify. To do this he used 'naïve' coloured crayons: 'red for French, blue for Prussian, black for doubtful'. These cards he despatched 'religiously' to General Trochu, but never heard a word back in return.[16]

Nadar had established a simple rope cordon around the balloon site. Theoretically, this was to protect the *Neptune*, and the three bell tents where the Aérostiers team took turns to sleep and eat, from restive crowds. The weather had been cold and wet, and they had few provisions, but a local restaurateur known as Monsieur Charles provided them with cheap meals and wine. The patriotic young Mayor of Montmartre, future Prime Minister Georges Clemenceau (only elected on 5 September), after objecting violently to their commandeering the Place without his permission, suddenly sent them a cartful of hay to sleep in. To this he added a gift of five or six large dogs to keep the area secure, and if necessary to warm their bell tents.[17]

The Prussian encirclement of the city was so complete that it was immediately clear that balloon observation was practically superfluous. It was now that Nadar began to consider the much more perilous possibility of actually crossing the siege lines. The following day, 17 September, he wrote urgently to Colonel Usquin, at the Council for the Defence of Paris: 'Monsieur Cornu, whom we had the honour to see yesterday, spoke to us of the fabrication of *free flight* Aerostats, in the event that communications should be entirely cut. This concept is an excellent one, and all our personnel are immediately ready to volunteer for such missions, one after the other, as necessary.'[18]

He added, on a strictly practical note, that he only had one balloon, the *Neptune*, and that the Council should be aware that any new balloons would take him a minimum of two days to complete. However many specialists were employed, the canopies would always require 'three separate coats of varnish' to seal them, and each coat took at least twelve hours to dry properly.

Nadar's historic proposal for launching a first 'free flight' balloon out of Paris produced absolutely no effect for several days. This left him and Jules Duruof in a fever of frustration. All the time they could see – and smell ('insupportable stink of burning straw') – from the observation

position of the *Neptune*, four hundred feet above the Butte, that the Prussians were digging in and preparing their Krupp guns. They also knew that Louis and Jules Godard, based nearby at the La Villette gasworks, were making similar proposals to the Defence Council. As so often in war, there was growing rivalry between allies, as well as cooperation.

On 21 September, Monsieur Rampont-Lechin, the Director of Postal Services, finally issued a top-secret request to the Council for National Defence: 'It has become absolutely necessary and urgent to employ aerostats to re-establish communications with the provisional government in Tours, which is required for numerous pressing reasons of a psychological as well as military and political nature.'[19]

Top-secret or not, rumours of this decision quickly reached the place Saint-Pierre. Nadar's No. 1 Aérostiers were now desperate to be given the go-ahead, and fully rigged the *Neptune* for free flight. But still the Council delayed. Unknown to Nadar, another balloon was somehow procured by the Postal Services itself – possibly from the Godards – and clandestinely inflated at a hidden site in the Vaugirard gasworks, in the south-west of Paris, on the afternoon of 22 September. Its aeronaut was a professional, Gabriel Mangin. But the balloon was found to be in 'a hopeless condition', the pilot was unenthusiastic, and the launch was abandoned.[20]

At this point, on the evening of the 22nd, the *Neptune* was finally given secret instructions to prepare for a launch at dawn the following day. The task would be to carry an undefined amount of mail and government despatches to Tours. The westerly air current was favourable, but the Aérostiers had now been waiting in full readiness for over forty-eight hours. All this time, the old and worn-out balloon had been fully inflated but steadily leaking coal gas. Much battered after six years of Duruof's festival flights, and the recent regime of tethered observation ascents, the *Neptune* had brittle varnish, numerous small punctures and splitting seams. It was now having to be repaired not merely daily, but hourly.

Throughout the night, the Aérostier team of volunteers steadily painted fresh cow-gum on the seams, and stuck on cotton patches as each new hole appeared. Nadar was hoping that the gentle westerly airstream would hold good. Duruof was quietly wondering how much the mail would weigh. Both of them privately speculated about whether the balloon would simply fall apart in the first hundred feet. The *Neptune* was, they said fondly, 'a noble piece of wreckage'.

Finally, at 7 a.m., just after dawn on 23 September, the harassed Directeur des Postes, Monsieur Rampont-Lechin himself, arrived in a fiacre with several canvas sacks containing 125 kilos of despatches and public mail. This was the equivalent in weight of two extra passengers, for a balloon that had barely been able to lift a single observer. Watched by a grimly silent group of soldiers, Nadar and the Aérostiers team hastily loaded the mailbags, disconnected the municipal gas pipe, helped Jules Duruof into the basket, weighed off, and shook hands. Then they stood awaiting Duruof's final command: *Lachez-tout!* – 'Let go all!'

There is a photograph taken at this moment, just before the launch, from the grassy Butte just above the place Saint-Pierre.[21] It shows the grey light of an urban dawn, a bleak huddle of Paris rooftops and windows, and the unpaved square almost deserted except for a group of dark figures in heavy coats clustered anxiously round the balloon basket. The balloon itself appears large, dirty and undecorated, swaying above the military bell tents, and casting no shadow. The long, thick snake of the gas pipe is still attached to the balloon's mouth, so it was evidently being reinflated until the very last moment. To the left, a dispirited line of soldiers lounge as close as possible to Monsieur Charles's café on the

southern side of the Place. To the right, waiting behind Nadar's cordon, are a sparse and dejected audience of some twenty people, most of whom appear to be small boys sitting on the ground. Contrary to later legends, there are no flags, no cheering crowds, no military band. It is a bleak and unglamorous image of grim determination.

Just after 8 a.m. – no balloon launch ever happens on time – Duruof shouted the irrevocable command, 'Lachez-tout!', and Nadar stepped smartly back from the heavily loaded basket. It was at once apparent that Duruof had a special flight plan in mind. Contrary to standard procedure, and using exactly the technique he had employed with Gaston Tissandier in Calais, he immediately cut away a huge sack of ballast and sent the ancient, creaking *Neptune* leaping vertically into the air. Whatever the risk of rupturing the old and rotten canopy, Duruof was absolutely determined to gain height and clear the Prussian lines. He might fall from the sky, but he would definitely not be shot out of it.

The *Neptune* rose with impressive speed, effortlessly cleared the nearest line of rooftops (the photograph shows nine-storey apartment buildings), and sailed away to the south-west. To Nadar's amazement the scattered, apathetic group of soldiers suddenly began cheering, and the shouts and huzzahs were taken up by the small boys across the Place. Duruof's fellow aeronaut Eugène Farcot recorded proudly: 'Duruof took off in high style, shouting that he would see us all in Le Havre. He slashed away the trailing ropes, and emptied an entire sack of ballast, coolly setting out in the direction of Versailles, very much the old and practised *routier* of the air.'[22]

Wilfrid de Fonvielle, who also seems to have turned up for the launch, remarked admiringly: 'Duruof challenged all the fury of the Prussian guns in an old, small, clapped-out balloon – he was a modern Curtius – he simply flung himself into the clouds, and went at them neck or nothing [*à corps perdu*] ... He *shot* his balloon straight upwards like a shell from a mortar.'[23]

Duruof later left a detailed account of this first-ever balloon escape from the besieged city. Having reached a height of five thousand feet without disintegrating, the *Neptune* unexpectedly turned north-westwards, crossed the Seine, and floated slowly over the 'black ant-heap' of the Prussian lines beyond Mont-Valérien. Duruof could hear the sinister crackling of musketfire below, and hunched down low into his basket. He felt vibrations, but no direct hits. Only later was it established that 3,500 feet was the safe height, out of range of most standard Prussian weapons. But this had to be judged by eye, as few of the later siege balloons were equipped with anything as sophisticated and expensive as aneroid barometers.

On this occasion, by way of reply, Duruof threw out handfuls of printed business cards embossed '*Nadar Photographe*'. It was later said that each one had been marked personally by Nadar in the top right-hand corner: 'Compliments to Kaiser Wilhelm and Monsieur Von Bismarck'.[24]

Duruof noted that the Prussians attempted to elevate their big field guns to fire up at him (like the Confederate artillery in the American Civil War), but without apparent effect. It was more disconcerting to see a line of Uhlan cavalry setting out in hot pursuit. But eking out what little remained of his ballast handful by handful, and throwing out more business cards, he nursed the leaking *Neptune* as it continued to drift north-westwards along the meandering line of the Seine.

To his immense relief, all signs of pursuit gradually disappeared after Mantes, apparently obstructed by the many twists of the river. Three hours after launching, at 11 a.m., he landed successfully at Corneville, near Evreux. He had travelled twenty miles, about a third of the way to Rouen. He was greeted ecstatically by the villagers, who could not believe that he had come from Paris until they were shown the official mailbags. Duruof commandeered a cart and trotted briskly to the nearest railway station, where by good luck he was able to board a direct train to Tours.

He arrived at Tours, with all the mail intact, by 4 p.m., and delivered his despatches to the government in exile. These included a special address to the nation from Léon Gambetta, which was printed in the next day's newspapers throughout France. Gambetta announced that Paris was preparing 'a heroic resistance'. All Prussian disinformation should be ignored. All political parties were united. The city could hold out all winter. 'Let all France prepare herself for a heroic effort of will!'

By the next evening the news had begun to spread like wildfire. It was not just the arrival of private mail, or Gambetta's public message. It was the fact that the siege of Paris had been broken by a balloon. Prussian firepower had been beaten by French air power. Paris was airborne and alive. It was a technical, but above all, a huge propaganda, triumph. In celebration, the artist Puvis de Chavannes painted the figure of a defiant Marianne with fixed bayonet standing guard on the north-western ramparts, bidding farewell to a departing aeronaut. Completed in November 1870, the dramatic picture shows the balloon directly above Fort Mont-Valérien. It was made into a lithograph and widely distributed in Paris as a propaganda poster.

4

The propaganda moved beyond France. Among the mails Duruof carried was an open letter from Nadar to *The Times* in London, appealing for international support. It was copied by government clerks at Tours, and then immediately sent on by train to Le Havre, where it was transferred to a steamship, offloaded at Dover, and delivered by overnight Royal Mail express train to the Continental sorting office in London. Finally it was published by *The Times* in the first edition of 28 September 1870, a mere five days after it had left Nadar's bell tent on the place Saint-Pierre. It appeared on the editorial page under the dramatic title 'From a Balloon'.

The Times also chose to make this day a special siege edition. It took the almost unheard-of step of publishing on its sacrosanct front page a huge, half-page map of 'Paris and Environs'. Actually what it showed was 'Paris and its Defences'. It displayed all the central *arrondissements*, the inner ring of defences, the disputed outlying villages, the main redoubts and ramparts, the twelve perimeter forts, and all the incoming roads and railway lines (though of course these were now cut). The level of detail is extraordinary, and copies must soon have been pinned up in every Prussian officer's tent and mess room. Given *The Times*'s anti-French bias, this is quite possibly what was intended.

Nadar understood the business of publicity and propaganda as well as, perhaps better than, any politician in Paris. He was also aware of the probably hostile attitude of *Times* readers. So his historic balloon letter consisted of just three short paragraphs, which the newspaper left entirely in their original French. He omitted any of the expected Gallic melodrama, self-justifying rhetoric or flamboyant heroics. The tone was modest, down-to-earth, frank.

Nadar began by thanking *The Times* for the 'hospitality' of its pages. (He was of course providing it with a sensational balloon scoop.) On previous occasions the newspaper had been 'extremely severe' towards 'imperial France', and it had been largely justified in this attitude. He himself – '*moi Français*' – was indignant and ashamed at 'the deplorable example my poor, benighted country' had given over the last twenty years. It was undoubtedly imperial France's 'error' to declare 'this abominable war' against the Prussians.

But now he begged English readers, with their famous sense of fair play, to reconsider the position. Here was a new, young, idealistic French republican government in power. Its peace proposals had been 'disdainfully' refused by the Prussians. Its sovereign lands had been 'cruelly' invaded and despoiled, by an enemy which had become 'greedy and over-confident'. Most of all, the Prussian military hostilities were now being pursued not merely against the French army, but against the civilian population itself, the ordinary 'people of France'.

The whole temper of that people, the people of France and specifically the people of Paris, was profoundly altered. 'I could only wish, Sir, that you could bear witness to the sudden, unexpected sight of Paris transformed and regenerated, and now standing utterly alone in the face of supreme danger. The city of pleasure and frivolity has become silent, grave, and serious-minded.'

'Wars,' Nadar concluded pointedly, 'are not won with cannons and rifles alone – there is also the small matter of having right on your side.' Prussia had become an 'insatiable' enemy, too sure of itself, too self-justifying, too vindictive. Imperial France had been justly punished in the first place. But now it was Prussia's turn to receive just punishment, a punishment that it had brought upon itself: '*La Prussie va recevoir le châtiment qu'elle provoque.*' For readers of *The Times*, many of whom would have known Victor Hugo's poetry, Nadar's clinching sentence must have had a particular ring.[25] A version of this same letter also appeared in *L'Indépendance Belge* on the following day, 29 September. The communications blackout had been decisively broken.

Three more balloons followed in quick succession over the next week, on 25, 29 and 30 September. Their aeronauts were Gabriel Mangin (his second attempt), Louis Godard and Gaston Tissandier, and they took off from the gasworks at La Villette and Vaugirard. All three of these balloons crossed the Seine and landed safely to the west of Paris carrying mail. They also carried baskets of carrier pigeons provided by a patriotic 'columbine' society, L'Espérance, to see if replies to the despatches could be flown back into Paris from the sorting office in Tours. When several of the pigeons returned over the next few days, it was clear that a complete outward-and-return postal service was now possible.

The Defence Council now officially announced the formation of the Paris Balloon Post.[26] There were to be two kinds of delivery: *monté* and *non-monté* – by manned and by unmanned balloons. The first would take

proper private letters; the second would accept only official ten-centime postcards with standardised message boxes to tick. Naturally, only the first ever caught on with the Parisians.

The third of the balloons, *Le Céleste*, was manned by Gaston Tissandier, who landed near Dreux, seventy miles due west of Paris in the department of Eure-et-Loir. He broke his arm, but still delivered the mail. Tissandier was also tasked with setting up a communications centre at Tours, and investigating all the possible methods of getting messages back into Paris. His ingenious professorial brain came up with numerous ideas, including not only carrier pigeons, but also messenger dogs, river flotation bags, and even balloons flown back into the city. His concept was that, as in the American Civil War, balloons could be mounted on trains. They could then be rapidly deployed to positions precisely upwind from Paris on any particular day, inflated and immediately released.

It was a supremely hazardous undertaking, but morale was high. Tissandier wrote: 'The appearance of these first balloons in the provinces produced universal excitement and enthusiasm. In less than eight days,

literally tens of thousands of families had received precious news from their besieged relatives by means of the air.'[27]

He was not exaggerating. As the weight of each letter was limited to four grams, Duruof's original 125-kilo mailbag had held over three thousand letters. The next three balloons carried two or more mailbags each, totalling between them over nine hundred kilos of mail. This produced a grand total of well over twenty-five thousand letters delivered by the first four siege balloons in the last week of September 1870. These numbers would soon rise dramatically.

Victor Hugo wrote to Nadar: 'One would have to be a pinhead not to recognise the huge significance of what has been achieved. Paris is surrounded, blockaded, blotted out from the rest of the world! – and yet by means of a simple balloon, *a mere bubble of air*, Paris is back in communication with the rest of the world!'[28]

5

The new Defence Council decided on a propaganda coup. On 7 October the fifth and sixth balloons, christened the *Armand Barbès* and *George Sand*, were launched simultaneously from Nadar's place Saint-Pierre. The dual launch was partly intended to confuse the Prussians, for aboard the *Armand Barbès* was the key member of the new republican government, Léon Gambetta. A radical lawyer and journalist, already known and admired as the dynamic Minister for the Interior, Gambetta had become

the hope of all France. Still in his thirties, he was renowned for his populist sympathies and fiery, patriotic speeches.

The council had appointed him Minister for War, and given him the
crucial task of energising and reorganising the provisional government
in Tours. His instructions were to mobilise the population, recruit fresh
troops, and put the army of the Loire on a renewed and aggressive foot-
ing. In short, he was to inspire 'the sentiment of resistance' throughout
the south and west of France. With Gambetta went the editor of the leading newspaper *La République Française*, whose parallel job was to revital-
ise political publicity for the provisional government.

Altogether, it was a brilliant but risky initiative. The *Journal des débats*
called it the launching of the 'Balloon Government'. But some critics
believed it was, rather literally, putting all their eggs in 'one basket'. In
the event of their deaths, or – even worse – their capture alive, the propaganda coup would turn devastatingly in favour of the Prussians.[29]

Victor Hugo later recorded in his vivid diary, *Choses vues*, how he
happened to be out 'wandering the boulevards' that morning, and turning his steps towards Montmartre, chanced to witness the last few
minutes before the launch. In fact he had almost certainly been alerted
by the ever-resourceful Nadar. Hugo found quite a crowd in the big, bleak
square: a detachment of General Trochu's infantry, a cluster

of bemedalled officers, a whispering group of Parisian workers, the assembled Aérostiers, and Nadar looking pale and exhausted. Then there were the two balloons, one dirty yellow – the *Armand Barbès* – and one dirty white – the *George Sand* – neither in the least impressive.

The white balloon was festooned with limp tricolour flags, and was clearly intended as the decoy. Nonetheless the *George Sand* carried a full payload of mail, and also two dauntless American businessmen, a Mr Reynolds and a Mr May, who were going to arrange a huge arms deal for Gambetta. Besides, Hugo noted, this was a *literary* balloon and a *lady's* balloon as well: the novelist's name alone would certainly put the philistine Prussians in their place.[30]

Hugo's diary recorded tersely:

There were whispers running through the crowd: 'Gambetta's going to leave! Gambetta's going to leave!' And there, in a thick overcoat, under an otter-fur cap, near the yellow balloon in a huddle of men, I caught sight of Gambetta. He was sitting on the pavement and pulling on fur-lined boots. He had a leather bag slung across his shoulders. He took it off, clambered into the balloon basket, and a young man, the aeronaut, tied the bag into the rigging above Gambetta's head. It was 10.30, a fine day, a slight southerly wind, a gentle autumn sun. Suddenly the yellow balloon took off carrying three men, one of them Gambetta. Then the white balloon, also carrying three men, one of them waving a large tricolour flag. Under

Gambetta's balloon was a small tricolour pennant. There were cries of
'Vive la République.'[31]

This historic moment was also commemorated in several popular prints
and engravings, some more imaginative than others. The main difference
between them lies not in the depiction of the balloons, but in the attitude and mood of the watching soldiers and Parisian crowd. In some, the
mood is evidently tense and sober, even sceptical. In others the crowd is
packed, wildly supportive and enthusiastic.

One possibly genuine photograph has survived, which shows the
moment after lift-off in the place Saint-Pierre. The *Armand Barbès* is about
twenty feet up, swinging in what is evidently a high wind (not Hugo's
gentle breeze). There are no tricolour pennants attached anywhere.
Gambetta, looking tense, has grabbed the edge of the basket to steady
himself, and is dramatically holding out his fur hat to bid farewell to a
small group of soldiers and civilians below.

But the mood is subdued. On the left, some soldiers raise their képis,
one gentleman doffs his top hat, a guardsman stands impassively leaning
on his musket with fixed bayonet. To the right, near the solitary gas lamp,
the tall, spindly figure of Nadar can be seen standing alone, still and
solemn, his right arm raised in a silent salute. Behind him, near the bell

tents, are a loose circle of about a dozen watching militia men, either standing or sitting. Not one has either lifted his hat or raised his arm. Compared to the popular prints, the square is shown as largely empty, and the ground scattered with hastily discarded ballast sacks. Even if some of the figures have been improved, the authenticity of the bleak photograph is striking.

Owing to an unexpected wind shift, Gambetta's balloon suddenly turned almost due north, passing over Saint-Denis and Le Bourget, an area known to be infested with Prussians. Nadar watched it with binoculars from the Butte, ice in his heart. Worse, the *Armand Barbès* could not gain sufficient height before crossing the siege lines. A brisk fusillade of musketfire rose up towards them, some balls hissing past, and several striking the base of the willow basket with a sharp crack. Then some actually pierced the canopy, causing the support ropes to vibrate above the passengers. It was a moment of terror. But as the American Civil War aeronauts had discovered, these produced only tiny, neat punctures and failed to ignite the gas.

But the position was still perilous. Gambetta was grazed in the hand by a musketball, and fell back stunned against the sacks of mail. The balloon lost buoyancy, sank, and actually came down in a field near

Chantilly that had just been vacated by a squadron of Uhlan cavalry. Alerted by warning shouts from field workers, the pilot hurled out ballast and the *Armand Barbès* lurched back into the air again. The Prussian cavalry, in full cry, pursued them cross-country for several miles, as in some nightmare dream sequence. When they crossed the village of Creil, twenty-seven miles from Paris, they were still only seven hundred feet above the ground, well within musketshot.

Finally, after about three hours, they skimmed over the heavily wooded region of Compiègne, near Epineuse, thirty-eight miles from Paris. Here they seemed to have temporarily lost their pursuers, and the pilot decided to risk a quick emergency landing. But he missed the open ground, and crashed into an oak tree, where the balloon hung suspended in the branches for several minutes, helpless should the Uhlans arrive. Eventually some local villagers climbed up and pulled them clear, bundling them and the mailbags unceremoniously into a hay cart. It is not clear whether Gambetta was recognised at this point, or even if the villagers realised that they were dealing with Parisians rather than Prussians. It was also said in some reports that the bearded man had been found hanging upside down from the anchor rope.[32]

Luckily, the mayor of Epineuse did recognise Gambetta, and realising the imminent danger from Prussian troops, transferred the whole party into his private coach and hurried them northwards towards Amiens. Halfway there, at the village of Montdidier, Gambetta was bandaged and revived with stiff tots of *eau de vie*. He sent off a suitably clipped and optimistic message by carrier pigeon:

> Arrived after accident in forest at Epineuse. Balloon deflated. Escaped Prussian rifle fire thanks to mayor of Epineuse and reached here Montdidier whence leave one hour for Amiens then railway to Le Mans or Tours. Prussian lines end at Clermont, Compiègne, Breteuil in the Oise. No Prussians in the Somme. Everywhere the people are rising. Government of National Defence acclaimed on all sides. – Léon Gambetta.[33]

At Amiens, they spent the night recovering, and by the following evening they – and the mailbags – had safely reached Tours via Rouen. The crew of the *George Sand* were already there. At the railway station, lined with National Guards, Gambetta climbed up on a porter's trolley and gave one of the great fighting, patriotic speeches with which he would rouse the nation: 'If we cannot make a pact with Victory, let us make a pact with

Death!' The republican government in exile proclaimed a mighty propaganda coup.[34]

From then on, the National Council for Defence was committed to balloons. It was clear that they were fully capable of breaking the Prussian siege, and that combined with the railway network, they could outwit and outrun the Prussian occupation forces beyond Paris. Two non-stop balloon-manufacturing centres were quickly established: one at the Gare d'Orléans (now Austerlitz), under the Godards; and the other at the Gare du Nord, under one of Nadar's original Aérostiers, Camille Dartois.

The high-ceilinged stations were empty of trains and so provided huge and convenient open-plan buildings where balloons could be mass-produced. The vast stretches of balloon material were spread out on huge trestle tables, where they were cut, sewn and varnished by hundreds of volunteer dressmakers. They could then be hung from the overhead iron girders above the tracks, inflated with air pumps, dried, and moved rapidly to their launch sites at the La Villette and Vaugirard gasworks.

The plan was to mass-produce a standard siege balloon of seventy thousand cubic feet, made of cheap calico, capable of taking two men, a cage of pigeons, and at least three hundred kilos of mail. They were essentially 'disposable', designed to last for a single flight. The Dartois balloons from the Gare du Nord were plain, no-nonsense white – like button mushrooms in the sky, it was said; while Godard's balloons at the Gare

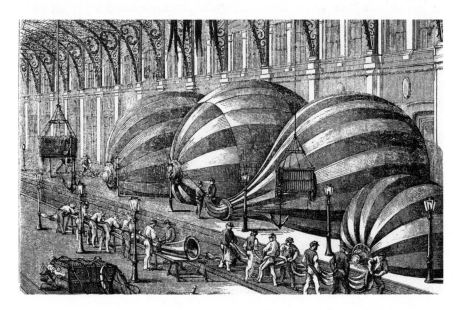

d'Orléans were all candy-striped, suggesting a certain teasing spirit of levity in the face of Prussian boorishness.[35]

Crucial to their propaganda value were the names assigned to each balloon. Knowing that a thousand Prussian telescopes and field-glasses would be furiously trained on every canopy as it floated slowly overhead, the Defence Council launched above the Prussians a veritable checklist of French genius. But admirably, and in the true Enlightenment tradition, they did not limit themselves exclusively to French citizens. In effect they sent over an airborne *cours de civilisation*.

The balloon names naturally included many inspiring soldiers and statesmen: *Lafayette, Armand Barbès, Gambetta, Louis Blanc, Washington, Garibaldi* and *Franklin*. But equally there were many scientists: *Archimède, Kepler, Newton, Volta, Davy, Lavoisier*. There were some inventors: *Daguerre, Niepce, Montgolfier*; but surprisingly only two writers: *Victor Hugo* and *George Sand*. There were a number of patriotic salutes: *La Ville de Paris, La Ville d'Orléans, La Bretagne, La Gironde, L'Armée de la Loire*. Finally there were several well-chosen political watchwords for the Prussian soldiery to consider: *La Liberté, L'Egalité, La République-Universelle, La Délivrance*. There was also the occasional direct provocation: on 17 December they sent up a balloon named *Le Gutenberg*.[36]

Of course all these names also heartened the French patriots below, not least in Paris itself, and served to stiffen the 'sentiment of resistance'. It is true that later, when night launches were adopted, this propaganda became much less visible, at least to the Prussians. Yet these balloon names continued to be circulated in military reports, newspaper editorials and news telegrams on both sides; and eventually had their rippling propaganda effect right across Europe and as far afield as Scandinavia and America.

Hugo was delighted when Nadar arranged for balloon No. 13 to carry his name skyward. The *Victor Hugo* was launched on 18 October, and piloted by a member of the No. 1 Aérostiers, Jean-Pierre Nadal. In an exceptional gesture, the launch site was fixed in the gardens of the Tuileries, to achieve maximum publicity. The balloon rose amidst cheering crowds, carrying besides its cargo of postbags and pigeons several thousand copies of a hastily-written propaganda letter by Hugo addressed 'To the Prussians'. It urged them, in the most magniloquent terms, to sign an honourable armistice, leave French soil and march peacefully home. 'I believe its effect,' confided Hugo, 'will be incalculable!'[37]

From 7 October 1870, depending on wind direction and weather conditions – and it was becoming a bitter winter – postal balloons were launched regularly two or three times a week. Night launches began on 18 November. A total of over fifty balloons had been launched by the end of December. The Prussians were not amused by all this, and Bismarck was recorded as remarking drily, '*Décidément, ces diables de Parisiens sont bien ingénieux.*' He was pleased when three balloons, the *Normandie*, the *Galilée* and the *Daguerre*, were unexpectedly captured in November, having landed behind enemy lines in bad weather. Both the crews and the mailbags sinisterly disappeared.

Bismarck wrote to the American Ambassador shortly after: 'I take this opportunity of informing you that several balloons sent out of Paris have fallen into our hands and that the persons travelling in them will be tried by a court martial. I beg you to bring this fact to the French government's notice, adding that any person using this means of transport to cross our lines without permission, or to engage in correspondence to the detriment of our troops, will be subjected, if they fall into our power, to the same treatment.'[38] This was a fairly explicit warning that balloonists would not be treated as regular combatants, but would be summarily shot as spies.

After the capture of the three balloons, launches from La Villette and Vaugirard tended to be at night, allowing them several hours of darkness to clear the Prussian lines. But this made navigation even more haphazard. Uncertain of their line of flight or their location, fearful of being shot as spies if they were captured, aeronauts were inclined to press on as far as possible before they came down low enough to establish their whereabouts. This resulted in some fantastic long-distance flights, but also several tragic disasters.

La Ville d'Orléans took off on the night of 24 November, ran into a storm, and landed fifteen hours later on a snow-covered mountainside in Norway. It was a record distance of 840 miles, at an average speed of just under sixty miles per hour, passing through air temperatures as low as –32 degrees Fahrenheit. Astonishingly, the two-man crew managed to hike through snowdrifts to Christiana (present-day Oslo), and all the mailbags but one were safely retrieved by Norwegian farmers.[39]

Le Jacquard, launched on 28 November, disappeared in another direction, over the Irish Sea, and was never seen again. Its one-man crew, a twenty-seven-year-old sailor named Alexandre Prince, had given a number of rousing patriotic interviews before his departure, and became

the subject of several poems and posters after his mysterious disappearance.^Ψ *La Ville de Paris*, launched on 15 December, landed in Wetzlar,
Germany, where its mail was immediately seized and its crew probably
shot. *Le Général Chantzy*, launched on 20 December, went as far east as
Bavaria, and met a similarly unknown fate.

Despite these tragic failures, the Parisian enthusiasm for balloon
post did not falter, nor the supply of aeronauts to deliver it. In fact
balloons swiftly became vital to Parisian morale, a heroic part of the siege
mythology. 'The wind was our postman, the balloon was our letterbox,'
recalled the poet Théophile Gautier. 'With each departing aeronaut, our
deepest thoughts also took flight, our hopes and fears, our wishes for
absent loved ones, our heartaches and our longings, everything that was
good and fine in the human spirit ... took to the air.'[41]

Ψ The mystery of *Le Jacquard* throws some light on the spirit of the siege balloonists.
Prince had never been in a balloon before, but in the last-minute absence of a regular
aeronaut, he volunteered to fly from the Gare d'Orléans on a night launch. 'I reckon
to make a good long flight,' he said simply. 'People will talk about my trip.' His instructions were basic. He was not to come down anywhere before dawn. He must be absolutely certain that he was completely clear of the Prussian lines. The mail must only
be delivered into friendly hands. He was last reported by a British fishing boat, thirty
miles due west of the Scilly Isles, and flying exceptionally high. However, some of his
mailbags were later found among the rocks at the Lizard, the most south-westerly
point of England. What had happened? His flight path on a marine chart indicates a
roughly straight trajectory between Paris, Cherbourg, the Lizard and the Scilly Isles.
While still flying low, Prince probably saw the tip of the Lizard below, and realised it
was his last possible point of landfall before the Atlantic Ocean. So he deliberately
threw out his mailbags in the desperate hope that they would be found and delivered,
even by the English. (In fact some of this mail actually was later delivered.) But in
doing so he was throwing out a large part of his remaining ballast. As he must have
known, *Le Jacquard* was doomed to gain massive height and to continue ever westwards over the Atlantic Ocean. Alexandre Prince's story – the lone figure in the high
balloon headed west into the sunset – had a peculiar power to haunt Parisians. His
name can be found in golden letters on a solitary memorial in the Salle des Pas
Perdus, in the present Gare d'Austerlitz.[40]

6

With thousands of private letters going out each week from Paris, the increasingly urgent question was how to organise a postal reply system. Albert Tissandier – not to be outdone by his younger brother – had successfully flown out in the *Jean-Bart* on 14 October. He now joined Gaston in his attempts to establish a balloon service, flying back into Paris with sacks of return mail. After taking meteorological advice about the prevailing autumnal winds, the brothers set up Rouen, on the river Seine, as their centre of operations. They were sixty-eight miles north-west of Paris, potentially a four-hour flight back to the capital. With the approval of Tours, a temporary mail-sorting bureau was established at the Rouen post office, with letters from all over France being brought in by rail. On 7 November they took off in the *Jean-Bart* from Rouen gasworks, carrying 250 kilos of return mail on a promising south-easterly air current.

Their plan was to follow the meandering line of the Seine all the way to Paris. By evening they had gone twenty miles in precisely the right direction, and when the wind dropped at dusk, they came down stealthily outside the little town of Les Andelys on the Seine. They were now only forty-nine miles from Paris, and had every hope of reaching their destination the next day, although they were dangerously close to the Prussian lines. But the next morning the wind had backed and was blowing to the north. The Tissandiers bravely relaunched, hoping to find a southerly wind at a higher altitude. They spent the entire day in freezing clouds at ten thousand feet, unable to see anything beneath them, and experiencing temperatures of 14 degrees below zero, but calculating that if they could only hold on, they should be over Paris by dusk.

When they finally came down, they found themselves suspended above the middle of the Seine and surrounded by steep cliffs and dark, impenetrable woods. They were in the forest of La Bretonne, at Heurteauville, far to the north of Rouen. To their dismay, they realised that they had gone backwards, and were now eighty-one miles from Paris. They were only saved from drowning when some villagers rowed out under the light of the moon and succeeded in towing the *Jean-Bart* to the shore. The whole attempt was then abandoned. Afterwards, Albert Tissandier insisted that it was not an entire failure, as it gave him the subject for a wonderful engraving of the moonlit rescue.[42]

Gaston now considered other fantastic schemes to get mail back into Paris. Some were proposed by the Académie des Sciences, some by fellow aeronauts, some by more eccentric members of the public. Among the most imaginative suggestions sent in were a plan to harness the eagles from the Jardin des Plantes; a blueprint for a small steam-powered dirigible balloon; the concept of floating the mail down the Seine in camouflaged containers; and the almost surreal proposal to suspend a seventy-mile aerial telegraph line, on a chain of free-flight balloons, all the way up the length of the Seine from Rouen to the capital.[43]

But pigeon post remained the tried and obvious method. Baskets of homing pigeons had gone up with every balloon after the *Neptune*. The problem was that their return flights were unreliable, and besides, each pigeon could only carry half a dozen brief messages, laboriously copied in tiny handwriting, and slipped into a leg ring. What was required was some ingenious way of hugely increasing each pigeon's postal payload. Each bird needed to carry not a dozen messages, but several *hundred*. Hence the fanciful idea of employing eagles.

The crucial technical discovery depended not upon eagles, but upon photography. Once again, the idea seems to have been conceived by Nadar, although it was actually pioneered by another commercial Paris photographer, René Dagron.[44] The key idea was the use of microfilm. Throughout October, working quietly in his Paris laboratory at the rue Neuve des Petits-Champs, Dagron invented a simple but revolutionary system of photographing letters and then reducing them to miniaturised film negatives. His method was this. He mounted hundreds of letters at a time onto huge flat boards, and then photographed them with a single exposure taken from a fixed camera. These letter-boards could be photographed as fast as they could be mounted, the letters simply being placed side by side under a retaining sheet of glass. Each photograph was then reduced to a single tiny negative, on part of a roll of collodion film. The process was very rapid, very economical, and easily repeatable. The result was a single roll of collodion microfilm, no thicker than a roll of cigarette paper, which could easily be inserted into a goose quill. The quill could then be attached to a carrier pigeon's tail feathers with waxed silk thread.[45]

Astonishingly, each collodion roll contained enough negatives to record well over a thousand two-page letters. As each goose quill could take four or five collodion films, one rolled tightly inside the other, a single pigeon could carry up to *five thousand* letters. Moreover, each roll could be reproduced scores of times, so that many copies could be sent by many different pigeons. Thus the chance of at least one safe home-run was greatly increased.[46]

It was Nadar who first heard about Dagron's work, and introduced him to his old contact Monsieur Rampont-Lechin, head of the Bureau de Poste. Nadar wrote Rampont a beautifully clear, technical letter describing the whole process, as precise and detailed as a patent claim, and recommending Dagron to lead the project. He was undoubtedly shocked when Dagron negotiated a large fee with the government, and, when asked to fly out to Tours, added twenty-five thousand francs of danger money. This was hardly in the spirit of the Aérostiers.[47]

On 12 November, four days after the failure of the *Jean-Bart* expedition, Dagron flew out of Paris aboard *Le Niepce*, carrying a precious cargo of cameras and microfilm equipment, weighing six hundred kilos. His secret mission was to explain and install this top-secret message system at Tours, and then at Clermont-Ferrand, well clear of the Prussians. The balloon had a difficult flight, descending over the Marne, and Dagron and his weighty photographic kit barely escaped capture after a rough landing and cross-country pursuit. But by the end of November the first microfilm pigeon posts were flying in.

Meanwhile, the Bureau de Poste in Paris had organised a mass-distribution system. Whenever a pigeon returned to a rooftop roost anywhere in the city, it was inspected by a duty pigeon officer. Certain pigeon roosts or *colombiers* became famous for their successes, but all roosts were manned twenty-four hours a day. The precious rolls of collodion film were immediately extracted from the pigeon's quill, and rushed over to the Bureau within minutes of their arrival. Here the films were carefully unwound and cleaned, by soaking them in a mild ammonia solution. They were then cut up into separate strips of negatives, and slipped into a battery of magic lanterns. These projected the negatives onto large screens, permanently installed in a series of darkened transcription rooms. Once magnified through the lanterns' special lenses, the individual letters and despatches became as large as posters, and were easily legible.

A team of clerks, also on continuous twenty-four-hour duty, sat in front of the large, luminous screens, transcribing the letters and despatches back onto paper. This, of course, was the slowest, most exhausting and least reliable stage of the process. The hot, dark, smoke-filled transcription rooms were places of tension and high drama. The

clerks saw fragments of history in the making, but were also privy to intimate family business, passionate lovers' letters, and the endless heartbreaking private tragedies of war. There seems to have been little censorship. Government despatches got priority, but all post was eventually sorted and delivered to its destination throughout Paris, usually within a week. By this technique a single pigeon could carry enough written material to fill the pages of an average-size novel. (Indeed, it is odd that these exceptional circumstances did not actually produce a novel.)

The Prussians naturally took counter-measures, including regular shooting patrols using buckshot cartridges, and, most effectively, teams of hunting hawks and trained falcons. But probably the winter weather took the greatest toll. Between September 1870 and February 1871 around 360 carrier pigeons were released, but only fifty-seven made it home to Paris, a success rate of about one in six. But because of the system of microfilm and pigeon duplication, the actual success rate for message delivery was far higher. Some ninety-five thousand individual letters and despatches were sent from Tour and Clermont-Ferrand. Of these it is estimated that more than sixty thousand items were finally delivered to their addressees in the besieged city, a success rate of more than one in two.[48]

The arrival of pigeons became as significant to Parisians as the departure of balloons. Both became part of the psychology of the city's resistance, and eventually its mythology. Just as the named balloon

launches were announced beforehand, so the news of the latest pigeon-
post arrivals would be officially advertised in the press. The pigeons too
were given names: *Gladiator, Vermouth, Fille d'air*. Households all over
Paris would wait in anticipation for their arrival. International news was
vital to morale, and was quickly printed in the wartime editions of *Le
Moniteur, Le Journal des débats* and other Paris papers. But family news –
health, money, food, children, domestic animals, gardens – was at a
premium. And there was always the shadow of the Prussians. 'In all
history, there will never have been a more beautiful or more touching
legend than that of these saviour birds,' wrote the journalist Paul de
Saint-Victor. 'They brought back to Paris the promise of distant France,
the love and memories of so many separated families.'[49] Once again Puvis
de Chavannes produced a poignant picture, *The Pigeon*, showing the now
emaciated figure of Marianne on the Paris ramparts. This time she is
shielding a carrier pigeon, and warding off a Prussian hawk, with the
belfries of Notre Dame symbolically in the background.

Letters between wartime lovers were particularly important and
intense. Even the ageing Théophile Gautier tried to keep in touch with
his mistress, the beautiful dancer Carlotta Grisi, who was safely
ensconced in her manor house in Geneva. Like so many others, he
despatched a fortnightly stream of duplicated letters, faithfully sent by
each departing balloon post. He numbered each of the letters (over

seventeen are known) so that lost ones could be identified. He also dated
them, like many correspondents, according to the number of siege days
that had elapsed.

On 30 November 1870 Gautier wrote duplicated letter No. 7, 'on 74th
day of Siege':

> *Darling Carlotta ... This morning I regaled myself with a rat pâté which
> wasn't at all bad. You will understand the sadness of our life. The rest of
> the world no longer exists for us ... Ah! My poor Carlotta, what a wretched
> year, this 1870. What events, what catastrophes! And all without the solace
> and sweetness of your friendship ... I imagine that my dear ones might be
> ill, unhappy, or what would be far worse, forgetful. A balloon is leaving
> tonight: will it be more fortunate than the earlier ones that were captured?
> ... Be assured that if I do not come and see you, it is merely the fault of
> 300,000 Prussian soldiers.*[50]

Victor Hugo put both the balloons and the pigeons into many of the
forty-five poems collected in his remarkable month-by-month verse journal of the Paris siege and the Commune, *L'Année terrible*. They became
explicit, airborne symbols of Parisian resistance to the Prussians. Again
and again Hugo uses the lyrical image of the dawn light in the east, with
the silhouette of a departing balloon or the tiny flicker of an incoming
pigeon: 'the ineffable dawn where fly the doves'.[51]

Perhaps the most unusual of these resistance poems is his '*Lettre à
une Femme*' – 'Letter to an Unknown Woman'. It is dramatically subtitled
'*Par Ballon Monté, 10 Janvier*'.[52] January 1871 was the final month of the
siege, and this is central to the bravado of Hugo's poem 'sent by manned
balloon' (probably the *Gambetta*, which left at 3 a.m. on 10 January). By
this time General von Moltke had finally overcome Bismarck's scruples,
and the Prussians had begun to bombard the centre of Paris five days
before, on 5 January. This was the first time in modern history that a
great European capital, and its overwhelmingly civilian population, had
been bombarded. It caused profound shock. Over twelve thousand shells
fell without discrimination, mostly on the Left Bank, striking the
Panthéon, the Salpêtrière hospital, the Montparnasse cemetery and the
Sorbonne.[53]

Ignoring these horrors, Hugo summons up, with a surprisingly light
touch, the more banal daily discomforts of the siege. He refers to the
sight of the raw tree stumps all along the Champs-Elysées, the icy chill of

the unheated apartments, the huge queues outside the food shops, the constant thump of incoming Prussian shells, the drunken, desperate soldiers singing in the streets at night, and the awful massacre of the animals ('We consume horse, rat, bear, donkey ... our stomachs are like Noah's ark'). There are also the small, insidious, personal privations: no clean shirts, no gaslight to work by, no white bread, the reduced diet of vegetables ('Onions are worshipped like gods in ancient Egypt'). Above all, there is the awful gloom of the early-night-time blackout in the erstwhile City of Light: 'at six in the evening descends the dark'.

But the message of Hugo's resistance poem is that the old pleasure-loving Paris has been transformed, just as Nadar had claimed in that very first balloon letter to *The Times*. Paris has learned to accept all these hardships, without complaints, in the name of France. She will never surrender. Paris can take it. This image of Paris as a beautiful, stoic, unconquerable woman is the one that Hugo most wants to project. But she is also male in her determination – '*un héros ... une femme*'. While Moltke bombards the Parisians, and Bismarck starves them, this strange, smiling, symbolic figure stands defiant on the ramparts. She – or he – gazes upwards, 'at the balloon that sails away, at the pigeon that flutters back'.

Soit. Moltke nous cannone et Bismarcke nous affame.
Paris est un héros, Paris est une femme!
Il sait être vaillant et charmant; ses yeux vont
Souriants et pensifs, dans le grand ciel profond,
Du pigeon qui revient au ballon qui s'envole!
C'est beau – le formidable est sorti du frivole.[54]ⴅ

In fact Paris was bombarded into submission within three weeks. The city capitulated on 28 January 1871, in the nineteenth week of the siege. The Prussians marched briefly down a deserted Champs-Elysées on 1 March, exacted enormous punitive reparations, and the Commune of Paris erupted in April. The penultimate siege balloon, *Le Richard Wallace*, named after the British philanthropist who paid for the municipal fountains of Paris, had left from the Gare du Nord at 3.30 a.m. on 27 January. Manned by the last of Nadar's original Aérostiers, Emile Lacaze, it flew fast and sure, almost due west, covering over 350 miles and reaching the large naval port of La Rochelle on the Atlantic coast in the late afternoon.

Sailors spotted Lacaze, flying low, and shouted for him to valve gas and come down. Witnesses later said that he waved to people on the quayside, and then unaccountably threw out sacks of ballast, gained height and sailed on over the bay of Arcachon, and out into the open Atlantic. He also threw out his mailbags, since several of them were later washed up on the shore.[56] No one knows exactly what Emile Lacaze had in mind. Perhaps his release valve had jammed, like Major Money's long ago. Perhaps, like Alexandre Prince, he deliberately sacrificed his own life for the mail. Or perhaps he simply could not bear the thought of Paris surrendering. Perhaps the last of the Aérostiers was determined to head

---

ⴅ Hugo's other heroic siege poems include 'Paris bloqué', 'Du Haut de la muraille de Paris', 'Les Forts' and 'Le Pigeon'. But his letters and prose diary suggest that the siege sometimes rather suited him. As well as being cosseted by his ageing but ever-faithful mistress Juliet Druot, and provisioned by generous well-wishers, his diary shows that he was also privately visited by a stream of late-night female fans and acolytes. Most brought small, innocent offerings, *billets-doux*, sweetmeats or poems. But some were anxious to offer various forms of sexual favour to the great man. Among these were Gautier's unhappily married literary daughter Judith Catulle-Mendès, and none other than the future Communard leader, the fiery and headstrong Louise Michel.[55]

on out to America, the land of the free, three thousand miles away. At all events, neither Lacaze nor *Le Richard Wallace* was ever seen again.

7

What was truly significant about the Paris siege balloons can never be reduced to statistics. Nonetheless, the bare statistics are astonishing. Between 20 September 1870 and 28 January 1871, a period of 130 days, a total of seventy-one free-flight balloons were organised within central Paris. Of these, sixty-seven were successfully launched and overflew the Prussian siege lines, at various heights and at various times of day or night. Several of them flew as far afield as Bordeaux, Brittany, Cornwall, Belgium, the Netherlands, Norway and Germany, although, despite rumours to the contrary, none ever crossed the Alps.[57]

Amazingly, only five balloons were ever captured. Two were trapped on the ground just south of Paris (near Chartres and Melun), brought down by a combination of bad weather and Prussian musketfire. The other three were caught helplessly in powerful easterly air streams, and

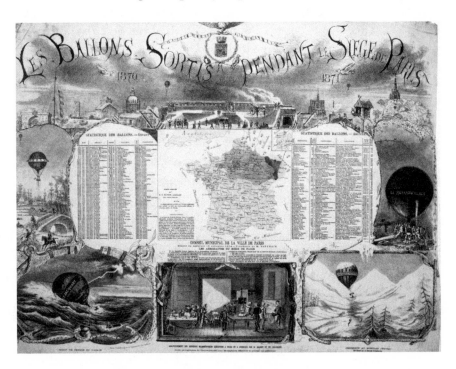

flew several hundred miles into Germany or German-occupied territory. They landed outside Metz, Koblenz and Stuttgart (a distance of three hundred miles), their crews still believing they were in French territory. The fate of these men is not certain: some were imprisoned and later released, others disappeared and were probably shot out of hand.

Although not captured, several balloonists were badly hurt on landing, and the pilot of the *Fulton* was killed. Three were missing presumed drowned – two over the Atlantic, one over the Irish Sea. It is extraordinary that there were not more casualties. No balloon was ever brought down by the much-feared Krupps field gun, and no counter-attacking Prussian balloon corps ever materialised – though it is an interesting indication of English attitudes that the aeronaut Henry Coxwell volunteered to organise one.[58]

All the balloons carried mailbags and baskets of carrier pigeons, but some had other types of cargo. The *Niepce* carried photographic equipment, the *Steenackers* boxes of dynamite, the *Armand Barbès* Léon Gambetta, the *Victor Hugo* several thousand propaganda pamphlets, the *Général Faidherbe* messenger dogs, and *La Volta* – with magnificent insouciance – special telescopes and the astronomer Pierre Janssen from the Académie des Sciences, to observe a rare eclipse in the clear skies of the south of France. Many later balloons also carried the first airmail newspaper, *Le Ballon-poste*, pioneered by *Le Figaro*.[59]

Fantastic efforts were made to deliver the mail safely. The two balloonists Rolier and Bézier, who made the landing on the snow-covered Norwegian mountainside, walked for two days through freezing drifts until rescued. They only knew where they were when a Norwegian forester showed them a box of matches with a picture of Christiana on it. They then discovered that the remnants of their balloon had been blown into the next valley, where three of the four mailbags – the third split open – were recovered by Norwegian farmers, and brought back to the capital. Even more remarkably, the farmers faithfully restored to them their two telescopes, a Scottish plaid, a cooked goose, two baguettes and a bottle of brandy. Travelling back by train and boat via London and Saint-Malo, Rolier and Bézier delivered the mail to Tours on 8 December, a mere two weeks late.[60]

The final balloon statistics can never be certain, and may always owe something to French propaganda against the Germans (just like the Battle of Britain statistics seventy years later). But it is generally agreed that between sixty-four and sixty-seven balloons were successfully launched; that fifty-eight or fifty-nine landed safely in friendly territory; that 102 passengers (not including the aeronauts) were safely transported, together with four hundred pigeons and five dogs; and that not thousands, but *millions* of letters were successfully delivered.𝖄 And not necessarily in France: one of the Norwegian mailbags contained a letter sent to Africa, while others were delivered to addresses in Sydney, Australia, and San Francisco, USA.

At all events, it was the first great and successful airlift in European history. If it did not win a war, it saved the morale and even the soul of a nation. It was the finest hour of the free-flight gas balloon, and the final justification of the Montgolfiers and Dr Alexander Charles.

8

Yet, surprisingly, there was no immediate official recognition of what this handful of balloonists had achieved. Perhaps this was because of the subsequent Commune, and the terrible divisions produced by its bloody suppression, when over twenty thousand people were killed on the barricades or subsequently executed by the Thiers government. There was little inclination to revisit recent Parisian history. Instead, various heroic but apocryphal stories began to circulate after the war, forming a kind of aeronautical folklore. With some justification, one of the best concerned Nadar himself.

The encampment at the place Saint-Pierre had long since been abandoned, and the No. 1 Aérostiers disbanded – though not without murmurings and 'a certain bitterness' that the Godards had stolen Nadar's thunder. Nevertheless, as Nadar himself wrote, no one could take away from him 'the honour of having first created the balloon-post service', or

---

𝖄 The cumulative figure must obviously be a broad estimate. But over a decade later, Henry Coxwell, not necessarily a friendly witness, put the figure at around three million letters. Glaisher and Flammarion put it at '2,500,000 letters weighing nearly ten tons', and L.T.C. Rolt agrees, while the modern French historian Victor Debuchy presents it characteristically as: '10,670 kilos of mail ... at 20 centimes per 4 grams per letter ... generating a gross income of 533,500 francs for the Bureau de Poste ... or a net sum of 294,150 francs'. Debuchy's figures, incidentally, when translated, would give a precise total of 2,667,500 letters delivered, which would surely satisfy even a Prussian accountant.[61]

of having successfully despatched Léon Gambetta by air to Tours in the *Armand Barbès*.[62]

This was perhaps the most significant single balloon flight of the whole siege, and it was around this flight that several legends formed. It began to be said that Nadar had not merely organised the *Armand Barbès*, but had actually accompanied Gambetta in the balloon. Having assured Gambetta's safe arrival at Tours, Nadar then immediately commandeered another balloon and – so the story went – daringly flew back into the besieged city. In fact, despite Gaston Tissandier's efforts, no balloon ever succeeded in flying back into Paris. But this account was widely believed, and was even reprinted in a popular history of ballooning by Fulgence Marion published three years later in Paris and New York, in 1874. According to the report, Nadar not only heroically returned, but actually engaged in an aerial dogfight with a Prussian balloon 'above the Fort of Charenton'.

The tall tale went like this. Nadar's mythical balloon was called (perhaps with a nod to Thaddeus Lowe) *L'Intrépide*. It was sighted just after dawn one late-October morning, floating in low from the west. As it approached the Charenton fort, a second balloon, also carrying a tricolour, rose from a nearby wood and rapidly drew alongside *L'Intrépide*, apparently intending to guide Nadar safely into Paris. (Such a manoeuvre, given the navigation limitations of a balloon, would in fact have been virtually impossible to perform.) But once close enough, in a piratical gesture worthy of Jules Verne, the helmeted pilot of the second balloon suddenly lowered its tricolour, hoisted a Prussian flag, and treacherously opened fire with a musket, puncturing the top of Nadar's canopy.

In an increasingly fantastic sequence worthy of Baron Munchausen, Nadar was reported to have performed the acrobatic feat of crawling up the outside rigging of his balloon, staunching the escaping gas with his scarf, and then clambering back down into his basket. He was next observed to produce a gleaming duelling pistol from his shoulderbag, coolly take aim at his adversary, and bring down the Prussian balloon with a single shot. Nadar then threw out a last bag of ballast, and safely sailed over the Paris ramparts to land triumphantly in the middle of the Champs-Elysées. In fact, of course, he never left Paris at any time during the whole siege.[63]

Nevertheless, Nadar's real-life career as publicist remained as remarkable as ever. While Paris struggled to rebuild its battered boulevards and burnt-out monuments (including the whole of the Palais du Louvre), and to recover its identity as the international capital of culture, Nadar turned his attention from aeronauts to artists. In April 1874 he organised the historic First Exhibition of Impressionist Painters, hung at his own gallery attached to his photographic studios at 35, boulevard des Capucines.

The definitive roll-call of contributors included Monet (with his famous signature canvas *Impression, soleil levant*), Renoir, Degas, Cézanne, Sisley and Pissarro, all modestly grouping themselves under the title 'Société anonyme des artistes peintres'. The provocative 'anonymous' was almost certainly Nadar's simple but brilliant marketing idea. The concept for this hugely influential exhibition had grown out of his friendship with Manet, who had earlier dedicated his painting *Jeune femme en costume Espagnol* to Nadar. In 1878 Nadar also helped to organise the first retrospective exhibition of Honoré Daumier's work. From this time on, the reborn idea of Paris as the luminous home of the Impressionist painters, and all the cultural glitter of the coming *Belle Epoque*, steadily took hold across Europe and America, and the city's renaissance began. ♆

The true history of the siege balloons emerged only slowly. In summer 1871 Camille Flammarion collaborated with James Glaisher in England to produce a second English edition of *Travels in the Air*. In this Glaisher added a short Preface on the aeronautical history of the siege, in which 'the balloon has proved itself so great an assistance to the French Nation'. He also printed a basic inventory of all the flights that left Paris – at this date given as numbering only sixty-two. Glaisher's marginal notes lamented the 'scourge of Prussian occupation', but also seemed to blame

---

♆ Nearly a decade later, in 1886, now aged sixty-six, Nadar achieved yet another publicity coup when he staged the first ever photo interview, or 'talking head'. This was a long animated conversation, held across a dining-room table, with the distinguished French chemist Michel-Eugène Choiseul, the inventor of margarine and an early expert on gerontology. The spry, twinkling Choiseul, with his wild mop of white hair and reputation as a scientific *enfant terrible* (very much to Nadar's taste), was celebrating his hundredth birthday. The dialogue was recorded by a stenographer, and published alongside a running series of 'candid' photographs of Choiseul, continuously 'snapped' in the act of talking and gesticulating. It produced something between a strip cartoon and the first filmed interview, and was catchily entitled 'The Art of Living to One Hundred Years Old'. Nadar himself lived to be eighty-nine. Less than a year before his death, in July 1909, he telegraphed Louis Blériot to congratulate him on having successfully flown the Channel: 'Heartfelt thanks for the joy your triumph has brought this antediluvian supporter of Heavier-than-Air – Nadar.'

republicans like Fonvielle for the destruction of the Second Empire: 'It is hard to forgive all the agitators who, by their writings, have helped to dethrone the best of Emperors and to bring France to her present terrible condition. Vive l'Empereur!'[64]

Fonvielle himself stood unsuccessfully as a Deputy in 1871, and published a series of inflammatory accounts of the Paris Commune, notably *La Terreur, ou la Commune de Paris dévoilée* (1872). He also wrote a thrilling account of his own flight out of Paris in the balloon *L'Egalité*, which after several terrifying brushes with the Prussians had landed successfully in Belgium on 24 November 1870. Soon afterwards he gave up both politics and ballooning altogether, and concentrated entirely on popular-science writing.

Gaston Tissandier was made Chevalier of the Légion d'Honneur, but less for balloon flying than for his organisational work for the provisional government at Tours. In 1872 he and his brother Albert founded an important popular-science journal together, *La Nature: Revue des sciences*, with Gaston writing and editing the articles, and Albert producing most of the illustrations. Fonvielle also contributed to this. The luminous

REVUE DES SCIENCES

ET DE LEURS APPLICATIONS AUX ARTS ET A L'INDUSTRIE

JOURNAL HEBDOMADAIRE ILLUSTRÉ

RÉDACTEUR EN CHEF

GASTON TISSANDIER

VINGT ET UNIÈME ANNÉE
1893
DEUXIÈME SEMESTRE

PARIS

G. MASSON, ÉDITEUR
LIBRAIRE DE L'ACADÉMIE DE MÉDECINE
120, BOULEVARD SAINT-GERMAIN, 120

globe on its cover could be mistaken for an ascending balloon, but in fact it was a rising sun.

On 31 July 1873 Jules Duruof and his wife Caroline made an epic twenty-two-hour flight in his new balloon, the *Tricolore*, over the North Sea. Once again, they launched from the place d'Armes in Calais, and once again in dangerous storm conditions. They had intended to postpone the flight, and it had been banned on safety grounds by the mayor of Calais, but they were surrounded by a hostile crowd, who mocked Duruof's claims to be a siege balloonist, and jeered at him as a 'patriotic coward'. This was too much for the erstwhile Aérostier. 'Let us show them we are not afraid to die,' he is reported to have shouted, somewhat stagily, to Caroline. He promptly cut away the retaining ropes, and they were whirled away across the square and straight out to sea. It might have seemed an almost suicidal gesture, but it was also well-calculated in the grand Duruof style. The next afternoon they were pulled out of the sea by a British fishing boat in the Skagerrak.[65]

The incident was widely reported in the French press, and letters began to pour in with personal memories of how Duruof and the other balloonists had brought Paris hope in those dark days. These stirred up enough publicity for the siege balloonists as a group to receive belated recognition. A special medal was struck for them, commemorating the '*Emploi des Aérostats pour La Défense de Paris*'.[66]

A national monument was also designed, a bronze balloon with various mythological figures clinging around the edge of the basket. A bronze homing pigeon was ingeniously incorporated, apparently fluttering around the rigging. It was installed on a large stone plinth at the

rond-point at the Porte de Neuilly in 1874, where it was much favoured by small boys, just as Nadar's original *Neptune* had been at the place Saint-Pierre. It also appealed, of course, to the local pigeons. French critics thought the balloon too small and heavy in appearance, out of scale with the figures beneath, and giving the impression of weighing them down to earth rather than lifting them up to glory. German critics were much more severe: they destroyed the entire memorial the moment they marched back into Paris in the summer of 1940.

In April 1875 the reputation of the balloonists' courage was publicly reinforced when Gaston Tissandier joined the three-man scientific team to make the high-altitude ascent in the *Zénith*. This was a deliberate attempt to break the Glaisher–Coxwell record, using a new oxygen breathing apparatus designed by the other two crew members, Théodor Sivel and Joseph Croce-Spinelli. Starting from the La Villette gasworks, with all its associations with the siege balloons, they reached around twenty-two thousand feet safely, and began using oxygen from small rubber bladders. Shortly after, they all collapsed from asphyxia.

When Tissandier, the only crew member with extensive balloon experience, recovered consciousness, the *Zénith* was hurtling downwards. The other two men were slumped on the floor of the basket, with blood

streaming from their mouths and ears. Somehow Tissandier managed to control the final descent, but both the others were found to be dead on landing. Tissandier was the sole survivor.[67] The French government was by now much more disposed to celebrate its aeronauts. Sivel and Croce-Spinelli were buried in an elaborate tomb in Père Lachaise, close to Sophie Blanchard's, dramatically showing their draped, life-size figures lying like chivalric knights of the air.

In 1876, partly as recognition of his heroism during the disastrous *Zénith* flight, Gaston Tissandier received the annual Gold Medal from the Société Française de Navigation Aérienne, and was elected its president. His ballooning experiences were now legendary, and over the next decade he published nine editions of his *Histoire de mes ascensions* (1878–88).

Like Nadar, Tissandier became increasingly concerned with the old problem of steering aerostats. His answer was mechanical power, and he worked on various ideas for fitting balloons with engines – the forerunner of the airship. In 1881 he exhibited a working model of an electric-powered, propeller-driven dirigible at the Exposition d'Electricité. Two years later, in October 1883, with his brother Albert he successfully piloted the first electric-powered dirigible, *La France*. (Its engine, incidentally, was manufactured by the German firm Siemens.) It was a brilliant

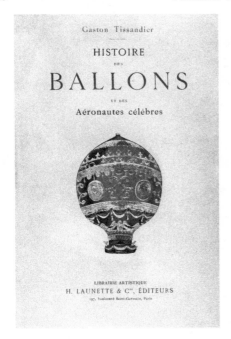

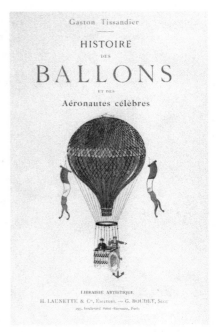

prototype of things to come, but was too clumsy and underpowered to cause great excitement. It did however greatly interest that Prussian officer Count Ferdinand von Zeppelin. Tissandier published the results of his researches in *Les Ballons dirigibles: Application de l'électricité à la navigation aérienne* in 1885.

Tissandier now regarded himself as retired from active ballooning, and, reverting to his professorial *métier*, worked steadily to put together a definitive history of the entire field. This appeared in 1890 as his superb two-volume *Histoire des ballons et des aéronautes célèbres*. Though he was appointed to the Commission for Military Aviation at the Ministry for War, and to the Commission for Civil Aviation at the Ministry of the Interior, he did not quite live to see a genuine aeroplane. He died in Paris on 30 August 1899, aged only fifty-five. His huge aeronautical archive, the finest single record of the Paris siege balloons, was eventually bequeathed to the Library of Congress in Washington, DC. Perhaps it was finally delivered by Emile Lacaze in the *Richard Wallace*.

The official French celebration of the greatest aeronautical triumph in history was symbolically crowned by Henri Giffard's giant captive balloon at the Paris Exposition Universelle of summer 1878. This huge multicoloured aerostat – more an aerial cathedral than a mere balloon

– was gloriously tethered at the very heart of Paris, in the place de la Carousel opposite the rebuilt Louvre palace. It was nearly nine hundred thousand cubic feet, well over ten times the capacity of the average siege balloon, and four times that of Nadar's *Le Géant*. It stood 220 feet high, and would turn out to be the largest conventional balloon ever constructed. It had a massive lift of twenty-seven tons – nearly three times the total weight of all the siege airmail taken out of Paris, and could carry over fifty people at a time. It would eventually take up over thirty-five thousand sightseers, and for the first time the concept of mass flying began to emerge in Europe.[68]

Yet the *Mammoth* also marked the end of the great balloon era. For all its size and power, it was a captive, a slave. It was raised and lowered under the tutelage of huge, hissing, iron steam winches, and controlled by the same gleaming, massive mechanism that powered factories and drove railways. The beautiful free-flight balloon had been turned into a monstrous, creaking, submissive, industrial elevator. The dream of flying had moved elsewhere. ◊

◊ But dreams of course persist. The famous Brazilian inventor Alberto Santos-Dumont (1873–1932) came to France and built a tiny dirigible 'No. 1 Airship' in 1898, regularly flying it around Paris from his apartment at 9, rue Washington. He frequently anchored above his favourite restaurant next door in the Champs-Elysées. In his 'No. 6 Airship' he finally flew around the Eiffel Tower in 1901. I lived at 30, rue Washington, opposite Santos-Dumont's apartment, for six months in 1994, and sometimes heard him taking off in the silent hours before dawn. I treasure the opening paragraph of L.T.C. Rolt's classic *The Balloonists* (1966), which begins: 'We were walking up the Champs-Elysées, and had reached the corner of the rue Washington when my friend Charles Dollfuss suddenly halted and, to the mild surprise of passers-by, struck the pavement a sharp blow with his stick. "Here," he announced, "Santos-Dumont landed in his balloon."' Later, even Santos-Dumont changed to lightweight monoplanes.

11

Extreme Balloons

———

1

From this time on, the dream of free flight was increasingly handed over to proponents of various forms of heavier-than-air machine. As Nadar and Tissandier had seen in France, Sir George Cayley in England, Count von Zeppelin in Germany and Samuel Cody in America, the future lay with the engine-powered airship; and very soon with the true fixed-wing aeroplane. The romantic age of the free gas balloon was passing. As Victor Hugo had predicted, the future lay with the bird, not the cloud.

Or, more strictly, the future lay with the bird's aerofoil rather than the balloon's envelope. Despite generations of would-be birdmen, it was not the flapping of birds' wings that ultimately held the clue to human flight, but the basic shape of their wing feathers. Birds' wings form a natural aerofoil. They are not flat or paddle-like, as one might think, but curved and concave. Amazingly, this basic aerofoil shape can be observed even in an individual wing feather.[1] This makes the upper surface area of each wing larger (or longer) than the lower one. In consequence, the air has to flow more rapidly over the upper surface, and more slowly over the lower surface. This produces a thinning or decrease of air pressure above the wing, and a corresponding build-up or increase of pressure beneath the wing. So, as a bird's wing moves through the air, it is in effect pushed upwards from below, and simultaneously sucked upwards from above. These combined forces produce aerodynamic lift, or flight. Moreover, this sort of flight, unlike balloon flight, is independent of wind direction. By adjusting the aerofoil curve of each wing separately, a bird can turn, climb and dive freely in three dimensions. Not even an airship could really achieve this.

Working airships would appear in France by the end of the 1880s. Charles Renard made seven flights out of Paris and back in an electric-powered airship in 1884–85.[2] In Germany, an experimental Zeppelin – with an aluminium body and a Daimler petrol engine – would fly over Lake Constance in 1900. The Wright brothers flew their aeroplane at Kitty Hawk, North Carolina, in December 1903; and Louis Blériot crossed the Channel in July 1909.

Meanwhile, aerostation itself began to seem old-fashioned, almost a form of antiquarianism. Within a decade it had declined essentially to a rich man's hobby, and fell largely into the hands of eccentric aristocrats and wealthy sportsmen. There was a great vogue for 'champagne ballooning', reaching its apogee in the Edwardian period, when the famous Gordon Bennett Annual Long Distance Balloon Race was inaugurated in 1906.[Ψ] Rules and clubs were formed, international cups and prizes competed for, birthdays and fashionable weddings were celebrated in the air. There was a glamorous ballooning 'season', as for racing, polo, or yachting. These rich amateur balloonists also enjoyed taking up literary and artistic figures on both sides of the Channel, like Guy de Maupassant or H.G. Wells, on what were essentially celebrity jaunts.

[Ψ] Interrupted by both World Wars, the Coupe Aéronautique Gordon Bennett was reinstated in 1983, and continues to this day. It is regarded as the premier free-flight gas balloon competition in the world. But it also remains perilous. In October 2010 I was at Albuquerque for the annual International Balloon Fiesta, when rumours of the disappearance of two experienced and much-loved local aeronauts, Richard Abruzzo and his co-pilot Carol Davies, began to circulate. They had won the 2004 Gordon Bennett Cup, and were the favoured crew in the 54th event, which that year launched from Bristol. They had flown southwards across the Bay of Biscay, over France, Spain and Italy, and then on the third morning turned east and started to cross over the Adriatic, between Brindisi and Serbia. Here, on 29 September, all radio contact had suddenly been lost, and no emergency beacon could be tracked. Over the next few days it gradually emerged that they had been killed in a thunderstorm when struck by lightning at five thousand feet, and had dropped like a stone into the sea. Their open gondola, still containing their bodies, was not recovered until December. I had flown and talked with some of their colleagues, and witnessed the consternation and soul-searching this terrible news caused. I was also shown the aluminium frame basket they had used on a previous prize-winning flight, proudly preserved in the Anderson-Abruzzo Albuquerque Museum. My notebook reads: 'The yellow panelling is torn where they were thrown out on a rough landing and Richard fell thirty feet and broke his ribs and pelvis.'

Maupassant went up in a balloon in 1887. He had printed invitation cards to the launch, as for a luncheon party, but its real purpose was to advertise the publication of his strange autobiographical novella, *Le Horla*. The balloon was named after the book, and the flight was an early form of publicity book tour from Paris to Belgium. *Le Horla* is a story of incipient madness, and Maupassant himself was already suffering from grave mental problems, from which he would die in 1893. Perhaps for that very reason the balloon flight seemed rapturous, a strangely releasing and therapeutic experience: 'The heady perfume of cut hay, wildflowers, damp green earth rose up through the air ... A profound and hitherto unknown sense of well-being flooded through me, a well-being of both mind and body, a feeling of utter carelessness, infinite repose, total forgetting ...'[3]

H.G. Wells ingeniously used the account of a runaway balloon flight to open his futuristic novel *The War in the Air* (1908). His protagonist, Bert Smallways, is accidentally swept away in a hydrogen balloon from Dymchurch Sands, on the Kent coast, and travels across the Channel and all the way to Germany. Initially Bert's sensations are euphoric: 'To be alone in a balloon at a height of fourteen or fifteen thousand feet – and to that height Bert Smallways presently rose – is like nothing else in human experience. It is one of the supreme things possible to man. No flying machine can ever better it. It is to pass extraordinarily out of human things.'[4]

But it is a sign of the times that Bert's balloon finishes up over a secret Zeppelin factory in Bavaria. He sees the future lying beneath him in the form of row upon row of 'grazing monsters at their feed'. These are 'huge fishlike aluminium airships', some over a thousand feet long, each capable of ninety miles per hour into a headwind, and fully equipped with guns, bombs and 'wireless telegraphy'. The fleet is commanded by an Admiral von Sternberg, who is described in terms of the Franco-Prussian War, as 'the von Moltke of the War in the Air'.[5] Having inadvertently observed all these modern secrets from his old-fashioned balloon, Bert Smallways is symbolically shot out of the skies by a volley of German gunfire.

What remained of serious free ballooning at the end of the nineteenth century was notable for increasingly extreme and quixotic flights. These were usually of two kinds: reckless and bizarre attempts to entertain local crowds, or else equally reckless attempts to establish some kind of world record. Such exploits were intentionally dangerous and controversial, and brought all kinds of drama and fatalities, usually accompanied by huge and sensational press coverage. Yet all but a few were quickly forgotten. It was, in a sense, the *fin-de-siècle* of ballooning: stylised, extravagant and gloriously picturesque, but ultimately as ephemeral as a breath of air. Yet among these strange latterday balloonists there were a small band who deserve to be remembered. They were often unearthly in their courage.

2

In Britain, this aerial champagne culture produced an extraordinary late fashion for female balloon acrobats and trapeze artists. This was a risqué tradition that had hitherto been largely confined to France, and the spectacular performances of the Garnerin and the Godard girls. Now it appeared in England, a striking demonstration of the *fin de siècle* of ballooning.

Dozens of celebrity female aeronauts and artistes began performing at county fairs and festivals across the land, executing acrobatics, releasing aerial firework displays, or doing parachute jumps, in a tradition that went right back to Sophie Blanchard. Many of them are only remembered by a few colourful posters that have survived in provincial museums, announcing their promised aerial feats, in sixty-point letterpress and garish red, green and gold illustrations.

Posters can still be found that announce 'Miss Marie Merton's ascent' at Wolverhampton fairground in 1891. Or newspapers that advertise Maude Brooks and Cissy Kent as 'the stars of Lieutenant Lempriere's Aerial Show'. A poster declares that on 2 June 1891 'Miss Maude Brooks will Drop from the Clouds' at the Cricket Grounds, Rotherham, in South Yorkshire: 'She will endeavour to alight within the Grounds, but in the event of not doing so, will return with all possible speed, appear on stage, and give an account of her Aerial Voyage.'

These performances were not as frothy and light-hearted as they seem. Maude Brooks was seriously injured when her balloon collapsed during an ascent from a Dublin garden party on 25 May 1893. She managed to release

her parachute, but fifty feet above the ground it tore and she landed heavily, breaking her legs and arms, and permanently damaging her spine. Such threats of death or injury hung over all of these aeronauts.[6]

Perhaps the most famous Edwardian balloon girl was twenty-year-old Dolly Shepherd. Flying regularly from the Alexandra Palace, and fairgrounds all over England, she popularised a truly terrifying balloon act in which she ascended several thousand feet hanging beneath a trapeze, then pulled a simple release cord and dropped back to earth by parachute. Dolly used no balloon basket at all, but hung fully exposed from her trapeze bar, dressed in a blue-trousered flying suit, with a jaunty cap and tight lace-up boots to show off her legs. Her only instrumentation was a tiny altimeter, which she wore as a silver bangle on her left wrist. She had many male admirers, and received several offers of marriage. But she had an even greater following among young working-class women, who regarded her as a portent of women's rights.

In 1908 Dolly gained a national reputation when she ascended on twin-parachute harnesses with her friend Louie May. When Louie's harness failed to release at twelve thousand feet, Dolly performed the

extraordinary feet of transferring the petrified Louie to her own trapeze, while still attached to the balloon. She then pulled her own parachute release, which worked, and with Louie's arms locked around her neck, brought them both safely back to earth on a single parachute. Louie was unhurt, but Dolly suffered severe back injuries which left her paralysed in a wheelchair for many weeks. Astonishingly, she recovered, and continued balloon parachuting for several years afterwards.[7]

It was an exceptional act of courage and, above all perhaps, of female friendship. Yet many felt that Dolly was being exploited by her balloon Svengali, a mysterious Frenchman known simply as 'Captain Gaudron'. He arranged all her flights, supplied her equipment (including her provocative uniform, and also the release mechanism that failed), but only paid her piecemeal, by each ascent, and certainly without any life insurance. Yet Dolly would always speak with a naïve rapture, and a certain nostalgia, of her balloon experiences. Her passionate attitude seems expressive of this late period of extreme risk-taking.

> I never lost that sense of wonderment and ecstasy whenever I floated alone in the awesome silence ... Every ascent renewed in me those same feelings of delight and contentment. When I soared upwards, above all earthly worries and discomforts, my mind was set free to wander at will and to absorb the sensations of gentle flight, and the beauty of everything around and below me. I never failed to marvel at my bird's eye view of the scenes below, whether rural or urban, forming an intricately woven tapestry above which I floated so effortlessly. In those days, flight in any form was an experience known to only a very few of us. Remember, no aeroplane flew in England until 1908.[8]

What was the appeal of these hugely popular and sensational displays? As the parachutes were still relatively crude, and the balloons increasingly old and ill-maintained, the performances were always far more dangerous than the sporting, fairground atmosphere suggests. Frequent injuries and regular fatalities occurred, as also happened in the Edwardian circus. According to Dolly, most of the parachutists with whom she worked, even the most glamorous ones such as Maude Brooks or 'Devil-may-care Captain Smith or handsome dashing Captain Fleet', somehow 'disappeared', as she put it in her memoirs.[9]

They may have been killed, but the more sinister possibility is that, like her, they suffered spinal or internal injuries as a result of a crash or a

The mysterious 'Universum', or 'The Pilgrim', an engraving made by
Camille Flammarion to illustrate his book *L'Atmosphère* (1888), in imitation of a
medieval woodcut. It shows the place where heaven and earth may meet.

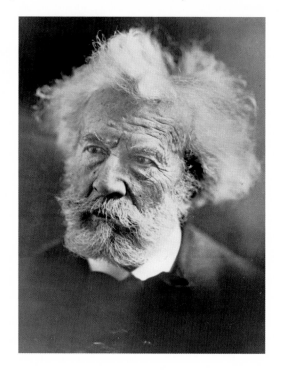

Camille Flammarion, the visionary
French balloonist, astronomer,
scientist and science-fiction writer,
looking every bit the part, at age
eighty-two in 1924.

WRITERS WHO TOOK TO THE AIR

Jules Verne

Victor Hugo

James Glaisher

H. G. Wells

David Hempleman-Adams

Ian McEwan

'The *Eagle* on the Polar Ice', a photograph by Nils Strindberg, 14 July 1897, which became a ghost-like symbol of the passing of the age of Romantic ballooning.

The *Eagle*'s crew. Left to right: Knut Fraenkel, Salomon Andrée and Nils Strindberg, before the departure of the second polar expedition, 1897.

'A Balloon Wedding in the Clouds'. An ultra-fashionable American wedding somewhere above New York, drawn for an Italian magazine in 1911.

Fanny Godard, a leading French female balloonist of the Belle Epoque, showing a lot of style and leg, photographed by Nadar in his Paris studio, 1879.

Dolly Shepherd, the most celebrated of the Edwardian balloon parachute girls, with her impressive uniform and high boots. In 1908 she saved her friend Louie May's life, when Louie's parachute failed at twelve thousand feet.

'Babar and Princess Celeste depart on their honeymoon for Paris'.
Cover of the first edition of *Le Voyage de Babar*, by Jean de Brunhoff, 1932.
Perhaps a gentle satire on Jules Verne's colonial attitudes.

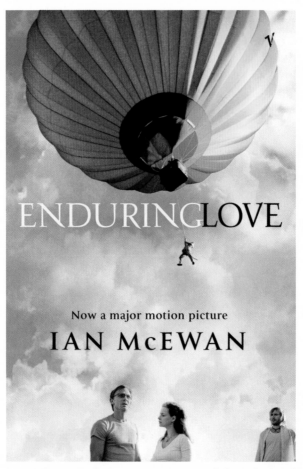

ENDURINGLOVE

Now a major motion picture

IAN McEWAN

Ian McEwan, *Enduring Love*, 1997, a novel about fate and fatal attraction, brilliantly defined by an unforgettable balloon incident in its haunting opening chapters, subsequently made into an equally haunting film, directed by Roger Michell. 'He had been on the rope so long that I began to think he might stay there until the balloon drifted down, or the boy came to his senses and found the valve that released the gas, or until some beam, or god, or some other impossible cartoon thing came and gathered him up' (from Chapter One).

'Airabelle', the heavenly cow, a regular favourite at the annual
Albuquerque Balloon Fiesta, New Mexico, a folksy icon of
American good humour and goodwill.

The renowned 'mass dawn ascent' at the Albuquerque Fiesta,
photographed by Richard Holmes, 2010.

'Earthrise', the famous environmentalist image, this time photographed from Apollo 11, July 1969. 'The dream of flight is to see the world differently.'

heavy landing. But unlike Dolly, they may have been paralysed or disabled for life. There was no attempt to regulate or license the displays, let alone to insure the lives of the performers, until the First World War brought such frivolities to an end. Nevertheless, like their contemporaries the suffragettes, many of them, such as Dolly, insisted that they were striking a blow for women's freedom.

These young aerial artistes, so dazzling in their courage and carelessness, were some of the last representatives of the great nineteenth-century tradition of ballooning as entertainment. They were both the stars and the victims of the show, and it would be difficult to judge how far they were liberated or exploited in their *métier*. Like beautiful sacrifices they would be 'offered up to heaven'; and like angels they would 'drop back from the clouds' for the edification of casual onlookers. Remarkably, such displays continued in America into the 1930s. But strangely, the ambiguity of their roles was mirrored in the other form of extreme ballooning.

3

The outstanding example of the extreme record-breaking balloonist was the Swedish engineer Salomon Andrée, and his fantastic efforts to reach the North Pole by balloon in 1896–97.

Andrée was born in 1854 in the tiny provincial township of Gränna, three hundred miles south-west of Stockholm, on the edge of Lake Vättern. He was brought up largely by his mother, Wilhelmina; his father, an apothecary, died when he was sixteen. As a boy Salomon ran wild, building rafts, sailing boats and on one occasion launching a fire balloon that set light to a local barn. He remained close to his mother all his life, calling her Mina, and sending her long letters confiding in her all his plans and secret ambitions. She seems to have given him an inner confidence and self-sufficiency that never left him. He grew up exceptionally tall, headstrong and adventurous, and defiant in his attitudes. He was strongly committed to the natural sciences, with a special fascination for engineering, meteorology and ornithology. By contrast, he disclaimed all interest in the arts and literature, claiming that concerts and art galleries bored him, and that he only liked adventure stories – notably the fantastic tales of Baron Munchausen.[10] Like a hero out of Jules Verne – or Nietzsche – his watchword became 'Mankind is only half awake!'

Andrée trained at the Royal Institute of Technology, Oslo, where he graduated with a first-class degree in engineering, and a passionate belief in the power of 'technology' to solve human problems. This engineering degree was itself a recent innovation, with particular attention paid to all forms of transport, including railways, engines and bridge construc-tions. Immediately on graduation, at the age of twenty-two, Andrée char-acteristically decided to visit the future, and travelled steerage to America, landing in New York with little money, no contacts and no work in prospect. Undaunted, he took the railroad south to the home of American science, Philadelphia. He arrived in time (probably as he had planned) for the Philadelphia International Exhibition of summer 1876, and enthusiastically toured all the stands, making notes of all the new mechanical inventions. To his delight he came upon a Swedish national stand, and at once succeeded in landing himself a job as a demonstrator and technical assistant for the duration. He also had his first glimpse of the importance of publicity and clever presentation in getting innovative projects 'off the ground' – a significant new American catchphrase.

But something unexpected occurred during these formative weeks. Andrée sought out not contemporary engineers and railroad designers, but the legendary old American balloonist John Wise, now retired (temporarily) at ground level on the east coast. They talked of the American dream of the Atlantic balloon crossing, and the theory of

prevailing high-altitude currents. They may even have talked of ballooning to the Pole, since Wise published a letter on this very subject three
years later in the *New York Times*.[11]

Young Andrée became fascinated by the technical challenge of
ballooning. Wise recommended the latest works on meteorology and
trade-wind patterns, and promised to take his young Swedish protégé on
an introductory flight once the Exhibition was over. Twice Andrée
climbed into one of Wise's balloon baskets, but twice Wise cancelled the
flight at the last moment due to bad weather conditions, a lesson in
prudence that Andrée did not perhaps fully appreciate at the time. To his
infinite frustration, Andrée never actually flew with John Wise in
America, though in later years he sometimes implied that he had, the old
American master handing on the aeronautical baton to the young
Scandinavian one.

Returning home, Andrée obtained a post in the Swedish Patent
Office, where he could study the development of every kind of mechanical
invention, but he remained restive and unfulfilled. In 1882 he made
another daring career leap, and volunteered for a two-year scientific
expedition to the bleak northern island of Spitsbergen. This was a
remote, icebound territory claimed by Sweden, lying inside the Arctic
Circle between 78 and 80 degrees north. Hitherto quite unexplored,
except by passing whalers, Spitsbergen was fast becoming the new
Swedish centre for the science of polar exploration. There was growing
rivalry with the Norwegians, who had discovered Franz Josef Land to the
east in 1873. At this time neither Pole had been reached, and numerous
expeditions – like that of Sir John Franklin to discover the North-West
Passage in 1845 – had been lost, always under terrible conditions,
including snow-blindness, limbs amputated as a result of frostbite, and
rumours of cannibalism. There was much speculation about the
unknown icy extremes at either end of the planet. Were they simply
frozen deserts, or were they inhabited by unknown tribes, or even by
unknown monsters? The great Norwegian explorer Fridtjof Nansen
wrote at this time: 'The history of polar exploration is a single mighty
manifestation of the power of the unknown over the mind of man.'[12]

The North Pole was particularly mysterious, with a powerful symbolic
presence in Icelandic literature, and in such works as Mary Shelley's
Frankenstein (1818), whose terrible dénouement takes place on the frozen
Arctic Ocean. No one knew if the ice pack eventually became land, and if

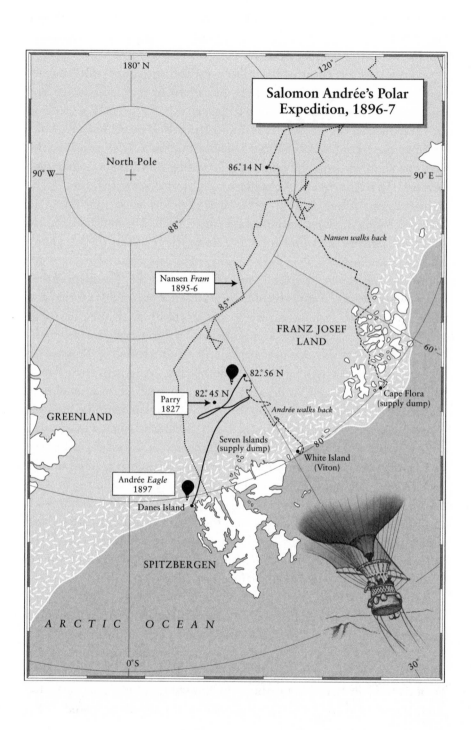

Salomon Andrée's Polar Expedition, 1896-7

180° N

120°

North Pole

90° W

86.°14 N

90° E

88°

Nansen walks back

Nansen *Fram*
1895-6

85°

FRANZ JOSEF
LAND

60°

82.°56 N

Parry
1827

82.°45 N

Andrée walks back

Cape Flora
(supply dump)

GREENLAND

Seven Islands
(supply dump)

80°

White Island
(Viton)

Andrée *Eagle*
1897

Danes Island

SPITZBERGEN

ARCTIC   OCEAN

0° S

30°

so what kind of creatures – besides the enormous and ferocious polar bear – might live there. Unlike the South Pole, the North Pole itself had no land mass or definable landmark, but was merely a geographical co-ordinate at 90 degrees north on the frozen ice cap. No expedition had reached further north than 83 degrees and survived to tell the tale. Using sledges, the English explorer William Edward Parry had got less than a hundred miles beyond Spitsbergen to 82.45 north in 1827, and the Royal Navy Commander Albert Markham had pushed to 83.20 north in 1876. But two American expeditions, led by Charles Hall (1871) and George DeLong (1881), had ended in disaster.

On Spitsbergen Andrée's determination and independence greatly impressed the expedition's director, Dr Nils Ekholm, who then held the important position of Senior Researcher at the Swedish Meteorological Central Office. Ekholm saw Andrée as a natural leader, with immense technical confidence. On his return to Sweden, Andrée began specialising in meteorology, and published several successful academic papers during the 1880s on electrical charges in clouds, and polar winds and weather patterns. He gained a reputation as a polar expert, and gradually the idea of mounting his own polar expedition from Spitsbergen began to possess his imagination.

Against this was the possibility of a more conventional, domestic future. Throughout his thirties Andrée had a long-term liaison with a married woman, Gurli Linder. She was deeply attached to him, and considered divorce; but perhaps her married status suited Andrée.[13] He used to say that 'marriage was too great a risk' for an explorer, and that his mother remained his closest confidante. He seems to have been curiously aloof and inexpressive in most of his friendships, although he had a natural gift with children, and would unbend and join in all their games with boyish enthusiasm, 'frolicsome and roguish'.[14] But more and more he became obsessed by finding a brilliant engineering solution to what he thought of as 'the challenge' of the North Pole.

The infinitely slow and wearisome traditional method of Arctic travel by dog sledge, or skis, or drifting boat, seemed absurd to Andrée. He thought of John Wise and the great American dreams of epic flight. Slowly the decisive project took shape. The modern engineering solution to the North Pole was clearly air travel. Surely it would be possible to fly there in a specially designed and engineered hydrogen balloon? He could 'conquer' the Pole simply by dropping from the skies, anchoring at 90 degrees north, and depositing a Swedish flag and a marker buoy. It would be the ultimate, planetary, record-breaking balloon flight, before the nineteenth century came to an end. Andrée felt he had taken on a national destiny. Marriage would have to wait. He spent the next six years totally dedicated to technical preparation, publicity and fund-raising.

Andrée took his first actual flight in a balloon surprisingly late, two years into his project, in summer 1892, having hired the Norwegian aeronaut Francesco Cetti, based in Stockholm, to teach him. Cetti described Andrée as 'disagreeably calm' when airborne, and impervious to the picturesque charms of ballooning. Instead he was excited by all the technical potential the balloon offered, notably the use of onboard cameras, and the possibility of mapping a large swathe of the unexplored Arctic with overlapping photographs. On the strength of such ideas, he managed to raise funds for his own first experimental balloon. This was a relatively small 37,230-cubic-feet aerostat, which he named *Svea*, after the national emblem, the fierce valkyrie Mother Sweden, tutelary goddess of the North. It was the first of many skilful publicity gestures.[15]

Between 1893 and 1895 he made eight short flights aboard the *Svea* in Sweden, all undertaken solo and without further training from Cetti. He proved a natural aeronaut, cool and resourceful, and was soon experimenting with various forms of baskets, instrumentation, sails and drag lines. On his third flight, in October 1893, he was caught in a violent storm and blown out to sea from Gothenburg, and right across the Baltic towards Finland. He should have been lost, but keeping his head, he skilfully crash-landed on an offshore island, jumped from the basket, and allowed the tattered remains of the *Svea* to blow away without him. They were eventually found fifty miles away on the Finnish mainland.

Because he was missing for forty-eight hours, Andrée attracted a great deal of publicity in the Swedish press, and the disaster was turned into a triumph. A crowd of three thousand people greeted him on his return to Stockholm aboard a Finnish steamer. His tall, Viking-like figure, with his thatch of blond hair and large, dashing moustaches became increasingly well known. On his last flight in the *Svea* he travelled 240 miles in little over three hours, and successfully used a rip-panel to land. He could now claim to be Sweden's leading aeronaut, although in reality his total flying experience amounted to about forty hours in the air.

Andrée published his flight reports in the *Royal Swedish Academy Journal*, and gained the powerful support of the leading polar scientist

A.E. Nordenskiöld. Building upon this, he shrewdly publicised his new method of steering the balloon, by combining a special type of drag rope with a newly designed sail. With such engineering innovations, Andrée caught the attention of the popular press, while simultaneously promoting scientific fund-raising. At academic sessions his gaunt, aristocratic appearance, sober and even melancholy, gave him natural charisma and authority.

He did not seem like an adventurer, though he was quick to spot opportunities. During one meeting of the Academy, Nordenskiöld raised the possibility of a 'photometric survey' of the Arctic from a fixed balloon, tethered at Spitsbergen. Andrée cleverly ran with this as a brilliant idea, though he had already conceived it himself, and merely added that it would be even better for the Academy to fund a *free* balloon, because then the survey could go as far as the Pole itself. Nordenskiöld was delighted with this response.[16] During the 1890s Andrée assembled serious scientific support for such a perilous and even quixotic expedition, and gained several wealthy patrons, including the great industrialist and arms manufacturer Alfred Nobel.

It helped him that polar exploration was increasingly in the news, and a question of national pride. In summer 1894 Nansen set out on his famous expedition to get as near as possible to the North Pole by drifting in the ice floes in his specially constructed boat the *Fram*. Nansen planned to overwinter on his journey, and somehow survive the six months of total polar night. He set out, amidst much Norwegian excitement, but failed to reappear in the summer of 1895. Nothing was heard of his expedition for the next twelve months. When he failed to return in the spring of 1896, a rival Swedish expedition seemed appropriate. Finally King Oscar II, monarch of both Scandinavian countries, expressed his approval of Andrée's attempt to 'conquer' the North Pole by the latest technical means, and made a substantial donation. Exactly as Andrée had planned, his balloon expedition had become a patriotic endeavour.

Andrée now hastened to put together a balloon crew. Shrewdly he first persuaded his erstwhile meteorological director, Dr Nils Ekholm, then in his late forties, to agree to accompany him. His professorial, bespectacled appearance somehow further increased the expedition's scientific standing. A second potential member of the crew was Ekholm's brilliant young assistant Nils Strindberg. Aged only twenty-four, Strindberg was a trained physicist and meteorologist, but also had wide interests in books and music. He drew, painted and played the violin. His family were prosperous and distinguished, and he was a nephew of the great dramatist August Strindberg.

This was another shrewd choice. Temperamentally the opposite of Andrée, Strindberg was a cheery and attractive figure, bubbling with life and fun, adding a good emotional balance to the team. He was disarmingly youthful in appearance, and soon attempted to cultivate an unconvincing Andrée-style moustache. Yet equally vital for the expedition, he had already made a name for himself as an open-air photographer, using the latest Eastman Kodak equipment. In this capacity he further enhanced the engineering credentials of Andrée's crew. Like Ekholm he knew nothing at all about balloons. But unlike Ekholm, he characteristically hurried over to Paris to take aeronautical lessons, which he regarded as a 'tremendous lark'.

Andrée formally presented his scheme, 'A Plan to Reach the North Pole by Balloon', in a long and masterful speech given first to the Swedish Royal Academy in Stockholm in February 1895, and then repeated to the Sixth International Geographical Congress in London the following

spring.[17] In his best dry, commanding manner he outlined the apparently overwhelming challenges posed by an Arctic balloon flight. In sum there were four huge problems: how to sustain the balloon in the air for at least thirty days; how to survive the extreme cold and the potentially fatal problems of icing; how to navigate the balloon on a continuous northerly course; and how to get home in the eventuality that the balloon came down on the ice.[18]

Then, one by one, he coolly analysed each of these formidable difficulties, giving his own precise technical answers. His central theme was that reaching the North Pole was no longer 'a purely scientific problem', or even a human problem. It had become a specifically technological problem, a straightforward 'task for the technologist' requiring a logical series of engineering solutions.[19] He could provide these with a package of brilliant inventions, ranging from an adaptable sailing rig, adjustable guide ropes, self-venting gas valves and ice-repellent balloon fabrics, to the smallest practical details, like an insulated cooking device, lightweight aluminium cutlery, and tinned condensed milk.

Many of Andrée's larger claims would turn out to be chimerical, but it was an extraordinary and captivating performance just the same. Not least because he sounded almost exactly like one of Jules Verne's fictional heroes, solemnly proposing 'An Aerial Excursion Across the Icy Continent' at some packed and breathless meeting of the Royal Geographical Society. It was as if the final destiny of nineteenth-century ballooning was to inflate fiction into fact (or indeed, vice versa).

Two great polar explorers, the American General Adolphus Greely and the British Admiral Sir Albert Markham (whose cousin Sir Clements Markham would later champion Scott's two expeditions to the Antarctic), were in attendance at Andrée's London talk, and gave a guarded welcome to the project. By contrast, Nansen's *Fram* scheme had been universally criticised when presented to the Royal Geographical Society three years previously.

Although he did not actually mention Verne, it is clear from later references in his papers and diaries that Andrée saw himself as fulfilling the destiny of the whole previous century of historic ballooning. He would mention Pilâtre de Rozier's pioneering ascents in France; Charles Green's night flight across Europe; John Wise's dream of crossing the Atlantic; James Glaisher's heroic attempt to explore the seven-mile altitude limit; the intrepid flights of the Paris siege aeronauts (and notably the flight that ended in Norway); Wilfrid de Fonvielle's bucolic five-day cross-country flights, anchoring the balloon each night; and the strange apotheosis of Henri Giffard's gigantic captive balloon at the Paris Exhibition, like some ancient heroic god shackled by a race of modern industrial pygmies.

Perhaps the most extraordinary blurring of fact and fiction occurred in the time frames that Andrée proposed for his polar expedition. His hydrogen balloon would be capable of staying aloft for at least a month, a feat never remotely approached by any previous balloon. It would be capable of carrying crew supplies and equipment that would last for at least three months, an equally astonishing boast. Moreover, it would be dirigible throughout the journey, both to the Pole and away from it, a highly contentious claim. Yet, paradoxically, the actual journey time from Spitsbergen to the Pole, a distance of some 660 nautical miles, would be amazingly short.

Andrée gravely proposed three possible time scales, all 'scientifically calculated' according to previous flight data. The first was based on the speed of the famous siege balloon of November 1870, that reached Norway from Paris in fifteen hours. Based on this balloon's velocity, Andrée projected that the North Pole could be reached from the Spitsbergen area in an astonishing six to eight hours. The second time scale, based on Andrée's own crossing of the Baltic in the *Svea* during the storm of 1893, would occupy 'little more than ten hours'.[20] Both of these figures made the crossing of the huge, fearful Arctic ice cap sound like a walk in the park. His audience was reduced to amazed silence.

The third projected schedule was perhaps a little more realistic. It was based on the meteorological records Andrée had himself taken at Spitsbergen during previous Arctic summers. He claimed that these revealed the existence of one of those largely regular and reliable 'oceanic air currents' of seasonal wind, which the great aeronaut John Wise and others had predicted. He believed that just such an oceanic current, a regular north-moving low-pressure cyclone, did indeed exist in the Arctic summer. It promised 'an average steady 16.2-mile-per-hour breeze' to the north, starting in June, which if unbroken would carry the balloon to the Pole in 'approximately forty-three hours'. This would be just under two days and two nights (except that in the northern summer there would be no nights). Andrée considered this time frame, between forty and fifty hours of travel, the most likely and also the most practical. It would allow proper time for observations, meals, sleep and carrying out a full photo-metric survey with '2 photographic apparatus and 3,000 plates etc'.[21] 𝕎

The one thing that Andrée's time projections did not include was how long, or which direction, the balloon would take to *return* from the Pole, although he did optimistically suggest that a journey from Spitsbergen directly across the Pole to the Bering Strait, a distance of 2,200 miles, might take a mere six days – 'that is, one-fifth of the time during which the balloon can remain in the air'.[22] He pointed out, with a rare smile, that once one had reached 90 degrees north, any direction was southwards, and therefore homewards: either towards Russian Siberia, or Canada, or Alaska, Iceland or even Greenland. Wherever one landed, he suggested, the Swedish pioneers would be greeted as aerial citizens of the world.

The balloon he presented was confidently christened the *North Pole*. Intended as the last word in aeronautical engineering, it was simultan-eously a kind of Vernian fantasy machine. Two hundred and twelve thou-sand cubic feet (about the same size as Nadar's *Le Géant*), standing ninety-seven feet tall and sixty-seven feet in diameter, it was constructed from three layers of hugely expensive double Chinese silk, and protected from ice by a special varnished cotton top canopy or *calotte*. Its venting valves were placed at the side of the balloon, rather than at the top, to prevent them icing shut. The conventional open neck above the basket was replaced by an automatic pressure valve, adapted from one designed by Giffard.[23]

𝕎 Andrée's idea of a 'photometric survey' of the Arctic was not ill-conceived. It even-tually led to continuous high-altitude surveys of the Arctic ice cap, beginning with NASA's 'Scanning Multichannel Microwave Radiometer' (SMMR) satellite in 1978. These first showed the huge seasonal expansion and contraction of the Arctic ice field, though not the thickness of the ice. It appears that the volume of summer Arctic ice has contracted by approximately 50 per cent since the year 2000, though the re-freeze of the ice cap in winter has remained roughly stable. These summer contrac-tions or meltings were particularly noticeable in 2007 and 2012. Model predictions suggest that there may be no summer ice cap at all by 2030. The cause of this may be part of a natural cycle (the end of the so-called '4th ice age'), or directly attributable to man-made global warming, or both. Either way, such shrinking, if continuous, would probably affect the Gulf Stream and the whole weather system of Great Britain and northern Europe. It would make it less temperate and more extreme, in storms, heatwaves and droughts. See 'Arctic Sea Ice' on NASA's *Earth Observatory* internet site. André's fellow Scandinavian scientists were already considering such possibilities.

The *North Pole* had an overall lifting capacity of 6,600 pounds (approximately three tons), of which three thousand pounds was free ballast in various forms. Of this, its complex system of three one-thousand-foot trail ropes, and eight shorter ballast ropes, provided 1,600 pounds – almost half – of the total adjustable ballast. So the ropes were crucial to its equilibrium. Altogether Andrée claimed to have designed into the balloon a large theoretical safety margin, allowing him to adjust altitude, and to respond freely to the expansion or contraction of hydrogen due to temperature changes in the Arctic air. Yet, apart from the *calotte*, he largely ignored the problems of moisture, Arctic fog and icing.[24]

The payload elements of the aerostat were mounted in three special sections, one above the other, suspended from the main balloon ropes. They consisted of a closed crew basket, then an open observation deck, and finally – *above* the balloon hoop – a conic or circular storage section. This arrangement had never been tried, or even tested before, but was intended to demonstrate all Andrée's technical skill and foresight.

At the base was the specially insulated and enclosed wicker basket, six and a half feet in diameter and five feet in depth, ergonomically designed rather like a yacht's cabin. Unlike conventional balloon baskets, it was sealed at the top with a flat roof, and accessed by a narrow hatch. Within, padded compartments were crammed with the latest scientific instruments, including chronometers, compasses, sextants, barometers, message buoys, and three pairs of Zeiss binoculars. Unusually for a balloon, it had a sleeping bunk, a galley and a night-stool. Typical of Andrée's ingenuity was a mobile spirit cooker and oven, which could be lowered beneath the basket to prevent the risk of fire, and lit by remote control.

Stores included three months' worth of tinned food, rifles and ammunition, fishing gear, a Swedish flag, and a reindeer-skin sleeping bag, large enough to take all three crew. This three-man sleeping bag was characteristic of the Scandinavian approach to Arctic exploration. It assumed teamwork, good fellowship, and the practical value of shared body heat. There were also marker buoys, thirty carrier pigeons, and numerous luxury items including champagne and Belgian chocolates.

The second section was the circular observation deck. This was effectively formed by the flat roof of the basket. Again, it was the equivalent of a yacht's cockpit, encircled by an adjustable canvas windshield or 'dodger', and protected by a chest-high wooden railing. The railing was an Andrée invention, known as the 'instrument ring', upon which a variety of observational instruments – cameras, barometers, ground-speed calculators – could be quickly bolted or unbolted as required. Such a deck, combined with the enclosed basket, meant that separate 'watches' could be kept, and the crew could take turns to go below to sleep or eat or write their journals, a vital consideration for morale on a long journey. Although of course the journey out was only intended to last fifty hours.

The third section was mounted, in another design innovation, above the balloon hoop, and accessed through the hoop by a rope ladder. It consisted of carefully selected packs of back-up supplies, stored in a system of canvas pockets and sealed compartments. Most of them were only intended for use if the balloon came down. Apart from further stores and ammunition, notable additions were a tent, a surprisingly large collapsible boat with paddles, and three self-assembly wooden sledges. These were Andrée's solution to any enforced landing on the ice.

Andrée's greatest ingenuity had been reserved for his special guidance system. This consisted of three sails, in combination with the series of heavy guide and ballast ropes. The sails were mounted on a horizontal bamboo boom slung from the balloon hoop, like the topsails of a square-rigged ship. The three main hemp guide ropes, each over a thousand feet

long, were slung from a hand-cranked winch that could pay them out or haul them back in again. When fully extended along the ice, they would drag and act as a kind of counterweight against the pull of the wind. In case they jammed in a crevasse or ice snag, Andrée had ingeniously fitted each section with exploding break points, and also with unscrewable metal disconnectors.

By slightly slowing the balloon down, the guide ropes radically changed its aerodynamics, making it behave like a kite held by someone running along the ground, or a sailing boat with its keel running through water. Thus Andrée believed that he had found a way of giving balloon sails a vital purchase on the airflow.

Additionally, in a brilliantly simple device, the angle of the trailing ropes relative to the balloon could be altered by running them through a heavy wooden swivelling block. Again, this was a design taken from sailing ships. The effect was to twist the balloon relative to the airflow, thereby automatically turning and altering the setting of the sails. Thus, by simply adjusting the angle at which the guide ropes left the balloon through the swivelling block, the sails could be turned to act like an airborne rudder. So Andrée believed he could redirect the course of his great balloon.

With this method, convincingly illustrated by fine engineering drawings, Andrée informed the Swedish Royal Academy that he could steer his balloon off the line of the wind by as much as twenty-seven, or even forty, degrees.[25] If his projected 'light Arctic breeze' deviated to the east or the west, he could bring the balloon back on a true northerly course with a simple adjustment of his swivelling block. Thus Salomon Andrée claimed at last to have solved a problem that had haunted aeronautics for almost exactly a hundred years. He had designed a self-sufficient, long-distance, dirigible free-flight balloon.

The North Pole was different from all previous free-flight balloons in one other crucial respect: it had a very narrow altitude band. In order to be dirigible, it always had to remain close to the ground, so its ballast and guide ropes would work. Andrée stressed: 'The weight of the balloon must be so balanced that when free it will stay at an average height of about eight hundred feet above the surface of the earth: viz. below the lowest region of clouds, but above the mists close to the ground.'[26] Accordingly, unlike conventional hydrogen balloons, the balloon envelope was fully inflated to keep it within this critical altitude band. It had little space for expansion. It would immediately vent gas through its automatic Giffard valve if it rose much beyond a thousand feet. So if, for any reason, it did rise higher, then gas would be lost and very large amounts of ballast would have to be abandoned to re-establish its equilibrium or balance. As ballast equalled flying time, it was a design innovation with unknown implications.

Of course, apart from some early prototype journeys in the little *Svea*, Andrée had hardly flight-tested any of these innovations. Most of them remained brilliant drawing-board ideas. Yet the whole project was modestly presented to the Swedish Academy as a logical exercise in practical engineering. Some critics wondered if this was not after all merely a version of techniques tried out many times before, and many years ago. Hadn't Charles Green tried guide ropes? And even before that, hadn't Blanchard tried sails? Was it an old fantasy, rather than a new technology?

Yet Andrée's calm authority, his 'scientific data', and perhaps his commanding moustache, quietly carried the day. Moreover, in his peroration he emphasised patriotic destiny, and gently mocked the attempts of the Norwegians, led by Nansen, who was still missing in the *Fram*: 'Who, I ask, is better qualified to make such an attempt than we Swedes? ... Is it not more probable that we shall succeed in sailing to the Pole with a good balloon, than that we shall reach it with sledges for transport ... or with boats that are carried like erratic blocks, frozen fast to some wandering masses of ice?'[27]

Accordingly the *North Pole* was funded, and swiftly built. Amid immense publicity, Andrée and his crew sailed to Spitsbergen aboard the *Virgo* in June 1896, accompanied by a small fleet of well-wishers, scientists and press. A crowd of forty thousand people saw him off from Stockholm docks. His mass of equipment, all of it proudly engraved or marked in red paint with 'Andrées Pol. Exp. 1896', was unloaded in a shingle cove on the north-western tip of Dane's Island. A huge wooden balloon hangar and a hydrogen-generating shed were swiftly constructed. Within four weeks the immense balloon was successfully inflated, and all the equipment prepared. The weather was fine and mild, perfect for a launch. But the wind blew steadily and provokingly *from* the north, not towards it.

They settled down to wait for Andrée's predicted light southerly cyclone breeze. It never came. Andrée gave endless press briefings, several tourist steamers came and went, the great balloon stirred uneasily in its wooden cage, and the Arctic air was suspiciously perfumed with escaping hydrogen. At the end of August, after two frustrating months, the whole expedition had to return to Sweden. Just before they left Spitsbergen, Nansen's ship *Fram* sailed quietly into the bay.

4

It was a strange and bitter anticlimax. Andrée stoically hid his disappointment, but was secretly devastated when Nansen himself triumphantly returned to Norway in September, having twice overwintered on the ice. In the first year the *Fram* had reached above 84 degrees north; and in the second, Nansen had set out with dog sledges from the *Fram* and had reached 86 degrees 14 minutes north, a formidable achievement, within a hundred miles of the Pole. Afterwards, Nansen and his colleague Hjalmar Johansen had succeeded in walking home together through the terrible pack ice, on the way surviving a second winter in a tiny ice-hut built on Franz Josef Island. It was a masterly demonstration of courage, comradeship and polar skills. Nansen recounted the trek in a superbly written travelogue, *Farthest North* (1897), which is still a bestseller.

Inevitably, Nansen stole much of Andrée's thunder with the Swedish public. Worse, he had inadvertently raised the bar for any future polar expedition, which would inevitably be regarded as a failure unless it penetrated well beyond 86 degrees north. Andrée briskly announced that he would try again in summer 1897, but support for the renewed expedition naturally wavered. Alfred Nobel continued his subsidy, and so did King Oscar, but there was growing criticism in the press. Was Andrée really a fantasist, a dreamer? Was his huge balloon a ludicrous anachronism?

Dr Nils Ekholm, Andrée's senior partner, now privately questioned him over the durability of the balloon fabric. He calculated that the balloon canopy, even while tethered in Spitsbergen, had been losing about 120 pounds of lift per day. In his view, this reduced the balloon's endurance in the air from thirty days (by the end of which time he projected it would have lost more than its entire lifting capacity of six thousand pounds) to an absolute maximum of seventeen days, and probably much less. Andrée agreed to increase the size of the balloon by sewing in new gores, but at Christmas 1896 Ekholm officially resigned from the 1897 expedition. He had lost faith in Andrée's dream.

Ekholm later published further reasons for resigning. They were revealing. He believed that Andrée had also lost confidence in the balloon's endurance, and its capacity to complete the whole expedition in the air. The vague talk of flying on over the polar ice cap to Greenland, Alaska or Siberia (depending on the wind direction) was a chimera. The balloon flight was a one-way ticket. He thought Andrée was secretly resigned to coming back the hard way, by sledge and boat. There was no 'engineering solution' to this. It would be a brutal, slogging, potentially suicidal journey of up to six hundred miles. Even the experienced and hardened Nansen had taken two years to accomplish it, and had been forced to winter on the ice. Ekholm, who was now forty-eight, did not think he was physically capable of such a trek, and he doubted if Andrée was either. He especially feared for young Strindberg. He summed up Andrée's dilemma as simply as possible. The balloon must have the scientifically proven capacity 'not merely to carry the expedition safely into the Polar area; but also safely out of it'.[28]℣

As for Nils Strindberg, his position was also altered. In October 1896 he had become engaged to his childhood sweetheart, a beautiful young woman called Anna Charlier. Both Strindberg's father and the Charlier family begged him to resign from the second expedition, as did his mentor Dr Ekholm. But Anna, deeply in love, understood Nils's desire to make his mark in science before he settled down to family life. So she supported his decision, though with deep secret misgivings.

℣ Ekholm always maintained his interest in aerial exploration, but he saw that the future lay with the heavier-than-air machine, and became the founding chairman of the Swedish Aeronautical Society in 1900. Nevertheless, he still believed, like Glaisher, that crucial meteorological data could be gathered from high-altitude balloons. In addition to his work at the Swedish Meteorological Office, he continued to publish scientific papers. His visionary paper 'On the variations of the climate of the geological and historical past and their causes' was published in January 1901 by the *Quarterly Journal of the Royal Meteorological Society*. It was one of the earliest academic papers to predict that increased CO_2 emissions, both natural and man-made, would eventually produce global warming. However, Ekholm argued that this would broadly be beneficial to mankind, and would ward off the threat of a new Ice Age. See 'On Variations of the Climate' on the *Nils Ekholm* internet site.

Meanwhile, Andrée purchased two of the latest Carl Zeiss cameras, reminding Nils that the 'photometric survey' of the Arctic from the air would be a scientific first, quite unlike anything achieved by Nansen. Much torn in his loyalties, both emotional and scientific, Nils eventually decided that honour required him to continue with the second expedition.

In spring 1897, after interviewing many candidates, Andrée replaced Dr Ekholm with Knut Fraenkel, a much younger man and a very different type. This choice may have confirmed Ekholm in his worries about the nature of the journey being undertaken. Like Andrée, Fraenkel was an engineering graduate of the Royal Technical Institute, but his main accomplishments were athletic. A youthful giant of immense strength and stamina, he was a fine gymnast, a mountain climber, an adventurer who had helped in his father's roadbuilding business in the north of Sweden. He was extroverted and good-natured, and turned out to be an excellent cook. At twenty-seven years old, he towered over the small, elegant Strindberg, and was much stockier than Andrée. But he accepted his authority, and evidently admired him far less critically than Dr Ekholm. Like Strindberg, he hurried off to learn ballooning in Paris, and despite some severe crash-landings, came back more enthusiastic than ever.

On 29 April 1897, just before the second expedition was due to sail for Spitsbergen, Andrée's beloved mother Mina died. The loss of this most important emotional tie for Andrée, now aged forty-two and still unmarried, may in some sense have loosened his last links with the earth. When Nansen wrote him a warm letter wishing him all luck, saying how much he admired what he was undertaking for Sweden, he also sounded a note of caution, quoting *Macbeth*: 'I dare do all that may become a man; who dares do more is none.' In the most tactful way, Nansen suggested that Andrée should not allow himself to be driven to take 'unreasonable risks' by patriotism, or any other influence: 'It is in drawing this boundary that true spiritual strength reveals itself.' Like a good mountaineer, Andrée should know when to turn back; or even when not to start.[29]

Andrée secretly added a codicil to his will, which included an ominous sentence: 'I write on the eve of a journey full of dangers such as history has never yet been able to show. My presentiment tells me that this terrible journey will signify my death.'[30]

The new expedition reached Dane's Island on 30 May 1897, repaired the balloon hangar on the bleak foreshore, and reinflated the balloon, which was now rechristened *Örnen* – the *Eagle*. Apart from this symbolic change – perhaps suggesting a triumphant flight, rather than a declared

destination – all Andrée's equipment, even his fine cotton handkerchiefs, remained emblazoned with the red insignia 'Andrées Pol. Exp. 1896'. For nearly six weeks they waited for the wind from the south. Finally, on the morning of 11 July, the barometer dropped and a cyclone arrived, bringing low grey cloud. It blew temptingly northwards across Virgo Bay, in a series of sharp blustery squalls. The final preparations took less than four hours, and were hurried. Andrée made a point of taking both Strindberg and Fraenkel aside, and asking them individually if they agreed to launch. Strindberg was notably impatient to leave, although there was some discussion about waiting for the wind to settle its direction and strength. But after two years Andrée was not to be restrained, and ordered the downwind side of the wooden hangar to be cut away with axes. Now there was no turning back.

At exactly 1.43 p.m. they shook hands, released a tangle of restraining ropes, and launched the huge balloon.[31] The *Eagle* rose with slow dignity, just cleared the hangar with a slight bump, then sailed magnificently out across the grey, choppy waters of the bay, heading a perfect due north. Everyone cheered and waved. The navigation sail was already billowing, though the complex cluster of trailing guide ropes made the balloon look curiously awkward. One of the journalists remarked that it was more like

a long-legged spider than a soaring eagle. A photograph records this iconic moment, and was later enhanced as a photo-illustration that was published by *Life* magazine, and around the world.

Then, in a few seconds, the mood shifted. It was noticed that the balloon was flying very low, at little more than sixty feet. The ropes dragged a broad wake of dark disturbed water behind it, but several of them appeared to be dropping away. What no photograph showed is what had happened in the first sixty seconds of the flight. As the *Eagle* rose, the lower sections of the guide ropes had begun untwisting their metal screw connectors. Even before the ends reached the water, they fell with a rattle onto the foreshore.

It was later found that the ropes had been coiled neatly on the ground outside the hangar, rather than stretched out straight along the shingle. They had simply twisted round and round as they were pulled into the air, and finally disconnected themselves. It was a classic case of a brilliant technical design failing at its first practical application. A workman saw this with a shout, but Andrée did not discover it until some minutes later, as he was already preoccupied with something else.

The *Eagle* was failing to rise any higher. Slowly it began to dip towards the surface of the bay. Within a few hundred yards the basket was

skimming and catching the water. Clearly the hurried 'weighing off'
inside the protected hangar did not correspond to the blustery conditions outside. Ironically, the huge, powerful *Eagle* had not been released
with enough initial lift. For a moment this seemed quite playful, and the
workmen back on the shore still cheered. But in a few more seconds the
basket was kicking up a bow wave, sinking deeper into the icy sea, and
looking as if it would actually submerge. Alarmed, Andrée ordered a
quick offload of ballast. In the emergency it seems that his two inexperienced crew, eager to obey their commander, immediately threw out four
bags of gravel each, far more than necessary.

The *Eagle* was seen to jerk sharply, then leap clear of the waves and
sail upwards, streaming water and trailing its shortened guide ropes high
into the air. The workmen cheered again. It was only then that Andrée
could see that he had lost most of his vital navigation system in the first
few moments of the launch. The balloon soon rose over the coastal hills,
and disappeared from sight, still climbing. Eventually it reached over two
thousand feet, far higher than Andrée had ever intended to fly, and the
automatic Giffard valve began releasing gas. Below them they could see
the first 'finely divided' ice floes and the 'beautiful dark-blue colouring' of
the Arctic Ocean. Just before they left the northernmost tip of Spitsbergen,
Strindberg dropped a departure message for Anna Charlier in a canister.
It was never found.

5

From now on the history of the expedition has four main written sources, which do not always tell quite the same story. The primary one is Andrée's official diary, which is in effect the captain's log. The next is Fraenkel's logbook, which consists mainly of meteorological observations. The third and fourth are Nils Strindberg's almanac, which mixes navigational with private records, and finally Nils's extended love letter to Anna, nine pages written in shorthand, which he intended to deliver on his return as a form of wedding present.[32] But there would be a fifth, unwritten source, perhaps the most eloquent of all: Nils Strindberg's Zeiss photographs.

At first the launch mishaps seemed comparatively minor, almost comic. The crew had drunk champagne, and were in high spirits. One of Nils's early almanac entries describes how he climbed into the hoop to admire the view, and the following cheery dialogue took place: 'Look out, Fraenkel!' 'What's up?' 'You'll get a shower-bath!' 'All right!'[33] Evidently Nils was happily urinating. The balloon found its new equilibrium at around a thousand feet, and stopped losing so much gas from its automatic valve. The wind was carrying them briskly north, though also several degrees eastwards. Still hoping eventually to adjust their course with the sails, Andrée had all hands splicing new full-length trail ropes from what remained of the originals. But he was relaxed enough to go below for an Arctic siesta.

All that first afternoon of 11 July the ballooning was smooth, sunlit and bucolic. They were thrilled by the sense of entering so swiftly and so easily into the unknown land of ice and snow. Nils recorded: 'Only a faint breeze from South East and whistling in the [Giffard] valve. The sun is hot but a faint breath of air is felt now and then. Andrée is sleeping. Fraenkel and I converse in whispers ... Ice is glimpsed a moment below us between the clouds. Course North 45 East magnetic. We are now travelling horizontally so finely that it's a pity we are obliged to breathe (as that makes the balloon lighter of course).'[34] It seemed almost magical.

It was only later, when he had time to make his calculations, that Andrée realised how much vital ballast they had lost. Adding together the eight hastily-emptied sacks of gravel (450 pounds) and the disconnected trail ropes (1,160 pounds), it amounted to rather more than 1,600 pounds, or one-third of their total ballast. As ballast equals flying time, this immediately shortened the balloon's possible endurance in the air by as much as half.[35]

There was also the problem of why the balloon had descended into the water in the first place. Possibly it had not been 'weighed off' properly. Possibly its sails had caught an unlucky downdraft of wind across Virgo Bay. Or possibly the balloon was leaking, and simply lacked lift, exactly as Dr Ekholm had feared. If so, this would require some radical rethinking of the journey. While Andrée lay below in his bunk, pondering these problems, Fraenkel prepared food, and Nils began photographing. He also started keeping a plotted chart of their course.

'Our journey has hitherto gone well,' Andrée entered briskly in his first recovered message, dropped by buoy at around 10 p.m. on 11 July. 'We are still flying at an altitude of 250 metres [830 feet] on a heading at first North 10 East, but later N. 45 East. We are now well in over the ice field, which is much broken up in all directions. Weather magnificent. In best of humours ... Above the clouds since 7.45 p.m.' It was signed as a team: 'Andrée, Strindberg, Fraenkel'.[36] Amazingly, there is no mention of the lost ballast ropes, or any comment on their increasingly eastward heading. Until midnight on 11 July the balloon continued to fly high and stable at around one thousand feet, but also continued to turn eastwards. In fact, by 1 a.m. on 12 July it was flying due east, no longer heading north at all. Nevertheless, up till then progress had been little short of astonishing. By the end of the first twelve hours they had covered some 244 miles, which was genuinely impressive. A sledge would have taken three weeks at best to cover the same ground.

At around 1.30 a.m. on 12 July, the loss of gas suddenly began to make itself felt. Entering a cloud, the balloon lost the warming influence of the sun and sank steadily into a new world: that of Arctic ground fog. From then on, conditions changed dramatically. Within four minutes the *Eagle* had dropped from eight hundred to sixty-five feet, and for the first time since Spitsbergen one of the shortened drag ropes touched the ice. The balloon would never again rise above three hundred feet.[37]

The crew did not know this, but their mood darkened. Nils noticed a huge, blood-red stain on the ice, where a polar bear had made a kill. It seemed ominous.[38] They could now see that what appeared relatively smooth and benign from a thousand feet was actually a surface of fearful irregularity, with twenty-foot humps and gullies of ice, sharp edges and rugosities, where the ice pack had stacked and compacted under huge wind and submarine pressure.

At the new low altitude, the balloon slowed to almost walking pace; worse, it gradually turned completely around and started to drift westwards. They were virtually retracing their steps. Throughout the day they threw out more ballast, including, at 4.51 p.m., the biggest buoy, originally intended to mark the Pole itself, or at least their furthest point north. Significantly, it was thrown overboard 'without communication'. This also suggests their change of mood. Immediately afterwards, the basket actually struck the ice forcibly, 'several times in succession'.[39] By 5.14 p.m. they had had 'eight strikes in thirty minutes'. This was menacing.

Throughout the remainder of 12 July, the *Eagle*'s speed and prospects steadily deteriorated. They were no longer advancing towards the Pole at all. Another of Andrée's key concepts, the 'steady summer breeze towards the Pole', had failed to materialise.

As they sank into the freezing fog, the sun disappeared altogether, and their horizon closed down claustrophobically, with visibility reduced to less than a mile in all directions. Their voices came back to them with a dull, muffled echo. It grew much colder. The ice no longer glinted and shimmered, its blue and white beauty replaced by a dreary, featureless grey. This gave Nils's cameras nothing to focus on, producing photographs without depth or scale. 'The snow on the ice a light dirty yellow across great expanses,' noted Andrée. 'The fur of the polar bear has the same colour.'[40]

Nils found himself waiting for the basket's next strike, each one shaking the wickerwork and vibrating up through balloon's entire rigging, making the canopy snap and creak overhead. He described this with an expressive term: 'the balloon's stampings'. It was as if the *Eagle* was putting its foot down, angrily demanding to kick clear of the hostile world of ice. Should they try to anchor, and wait for a better wind? Should they expend yet more precious ballast, and try to ascend above the fog? Uncertain what to do next, Andrée again went below to take stock and sleep on the single bunk. Nils and Fraenkel were left on watch.

At some point Nils climbed into the balloon hoop to be alone, and to write his letter to Anna. His thoughts went both into his almanac and into his letter, though with slightly different emphasis in each. In the letter he was determinedly cheerful: '12 July. Being up in the carrying ring is so splendid. One feels so safe and so at home. One knows that the bumps up here are felt less and this allows me to sit calmly and write to you without having to hold on ... Andrée is lying [below] in the basket cabin asleep but I expect will not get any proper rest. The sun vanishes in the fog.'⁴¹

Nils's later almanac entry gives an optimistic account of their progress during the earlier part of the day, when the sun still partially shone through the fog. But the misdating of the entry '12 June', instead of July – suggests an element of distraction in his thoughts.

12 June 21 hours 5 minutes o'clock ... This morning the height of car was 60 metres [190 feet] when the fog lightened enough to allow the sun to peep through. Every so often patches of blue sky. A refreshing sight after all the 'stampings' during the night. The carrying power of the balloon also increased finely. I wondered if we will make a high-level journey?

But by the time Nils wrote this entry, just after nine o'clock on the evening of 12 July, the chances of returning to a 'high-level journey' were slipping away. The balloon was growing heavier every hour with moisture from the freezing fog, and the lifting power of the remaining hydrogen was reducing as its temperature dropped: 'Hard and continuous bumps against the ground resulting from the fog that weighs us down.' Nils also noted that the wind direction had swung further round, '90 to 100 degrees', and was now threatening to blow them almost due west. Finally, at about 10 p.m., they stopped moving altogether – one of the drag ropes had become caught beneath a block of ice. At least temporarily, they were completely stuck – 'fastened very well', as Nils pointedly put it. Perhaps initially it was something of a relief. But now their progress did not look so impressive. After twenty-two hours' flying, they were supposed to be over halfway to the Pole, but they had not reached even 82 degrees North. They would remain stuck for the next thirteen hours.⁴²

From these entries of 12 July, it is possible to conclude that Nils had begun to have doubts about Andrée's balloon technique. He might have begun to wonder if the trail ropes and sails should be abandoned altogether. Perhaps they still had enough spare ballast to throw out, and let the *Eagle* fly freely into the Arctic sunlight? Once higher up, beyond one or even two thousand feet, the hydrogen would expand in the heat, the lift would rapidly increase, and the balloon would come to life again. By adjusting the Giffard valve, they could find a new ballast equilibrium and risk 'a high-level journey'. They could finally take the glorious chance of a free flight across the Pole.

Andrée was up on deck within an hour of Nils's almanac entry. He made a brief official report in his own journal, dated 12 July, 10.53 p.m.: 'Everything is dripping and the balloon heavily weighted down.' Then for about thirty minutes he obviously had a discussion about their prospects with Nils and Fraenkel. He noted what they all felt: 'the balloon sways, twists, and rises and sinks incessantly. It wishes to be off, but cannot.'[43] He then ordered them both down below to rest in the balloon car. They seem to have gone reluctantly. The question evidently discussed was whether they should cut the trail ropes, drop ballast and attempt a free flight above the fog before it was too late. The wind might take them west towards Greenland, or it might turn north towards the Pole again, or it might even carry them back southwards towards Spitsbergen. But at least it would be a flight. Though Andrée does not specifically say so, Nils – and probably Fraenkel – argued for this, 'the higher journey'. It was perhaps the most momentous decision of their expedition.

Andrée wrote in his journal soon after the others had gone below:

11.45 p.m. [12 July] Although we could have thrown out ballast, and although the wind [now blowing due west] might perhaps carry us to Greenland, we determined to be content with standing still. We have been obliged to throw out very much ballast today, and have not had any sleep nor been allowed any rest from the repeated bumpings, and we could not have stood it much longer. All three of us must have rest, and I sent Strindberg and Fraenkel to bed at 11.20 o'clock, and I mean to let them sleep until 6 or 7 o'clock [on 13 July] if I can manage to keep watch until then.

There was no hint in this entry of any dissension. The determination to 'stand still' seems to have been mutually agreed between the three of them, as a team. Yet the feeling that they 'could not have stood it much longer' suggests a certain tension. Andrée also added a curious reflection: 'If either of them should succumb it might be because I have tired them out.'[44]

Andrée was now alone on the deck of the stationary *Eagle* for several hours, until early in the morning of 13 July. It was a rare moment of solitary command, a historic pause: the heroic aeronaut aboard the greatest of the nineteenth-century free balloons. Using his Zeiss binoculars, he gazed around at the desolate, grey-yellow ice stretching to the horizon in every direction. The low fog and the utter solitude pressed down upon him: 'Not a living thing has been seen all night, no bird, seal, walrus or bear.'[45]

The balloon swayed slightly at the end of its trapped trail rope. At some point he sat down on the little wooden barrel they used as a seat, opened his journal on his knee, and made his longest and his only personal entry during the entire expedition.

Is it not a little strange to be floating here above the Polar Sea? To be the first that have floated here in a balloon. How soon, I wonder, shall we have successors? Shall we be thought mad, or will our example be followed? I cannot deny but that all three of us are dominated by feelings of pride. We think we can well face death, having done what we have done. Is not the whole endeavour, perhaps, the expression of an extremely strong sense of individuality which cannot bear the thought of living and dying like a man in the ranks, forgotten by coming generations? Is this ambition? The rattling of the guide-lines in the snow, and the flapping of the sails, are the only sounds to be heard, except the creaking [of the wind] in the basket.[46]

As Andrée clearly intended, this is a historic statement in the development of ballooning. It poses the question: are they at the beginning, or at the end, of a great tradition of aerial exploration?[ψ] Yet the truly surprising feature of this entry is its philosophical resignation. There is no sense of planning ahead, of assessing their chances. Andrée, like his balloon, is stuck fast, psychologically immobilised.

There is no attempt to consider options, or practical alternatives. It is almost as if the whole expedition is already over – 'We think we can

---

[ψ] They did have successors, but not as Andrée had imagined. In 1907 and 1909 an American journalist, Walter Wellman, attempted to fly an airship from Dane's Island, but only got thirty miles out and crashed on a glacier. In 1923 and 1925, the great Norwegian Antarctic explorer Roald Amundsen made two attempts to fly over the Pole by twin-engined seaplane, one from Alaska and one from Spitsbergen, During the second expedition he successfully landed on the ice at 88 degrees north, thereby beating both Andrée's and Nansen's records, and successfully flew back three weeks later. But perhaps his most appropriate feat was to launch an airship, the *Norge*, from the site of Andrée's launch on Dane's Island in 1926, and overfly the Pole itself. However, the first successful free balloon, in deliberate emulation of Andrée, did not reach the Pole until 2000. It had much modern equipment aboard, and flew the 'high-level journey' up to fifteen thousand feet. It was piloted by the British explorer David Hempleman-Adams, and actually managed a brief landing on the ice at 89.9 north. But it did not quite fulfil Dr Ekholm's original stipulation, and fly back again. Both balloon and pilot were brought back in a helicopter.

well face death, having done what we have done.' For Andrée, after little more than thirty hours, death is now the most likely outcome. Yet clearly these were not the feelings of either Nils or Fraenkel, who had every reason to live, and to return to Sweden.

By mid-morning on 13 July everyone had slept, and the situation had changed again. The capricious wind had come round through nearly 180 degrees, and was blowing eastwards once more. At 11 a.m. the trail rope, pulled in the reverse direction, suddenly broke free from the ice, knocking them all off their feet. They were now sailing back eastwards, on an almost reciprocal course. They resorted to a hot meal washed down with several bottles of 'the King's Special Ale'.[47] Then Andrée released four pigeons bearing the same very brief message. He gave their latitude as 82 degrees 2 minutes North, and said they were making 'good speed' to the east. He added less than a dozen words: 'All well on board. This is the third pigeon post. Andrée.'

Ominously, there were no further details of their plans or prospects; no personal comments; and no signature from either Fraenkel or Strindberg. Probably Andrée simply did not want to admit the true position.[48] But it was clear. At 5 p.m. that day they crossed back over exactly the same point at which they had been twenty-five hours earlier, at 4 p.m. on 12 July. The balloon had simply performed a huge west–east dog's leg. They had covered a further two hundred miles over the ice, but got not a mile nearer the Pole.

By this stage the technical state of the *Eagle* was critical. It was beginning to bump on the ice again, and it was now clear that the fog had frozen yet more moisture onto the balloon canopy and network cords, adding hundreds of pounds to her weight. The failure of the sun to emerge throughout 13 July meant that this process of icing-up was ever-accelerating. It was a situation, despite all his analysis of Arctic 'data', for which engineer Andrée had provided no 'design solution'.

As the afternoon of 13 July wore on, the bumpings made things increasingly difficult in the basket, and it became colder still. Andrée seemed lost in thought, Fraenkel let the cooker catch fire, and Nils started to feel ill: 'I tried to lie down in the car at 7 o'clock but in consequence of the bumping I became seasick and spewed.' He went up alone into the ring, pulled on 'a pair of balloon-cloth trousers, and an Iceland jersey', and read Anna's last letter. 'It was a really enjoyable moment.'[49]

At 8 p.m. on 13 July, probably in response to Nils's urgings to return to a 'high-level journey', Andrée ordered a major dump of ballast. They threw overboard six more marker buoys, the winch, the night-stool, and most of the remaining sacks of gravel. In total this amounted to 550 pounds, an enormous weight, which should have lifted them back well above the clouds. The *Eagle* stirred in response, rose to two hundred feet, and then stubbornly hung there, still shrouded in icy fog. By 10.30 p.m. it was down again, and striking violently against the ice.[50]

The only possible remedy was now so extreme that it would be a complete aeronautical gamble. Unless they threw out some of the equipment in the upper storage cone – which consisted of the tinned food, the spare ammunition, the sledges, the tent, the collapsible boat and the cooking fuel, all of it vital for survival down on the ice – the *Eagle* would never rise again. It was exactly the dilemma defined by Dr Ekholm. Did Andrée really trust the balloon alone to get them 'safely out of the Polar area', and back home? Would they commit themselves completely and finally to the air, rather than to the earth? They must have discussed this dilemma throughout the 'night' of 13 July, although there is no record in either Andrée's or Strindberg's journals of what may have been said. ⍟

6

Clearly Andrée concluded that it was too much of a gamble. At 6.20 the following morning, 14 July, he released further gas and started to bring

---

⍟ The physical and psychological stress that Andrée and his crew faced are vividly illustrated by the balloon flight to the North Pole made by David Hempleman-Adams in 2000. He used the latest propane burners, an autopilot, a GPS satnav, the most up-to-date survival kit, an Iridium mobile phone, a radio link providing a constant stream of updated meteorological data, and a helicopter back-up team. Even so, it took him five days to reach the Pole, and he nearly didn't make it. At one point he fell into exhausted sleep, hallucinated that he had landed, and awoke to find himself climbing out of the basket at thirteen thousand feet, believing the cloudbase to be solid, snow-covered ground. 'Then I wake up. I am standing in the basket, with one leg thrown over the side ... Only the harness is stopping me from jumping out, but I continue to jerk at the reins ... then I realize I am floating several thousand feet above the polar pack ice, one tiny step away from plunging out of the basket ... I feel frightened, really frightened, like no fear I've felt before.' See David Hempleman-Adams, *At the Mercy of the Winds* (2001).

the balloon down on what looked like a relatively smooth section of the ice floe. It was a fearful decision. They had been airborne for less than sixty-five hours, in a balloon designed to fly for thirty days. Now the *Eagle* would never be able to fly again. From this moment they were no longer aerial beings, but were committed to the ice, and to crawling painfully over one of the cruellest surfaces on the planet. It was not an easy landing, as the basket dragged for over an hour, banging, creaking and scouring up snow, until it finally turned over on its side. Nils recorded 'heavy shocks' until they came to a halt at 7.30 a.m.[51] Led by Andrée all three scrambled out, and stood bleakly on the frozen, featureless plateau.

It was a grim prospect. Their basket was no longer their snug home, their organised scientific laboratory, but a piece of wreckage, good simply for salvage. Their huge, proud balloon was a slack, half-deflated dome of tattered silk, crackling with frost and sitting brokenly on the ice, a mere piece of debris. Beyond, there was no sun and no clear horizon in any direction.

It was true that they had travelled 517 miles over the ice, and had landed without injury, and with most of their equipment and instruments intact. That was a genuine achievement. But they were still over four hundred miles short of the Pole. They had not reached Nansen's 'farthest north' point of 86 degrees, or the *Fram*'s at 84 degrees. They were not even quite at 83 degrees north, and barely beyond the point that Parry had reached in 1827. Moreover, they were 216 miles from the nearest land to the south, although they had four months left before the Arctic winter closed in on them. At first they all seemed stunned, and stood in silence, adjusting to their new world.[52]

Nils Strindberg seems to have been the first to react. He pulled the Zeiss camera off the balloon ring, walked clear of the landing site, and took the most memorable sequence of photographs of the entire expedition. The first is taken relatively close, from immediately behind the basket, which lies on its side, snow scooped up around it where it has dragged. Andrée, wearing a flat woollen cap, is in the centre of the frame, gazing down at the tangle of ropes, his back turned and his head bowed. Behind him, in darker clothes, Fraenkel stands with his hands thrust into his pockets, his back also turned but appearing to gaze intently at Andrée. There is a box, possibly a binocular case, flung down in the snow behind him.

The next photograph is taken from further back, and to the right. It shows the *Eagle* sitting upright and half inflated on the ice, a broken,

bulbous shape, very black against the dead glare of the Arctic fog. The gondola is dragged out behind it in a tangle of cords and cables. The sail hangs inert, apparently without its boom. Several guide ropes are still attached, indicating that Andrée never tried to jettison them. A few bits of equipment have now been unloaded, and dumped without order in the snow.

Andrée stands astride the storage section, as if beginning an assess-ment, and trying to reassert command. But Fraenkel still appears immo-bile on the ice, gazing at Andrée, stunned, as if he hasn't yet gathered the force to move. It is the unseen Nils, the youngest, the photographic witness, who has reacted with the most speed and energy. The bleak, almost abstract image that he captures, so poignant in all its human implications, becomes the defining picture of the expedition. It also becomes a larger symbol: the end of the romantic era of ballooning. ᛁ

The rest of their story belongs, as Dr Ekholm had feared and predicted, not to the air but to the ice. On 22 July, after eight days of preparation, they abandoned the *Eagle*'s gondola and the wreckage of the balloon, and began the long march south on the shifting ice floes. Andrée calculated that they had between two hundred and 250 miles to cover, depending on ice drift, and approximately 110 days of Arctic sunlight left in which to cover it. It was not impossible. If they headed south-west they might get back to the Seven Islands, off Spitsbergen; if south-east they might reach Franz Josef Land, where Nansen had overwintered. Both had

---

ᛁ During the rest of the expedition Strindberg took a further 240 photographic expo-sures, of which ninety-three have survived, about half in reasonable condition. His Karl Zeiss cameras worked at one hundredth of a second, and produced large 13×18cm Eastman-Kodak negatives, which could potentially produce brilliant images (like those from Shackleton's Antarctic expedition of 1914). But very few of the pictures appear to have been taken in sunlight. There are no defining shadows, no depth or details. Instead the dead, uniform glare from the ice field gives the images a curiously flat and ghost-like appearance, though this may partly be the result of water damage over thirty years. The collection does not constitute the projected 'photometric survey' of the Arctic. It is a tragic miscellany of survivor snapshots, from which Nils himself is mostly absent. The best are of routine, low-key subjects taken from the middle distance: Andrée standing on the upturned balloon basket; Fraenkel pushing a sledge stuck on a hump of pressure ice; Fraenkel and Nils standing over a shot polar bear; Andrée and Fraenkel breaking camp in the snow. None of the others have the drama of the first two balloon pictures. There are very few close-ups, except for a dead ivory gull pinned to a piece of canvas. There is also a picture of a set of three table forks, the third of which Andrée had laboriously constructed for Fraenkel. This last is perhaps another indication of Strindberg's mischievous humour. Most striking is the total absence of any surviving portraiture. We never see the balloonists' faces. Once they have returned to earth, they become virtually anonymous. The collection contin-ues to be subject to detailed analysis and technical improvement by the Swedish Aeronautical Society.

supply dumps. Andrée initially chose Franz Josef Land, but would later change course and head for the Seven Islands. They cut up a section of the *Eagle*'s silken fabric to protect their tent, and perhaps also as a memento of their great and disastrous flight. Then they mounted everything they could carry on the three sledges and the small collapsible boat, and finally turned their backs on the elusive Pole. But the Pole had not turned its back on them.

Initially they made good progress, of several miles a day. They hunted for food as they went, mostly shooting polar bear. A successful return seemed possible, provided they all kept well, the weather held, and there were no accidents on the perilous ice. Andrée's diary concentrated on the immediate logistics of the expedition, and on keeping his team together, which he did admirably. Fraenkel maintained a daily meteorological journal, stolid and reliable, but without adding much comment. Only Nils Strindberg's thoughts floated further afield and beyond the ice. For the first few weeks, he continued to write his tender letter to his fiancée Anna, promising his imminent return.

Occasionally his reflections turned back to the balloon. He even thought optimistically of a future airborne expedition. After nearly a month he wrote:

15th August. We made very good blood-pancake of bear fat and oatmeal
fried in butter and eaten with butter ... Proposals for alterations in the next
polar-balloon expedition. The drag lines to be sheathed in metal. The car
in the carrying ring. The gas to be somewhat heated by boiling water in the
car and condensing the steam in a sheet-iron vessel in the balloon. The
balloon to be of the same cloth [but] about 6,000 cubic metres [larger] in
volume.[53]

Nils's running letter to Anna was incomplete and undated. But gallantly
it maintains to the end the same light-hearted tone, as if describing some
ordinary holiday jaunt. He gives details of their day, the weather and the
things they talked about. He remains respectful and admiring of Andrée.
He makes no mention at all of the slogging trek, the exhaustion, the
growing terror, or the fact that he had recently slipped off an ice floe and
nearly drowned, and been left behind with the sledges to recover, while
Andrée and Fraenkel went on ahead to check their onward route.

Instead he drifts into a dream of Anna sitting at home in Sweden,
safe and sound. In fact, just like himself sitting on a sledge in the Arctic:
'The weather is pretty bad; wet snow and fog; but we are in good humour.
We have kept up a really pleasant conversation the whole day. Andrée has
spoken about his life, how he entered the Patent Bureau etc. Fraenkel
and Andrée have gone forward on a reconnoitring tour. I stayed with the
sledges, and now I am sitting writing to you. Yes, now [I imagine] you are
having an evening at home; and just like me, you have had a very jolly and
pleasant day.'[54]

But there was one fatal circumstance which only slowly became clear
as Andrée studied Fraenkel's irregular sextant sightings on the declining
sun, on its rare appearances. They were covering their daily distance over
the ice roughly as planned. Yet even as they walked away from the Pole,
they were being drawn back towards it. The Arctic ice pack was itself
drifting northwards almost as swiftly as they walked south. More than
once after a succession of hard days' marching (for example, between 31
July and 3 August), their sextant bearings showed that they had actually
moved several miles north. It was almost as if the North Pole, like a jealous god out of the Norse legends, having glimpsed the Eagle trespassing
on its horizon, was reluctant to let them escape from its icy grasp. They
were moving from science and technology back into the world of myth
and legend.

This agonising process, like trying to walk up a downward escalator, held them in thrall for the next two months. In an attempt to compensate for it, between 4 August and 9 September they angled their direction of march south-westwards, towards Spitsbergen. After covering eighty-one miles in thirty-five days, they found that the ice had also changed direction. They had been carried back virtually the same distance south-eastwards towards Franz Josef Land.[55] The net result was that they were headed almost exactly between their two possible landfalls, into the jaws of the open polar sea.

By mid-September, two months after their landing on the ice, the weather was closing in and the sunlight failing. But at last the Pole seemed to have relinquished its vengeful grip. The pack ice was drifting due south again, and with growing speed. They determined to conserve the last of their energy and provisions, and allow themselves simply to drift with it. Like Nansen before them, they would attempt to overwinter, but this time on the ice. All three of them were physically much weakened by the brutal trekking, the crude diet of bear meat, and the relentless sledge-pulling over the increasingly broken and treacherous ice. They all had frequent diarrhoea, bouts of snow blindness, and open sores on their feet. The athletic Fraenkel could no longer pull his own sledge unaided.

As Dr Ekholm had feared, they were all reaching the end of their tethers. Yet Andrée still thought their morale remained remarkably high: 'Our humour is pretty good, although joking and smiling are not of ordinary occurrence. My young comrades hold out better than I had ventured to hope. The fact that during the last few days we have drifted towards the south at such a rate contributes essentially, I think, to keeping up their courage.'[56]

In fact Fraenkel had completely stopped keeping his meteorological observations, while Nils Strindberg's almanac now consisted almost entirely of lists of meals. A last fragment of his letter to Anna recorded: 'Pull and drudge at the sledges, eat and sleep. The most delightful hour of the day is when one has gone to bed and allows one's thoughts to fly back to better and happier times, but now their immediate goal is where we shall winter ... It is a long time since I chatted with you ...'[57]

On 17 September something astonishing happened. They had their first sight of land since leaving Spitsbergen. It turned out to be tiny White Island, the most extreme north-easterly of the Spitsbergen archipelago, a mere seventeen miles long and dominated by an enormous six-hundred-foot glacier. They were still within the Arctic Circle, about 80 degrees north, and completely isolated in the polar sea. But the possibility of drifting south-westwards past White Island, and then all the way home, now became vivid. On the 18th they held a 'banquet' to celebrate. They ate seal steaks and 'gateau aux raisin with raspberry syrup sauce', drank a bottle of 1834 Antonio de Ferrara port given by the Swedish King, and sang the national anthem.[58] They were not behaving like doomed men.

In preparation for the last part of their journey, they reconnoitred a large ice floe several hundred metres across, and began to construct a snow hut or igloo at its centre. This was designed with much ingenuity by Nils ('our architect'), whose youthful sense of adventure had never faltered, and had become an essential dynamic in holding the team together.

What Nils designed, with a touch of genius, was a sort of imaginary balloon-gondola on ice. He drew it in both plan and elevation.[59] Of course it only had a single storey, and they crawled into it on hands and knees. But like Andrée's original balloon basket, it contained various compartments, for storage, cooking and sleeping. Characteristically, Nils christened the latter, designed for warmth, 'the baking oven'. Their new 'snug home' was scientifically equipped and appointed to last them, if necessary, through the winter until spring 1898. It had a vaulted roof, and even a chimney. It was as if the spirit of the *Eagle* would, after all, float them back to safety, but now in the form of an ice gondola floating on an ice floe.

On 1 October, Andrée sat in the entrance to their new base and made an unusually lyrical note: 'The evening was as divinely beautiful as one could wish. The water was alive with small animals, and a bevy of seven black-white guillemot youngsters was swimming there. A couple of seals too. The work with the snow-hut went on well, and we thought we would have the outside ready by the 2nd ...'[60]

But the polar spirit had not yet finished with them. As in Coleridge's nightmare ballad *The Rime of the Ancient Mariner*, written exactly one hundred years before in 1797, it continued to pursue them beneath the water. On the night of 2 October their ice floe suddenly started to break up. One crack passed directly beneath the side wall of their retreat, splitting it open and revealing a great chasm of icy black water below, which they only just prevented from devouring half their equipment. In the morning they repacked and prepared to take to their boat at a moment's notice. They were all deeply shaken. Andrée made a solemn entry in his diary: 'No one has lost courage. With such comrades one should be able to manage – I may say – under any circumstances.'[61]

The following day, 3 October, it became clear that the perilous condition of the pack ice would force them to land on White Island after all. They had journeyed approximately 190 miles across the surface of the ice from the *Eagle's* landing point. But owing to the ice pack's drift, their position was still about two hundred miles east of the nearest supply dump on Spitsbergen, and well over three hundred miles west of the Cape Flora dump on Franz Josef Land. In both directions lay open Arctic Ocean and lethal pack ice. White Island was now their only hope.

It was a bleak one. Their binoculars revealed a low, windswept, stony and utterly inhospitable shore, partially crusted with snow. Behind it in the ever-fading light reared the looming glacier. But it was still land. They were very weak, yet somehow they managed to transfer all three sledges and most of their remaining equipment, in what must have been a series of half-swamped boat journeys, from the splintering ice floe to the island. They had just enough strength to establish a camp, but it was only about a hundred yards up the beach. The best they could do was find a long, flat shelf of rock, about three feet high, parallel to the beach and facing out to sea. Beneath this they could shelter from the offshore wind, but only when sitting or lying with their backs against it. They were now reduced to the most elemental state of survival.

They arrived in a snowstorm, and the effort clearly exhausted them.[62] They did eventually manage to build a kind of makeshift hut, projecting just a few feet above the rock shelf. But much of their equipment remained scattered and disordered around the campsite. One of the sledges was never properly unpacked, and was left part of the way down the beach. Their written records effectively ended here. Fraenkel had already stopped keeping his scientific observations several days before they built the igloo on the ice floe, and never resumed them.

Andrée's diary is fragmentary, and finally incoherent, after the last great effort of reaching White Island. But there is one note that suggests he decided to name their last shelter after his mother, 'Camp Mina'. Another broken entry runs: 'bad weather and we fear ... we keep in the tent the whole day ... so that we could [work] ... on the hut ... to escape ...'

Perhaps the most strangely eloquent of all is an observation of the gulls which continually flew over their heads. Once these big birds had seemed symbols of freedom and escape: white, airborne and graceful spirits. But now they swooped low, and seemed vicious, noisy and hostile, like vultures: 'They fight, scream and struggle ... seem jealous of us ... no longer give the impression of innocent white doves ... but of outright birds of prey ... carrion [hunters] ...'[63]

Nils Strindberg took no more photographs, or at least none that have survived. Nor did he write again to his beloved Anna Charlier. But his indomitable almanac contains six last entries. They are all very short, but provide a final thread of narrative, and prove that the crew of the *Eagle* were all still alive in mid-October 1897.

> *2nd October. Our ice-floe broke close to the snow-hut*
>   *during the night.*
> *3rd–4th October. Exciting situation.*
> *5th October. Moved to land.*
> *6th October. Snow-storm. Reconnoitring.*
> *7th October. Moving.*
> *17th October. Home 7 o'clock a.m.*[64]

Probably only Nils, with his irrepressible youth and sense of adventure, could have described the sighting of bleak White Island, in the middle of the Arctic Ocean, as an 'exciting situation'. Or their exhausted wanderings around the beach as 'reconnoitring'. Yet he leaves the mystery of where they all were in the ten days between 7 and 17 October tantalisingly unexplained. Where were they 'moving' to? And why did they then come back 'home' at so precise an hour? Andrée's diary fragments throw no light on this either.

The word 'home' has a particularly haunting presence in Nils's almanac. It subtly changes its meaning throughout the four months of the expedition. It is first used to describe the *Eagle*'s gondola, and especially Nils's favourite position up in the balloon hoop, where he feels 'so snug and so at home'. Later, when they are down on the ice, it is used to associate the idea of Anna 'sitting comfortably at home' in Sweden with himself sitting alone on a sledge dreaming of her. Then, when they build the igloo in September, this also becomes 'home', the basket of the imaginary ice balloon that will eventually float them back to safety in the spring. Finally, 'home' seems to become the bare shelf of rock, looking out to sea, on White Island, Camp Mina. 'Home' – with its precise time entry – is the last complete word Nils ever wrote. This is where the last great romantic balloon expedition of the nineteenth century finally came to rest.

7

Yet exactly how the *Eagle* expedition ended only emerged a generation later, when modern aeroplanes already dominated the sky, and balloons seemed a half-forgotten relic of the past. No news had been received from the expedition after Andrée's early messages dropped from the balloon in July 1897, saying that all was well. A number of searches had been launched between 1898 and 1900, some financed by Swedish newspapers, but all without result. The crew's fate was a complete mystery, and with the upheaval of the First World War it largely faded from public consciousness.

Then, quite by chance, in 1930 a Norwegian whaling ship, the *Braatvaag*, anchored for twenty-four hours off the beach at White Island. A shore party casually wandering along the shingle where the snowline had retreated were astonished to come across the half-buried remains of one of the *Eagle*'s sledges. After further searches in the snow, two bodies and various documents were found. As with Captain Scott's doomed South Pole expedition of 1912, all journals, diaries and letters had been deliberately wrapped up and set aside, preserved by the Arctic cold. Most precious of all were Nils Strindberg's sealed tins of exposed photographic film. Through all their hardships he had never abandoned them.

Shortly after, a second expedition carefully uncovered, mapped and recorded the whole campsite. The scene they finally established was this. The skeletons of Andrée and Fraenkel, much pulled about by polar bears, lay within the driftwood frame, constructed beneath the shelf of rock, that Nils had described as 'Home'. Fraenkel lay flat, with open boxes of medical equipment scattered around him. Andrée was half-propped against the rock, a loaded rifle at his side, and their small kerosene cooker neatly placed on the section of rock shelf just above his left shoulder. The cooker still contained traces of kerosene. When pumped and lit, it burnt steadily.

Supplies of tinned food, fuel and ammunition littered around the camp showed that they had not starved. But much equipment had never

been unpacked. Death must have come quite soon, through exhaustion, illness or hypothermia, or a combination of all three. The disposition of objects suggested that Andrée had died last, having administered what medical aid he could to the dying Fraenkel, and then taken up his final half-sitting position as captain on his last watch.

But what had happened to Nils Strindberg? Further investigation of the rock shelf revealed a natural fissure about six feet long, thirty yards or so to the west of the camp. It had been carefully filled with large, flat stones to make a kind of cairn. Beneath them was the body of the third member of the *Eagle*'s crew. It was Nils Strindberg's grave. It had been lovingly and laboriously constructed, at what must have been a terrible cost to the dwindling physical reserves of his two comrades. Indeed, the effort may well have hastened their own deaths. From this it was clear that Nils, the youngest and most cheerful member of the expedition, had died first. What effect this must have had on Andrée and Fraenkel can only be imagined. It may have meant the end of all real hope of survival. But the immense care they took with his body and his personal effects showed how much they valued him to the end.

They wrapped up his letters to Anna, as well as a gold locket with her miniature, and his almanac. After what must have been a most difficult decision, instead of leaving Nils's engagement ring on his finger, they removed it and placed it with the letters. One of Salomon Andrée's last

acts was to put all these precious objects, together with Fraenkel's observation journal and his own unfinished diary, in meticulously sealed wrap-
pings, which he placed next to his own body where they might eventually
be found.

As the final remnants of their camp were collected and analysed in
1930, it became clear that the driftwood frame of their shelter had once
been covered by some sort of fabric. Closer inspection proved that part of
it was cloth from the *Eagle*. So the last thing Andrée's eyes rested upon
may have been the silken balloon canopy, billowing overhead in the Arctic
wind. He may even have dreamed that he was still flying.

Epilogue

———

My cluster of balloon stories appears to end here, down on the lonely winter ice, in gathering cold and darkness, with only death and failure, and falling hopelessly to earth. It might seem very far from those sunlit freedoms of the upper air, the glorious hope and 'hilarity' with which the whole dream began. The *Eagle*'s story certainly makes for a tragic conclusion. In its own way it might appear a strange modern replay of the original Icarus myth, with polar frost replacing solar heat, in Nature's revenge for mankind's eternal hubris.

I for one, surely a *hardened* biographer, have never been able to get Nils and Anna's heartbreaking story out of my head. Anna, incidentally, left Sweden for America, and although she eventually married, she left instructions that her heart should be buried separately with Nils's remains when they were found.[1] Indeed, the experience of the lovers and spouses of balloonists, those who remained on the ground, 'with their hearts in their mouths', has yet to be investigated and told.

But of course Andrée's expedition was very far from being the dead end to the overall ballooning story. Viewed from a proper aerial perspective, something revolutionary had been achieved. Over the brief hundred or so years between 1780 and 1900, through the extraordinary courage and recklessness of such men and women, the momentous idea of manned *flight itself* was at last established, after a hundred thousand years of human evolution. To fly, to inhabit the upper air, to claim our beautiful airy kingdom, could no longer be dismissed as a greedy human aberration, an unnatural trespass upon forbidden territory. It had become instead a proper dimension for the exercise of scientific genius and imagination, a new stage in our planetary evolution; and, one might hope, in our planetary self-knowledge. There truly was, as Félix Nadar put it with his exceptional gift for encapsulating and promoting new concepts, Le Droit au vol – 'The Right to Fly'.

The mechanical business of flight itself was certainly now handed over, via the airship, to the heavier-than-air machine, and within the next hundred years, to the rocket, the satellite, and 'ultimately' to the spaceship. Though neither the Apollo programme nor the current Mars missions are themselves the end, either. There will undoubtedly be further extensions in the forms of interplanetary – if not intergalactic – travel within the next hundred years or so; provided we do not burst our fragile planetary balloon in the meantime.

But the history of balloons has taken a different, and in some ways more subtle and provocative, path. Extreme ballooning, champagne ballooning, was itself only a phase, and dwindled away in the interwar years. But since the 1960s, ballooning has been reborn with the growing popularity of hot-air ballooning, and, it must be said, with its comparative cheapness and safety. Charles Green and the Tissandier brothers would certainly have approved of this development. It has emerged as a breathless form of tourist attraction, flying regularly over historic sites at dawn or dusk; and has become a major international sport, far outpacing the original Victorian 'recreational' coal-gas balloons; and like many modern sports transforming the old ideas of national rivalry.

Balloon fiestas in America have become as popular as book festivals in England; Colorado matches Cheltenham. Both attract increasingly large and knowledgeable crowds, and produce an infectious atmosphere of carnival. Indeed, it strikes me that there are several similarities. For example, the idea of the 'book launch' and the 'balloon launch' (events which Guy de Maupassant was perhaps the first to combine), with all their associated elements of excitement and possible disaster, have many factors in common. Perhaps balloons attract a more sportive, though not necessarily a more youthful, following than books; though both sets of devotees are impressively expert and enthusiastic.

On the other hand, ballooning requires no translation services, and major events take place annually all over Europe, and now increasingly in Africa and India. It might be said, even as Victor Hugo claimed, that the free balloon implies free airspace, an absence of hard borders, and therefore is democratic in its assumptions. Geo can belong to Demos, as François Arago pronounced with a flourish. There are now numerous international balloon fiestas all around the world, as at Velikie Luki, Russia; Rajasthan, India; or Lisburn, Ireland.[2]

Balloons are still used for many modern scientific observations, both peaceful and warlike. There are high-altitude weather balloons. There are balloons for surveying archaeological sites: 'buried' towns, or lost medieval villages, or Iron Age forts. There are military drone balloons, used for example by the US Army in Afghanistan. There are balloons or blimps used for advertising, such as the famous Michelin balloon, or Airabelle herself, who simply advertises milk. There are even modern tethered balloons used to study our environmental impact on the globe, such as those employing low-level automatic cameras over coral reefs.[3]

There is still a fascination with balloon records and 'firsts', often quixotically promoted as scientific research. In 1932, Professor Auguste Piccard, launching from a site near Zurich, rose to 53,149 feet in a pioneering form of pressurised gondola. Subsequent altitude and free-fall (jumping from balloons) records continue to be pursued. Colonel Joe Kittinger jumped from a balloon nineteen miles up over Florida in 1960. Most recently, in October 2012, Felix Baumgartner jumped from 127,852 feet, in a specially pressurised spacesuit, adding to the drama by having the whole event broadcast on live television and over the internet. He remained in free fall for over four minutes. One wonders what Jacques Garnerin would have made of that.

Horizontal records also continue to be challenged. The first crossing of the Atlantic by balloon took place, as we have seen, in 1978. The first non-stop round-the-world balloon flight, by *Breitling Orbiter 3*, succeeded, after many perils and disappointments, in 1999; the bright-red kevlar pressure-cabin gondola is preserved in the Air and Space Museum, Washington, DC. The first crossing of Mount Everest by balloon was achieved by Leo Dickinson in 2009, and produced the most awe-inspiring photographs.

But it should be clear by now that this book is not a conventional history of ballooning. In a sense, it is not really about balloons at all. It is about what balloons gave rise to. It is about the spirit of discovery itself, the extraordinary human drama it produces; and to this there is no end. It also explores how a single, counter-intuitive scientific discovery – that hydrogen can be *weighed* against atmospheric air – can have such huge social and imaginative impact. In this, it is about the meeting of chemistry, physics, engineering and the imagination.

My own lifelong fascination with balloons still puzzles me, ever since that original, childhood flight of fancy. I have written this book partly to find out why this is so. As indicated in some of my footnotes (and footnotes should be like little baskets of helpful provisions slung below the main machine), I have made a few hot-air balloon ascents myself. But this has always been as a passenger, never as a pilot. I think it is the idea, the upwards possibility, as much as the actual activity that attracts me. My heart leaps up when I behold *a dragon in the sky*.

I have mentioned some of my most memorable flights in France, America and Australia. But I also fondly recall the characteristic romance of a beautiful late-September flight, in the dusk over my home county of Norfolk, when the earth darkened below us, and the stars began to come out overhead; and we slipped down to land in a sweetly perfumed field. A field that turned out to be occupied by a large herd of shadowy, but distinctly inhospitable, prize Norfolk pigs.

A Chronological List of Classic Balloon Accounts, Both Fact and Fiction

————

FULL PUBLICATION DETAILS OF EACH ACCOUNT CAN BE FOUND IN
THE BIBLIOGRAPHY, UNDER THE AUTHOR'S NAME

The Marquis d'Arlandes, 'The First Mongolfière Ascent with Pilâtre de Rozier',
   21 November 1783

Dr Alexandre Charles, 'The First Hydrogen Balloon Ascent', 1 December 1783

Vincent Lunardi, 'My First Aerial Voyage in England', September 1784

John Jeffries, 'Across the Channel with Monsieur Blanchard', January 1785

Laetitia Sage, 'The First English Female Aerial Traveller', 1785

Major John Money, 'A Balloon Flight from Norwich to the North Sea', 1785

Tiberius Cavallo, FRS, A Treatise on Aerostation, 1785

Thomas Baldwin, Airopaedia, 1786

Jean-Pierre Blanchard, 'My First Balloon Ascent in America', 1794

Jacques Garnerin, Three Aerial Voyages, 1803

Rudolf Erich Raspe, The Surprising Adventures of Baron Munchausen, 1809

Mary Shelley, The Last Man, 1825

Jane Loudon, The Mummy!, 1827

Edgar Allan Poe, 'The Unparalleled Adventure of One Hans Pfaall', 1835

Monck Mason, Aeronautica, 1838

John Poole, 'Crotchets in the Air', 1838

Edgar Allan Poe, The Trans-Atlantic Balloon Hoax, 1844

Henry Mayhew, 'A Balloon Flight Over London', 1852

Jean Bruno, Les Aventures de Paul enlevé par un ballon, 1858

General George C. Custer, 'War Memoirs of 1862', 1876

Jules Verne, Five Weeks in a Balloon, 1864

Félix Nadar, Memoires du Géant, Paris, 1864

Gaston Tissandier, 'L'Histoire d'un ballon', 1870

James Glaisher, Travels in the Air, London, 1871

Gaston Tissandier, 'Les Ballons du siège de Paris', 1872

Thomas Wise, Through the Air, 1873

Wilfrid de Fonvielle, *Aventures aériennes*, 1876

James Glaisher, 'Address to the Young Men's Christian Association on a high-altitude balloon ascent with Mr Coxwell', 1875

Henry Coxwell, *My Life and Balloon Experiences*, 1887

Thaddeus C. Lowe, *My Balloons in Peace and War*, 1890

H.G. Wells, *The War in the Air*, 1908

Salomon Andrée, *The Andrée Diaries*, 1931

Jean de Brunhoff, *Le Voyage de Babar*, 1932

Anthony Smith, *Throw Out Two Hands*, 1963

Ian McEwan, *Enduring Love*, 1997

David Hempleman-Adams, *At the Mercy of the Winds*, 2001

Illustrations

'Mr Charles Green, the Aeronaut', painted by John Hollins, mezzotint by G.T. Payne, 1838. © National Library of Australia pic-an9548228

Mary Shelley, by Samuel John Stump, 1830. © Photo by Hulton Archive/Getty Images. The identity of this image has not been confirmed by the National Portrait Gallery, London

Jane Loudon, photograph reproduced in *In Search of English Gardens: The Travels of John Claudius Loudon and his Wife Jane*, edited by Priscilla Boniface (Lennard, Wheathampstead, 1987)

Edgar Allan Poe, woodcut by anonymous artist, from *A History of the United States of America* by Horace E. Scudder (Sheldon & Co., New York, 1897). © 2013 University of South Florida

Percy Bysshe Shelley, by Alfred Clint, after Amelia Curran, and Edward Ellerker Williams, oil on canvas, 1819. © National Portrait Gallery, London

Charles Dickens, by anonymous artist, c.1835. © Photo by Hulton Archive/Getty Images

Henry Mayhew, engraving after a daguerreotype by Beard, from an edition of *London Labour and the London Poor*, c.1865. © Photo by Hulton Archive/Getty Images

Félix Nadar, *Auto-portrait*, c.1855. © The J. Paul Getty Museum, Los Angeles

'Mr Glaisher insensible at the height of seven miles', engraving from *Travels in the Air*, edited by James Glaisher, 1871

'Paul takes off in the *Leviathan*', from *Les Aventures de Paul* by Jean Bruno (Bernardin-Béchet, Paris, 1869). Illustration by J. Desandr, 1869. © Courtesy, The Lilly Library, Indiana University, Bloomington, Indiana

'Le Ballon-Poste.' Paris siege poster advertising a weekly 'airmail' newspaper, the *Balloon Post*, containing 'a complete Journal of the week's events, and two columns of Private Correspondance' to be flown out by *Ballon Monté* (manned Balloon), for a subscription price of 20c. 'This week's edition of the *Balloon Post* gives clear and complete instructions on how to send and receive back Answer Postcards by which news and messages may be exchanged with all the departments of France.' With the kind permission of Grosvenor Auctioneers and Valuers

Le Ballon, by Pierre Puvis de Chavannes, oil on panel, 1870. © Mondadori Electa/UIG/age footstock

Major General George Custer, 1865. © Library of Congress, Prints and Photographs Division, LC-DIG-cwpbh-03216

'Thaddeus Lowe and his Balloon', by Mort Künstler, 1991. © Mort Künstler Inc, www.mkunstler.com

'In one bound we pass through the thick layer of cloud', woodcut from *Travels in the Air*, edited by James Glaisher, 1871

Albert (left) and Gaston Tissandier with their balloons *Zénith* (top left), *Jean Bart* (top right) and prototype airship below. © Library of Congress, Prints and Photographs Division, LC-DIG-ppmsca-02274

'Universum', or 'The Pilgrim', engraving imitating a medieval woodcut by Camille Flammarion, 1888, coloured by Hugo Heikenwaelder, 1998. © With the kind permission of Hugo Heikenwaelder

Camille Flammarion aged eighty-two, portrait photograph by Underwood & Underwood, 1924. © Library of Congress, Prints and Photographs Division, LC-USZ62-116545

Jules Verne, 1875 © Photo by Apic/Getty Images

Victor Hugo, 1880s. © Photo by Mondadori Portfolio/Getty Images

James Glaisher, 1875. © Photo by Hulton Archive/Getty Images

H. G. Wells, photo by Reginald Haines c.1908. © Photo/Hulton Archive/Getty Images

David Hempleman-Adams, courtesy of David Hempleman-Adams

Ian McEwan. © Eamonn McCabe

Left to right: Knut Fraenkel, Salomon Andrée and Nils Strindberg before the second polar expedition, by Gösta Flormans, 1897 © National Museum of Science and Technology, Stockholm

Photograph of Fanny Godard in her basket, by Nadar, 1879. © 2004/403/45 Musée de l'Air et de l'Espace, Le Bourget, France

Photograph of Dolly Shepherd, by A.E. Langdon, 1911. © IWM (Q 98454)

'Whimsical American style: a balloon wedding in the clouds', in *La Domenica del Corriere* by Achille Beltrame, 1911. © 2013. A. Dagli Orti/Scala, Florence

Cover of the first edition of *Le Voyage de Babar*, 1932. © Librairie Hachette, 1932/ Courtesy of Aleph-Bet Books

Film tie-in paperback cover of *Enduring Love* (London, Vintage, 1997), 2004. © Random House

The 2010 Albuquerque Fiesta, aerial photograph by Richard Holmes

TEXT ILLUSTRATIONS

4 A cluster balloon flight, Jonathan Trappe, 2010. © Barcroft Media

9 'The Perilous Situation of Major Money', from *Aeronautica Illustrata*, 1785. © Derek Bayes/Lebrecht Music & Arts

11 Nazca balloon flown in Peru by Julian Nott and Jim Woodman, 1975. © Photograph by Larry Dale Gordon

15 The Wetzel and Strelzyk families re-enact their escape, 1979. German Press Agency

17 Dr Jacques Alexandre César Charles receiving a wreath from Apollo, by E.A. Tilly. © Library of Congress, Prints and Photographs Division, LC-DIG-ppmsca-02190

19 'A Balloon Prospect from above the Clouds', from *Airopaedia* by Thomas
     Baldwin, 1785. © Science Museum/Science & Society

21 Philippe Lesueur's letter depicting the Montgolfier ascent from
     Versailles, with animal passengers, 22 September 1783. With the kind
     permission of the Anderson-Abruzzo Albuquerque International Balloon
     Museum

25 George Biggins's Ascent in Lunardi's Balloon, 1784. © Science and Society
     Photo Library/Science Museum/Getty Images

27 Engraving of Cyrano de Bergerac flying to the moon, from *L'Histoire comique
     contenant les états et empires de la Lune*, 1657

28 Earthrise viewed from Apollo 8, 24 December 1968. © NASA

30 'The Battle of the Balloons', printed by Bowles & Carver, c.1784. © The
     Royal Aeronautical Society (National Aerospace Library)/Mary Evans

32 The *Enterprise* at the Battle of Fleurus, 1794. Early flight collecting card
     issued by Romanet & Cie, Paris, 1895. © Library of Congress, Prints and
     Photographs Division, LC-DIG-ppmsca-02562

34 'Invasion Plans' featuring the *Thilorière* balloon, anonymous artist,
     reproduced in *Le Directoire* by Paul Lacroix, 1804. © Mary Evans Picture
     Library

36 'Monsieur Garnerin', drawn and engraved by Edward Hawke-Locker, from a
     sketch made on their aerial voyage, 5 July 1802. © Library of Congress/
     Science Photo Library

37 'An exact representation of Monsieur Garnerin's Balloons', engraving by H.
     Merke. ©The Royal Aeronautical Society (National Aerospace Library)/
     Mary Evans

38 Joseph Louis Gay-Lussac and Jean-Baptiste Biot at 4,000-metre altitude,
     1804. Early flight collecting card issued by Romanet & Cie, Paris, 1895.
     © Library of Congress, Prints and Photographs Division,
     LC-DIG-ppmsca-02561

40 Madame Blanchard, aeronaut, drawn by Jules Porreau. © Library of
     Congress, Prints and Photographs Division, LC-USZ62-97340

43 'Madame Blanchard during the balloon flight in Milan in the presence
     Napoleon on 15 August 1811', drawn by Luigi Rados. © Library of Congress,
     Prints and Photographs Division, LC-DIG-ppmsca-02180

45 Death of Madame Blanchard, 1819. Early flight collecting card issued by
     Romanet & Cie, Paris, 1895. © Library of Congress, Prints and Photographs
     Division, LC-DIG-ppmsca-02561

46 Sophie Blanchard's grave in Père-Lachaise Cemetery. © Gede

49 Death of Harris, 1824. Early flight collecting card, issued by Romanet &
     Cie: Paris, 1895. © Library of Congress, Prints and Photographs Division,
     LC-DIG-ppmsca-02561

272 *Le Ballon-Poste*. Paris siege poster advertising a weekly 'airmail' newspaper, the *Balloon Post*, containing 'a complete Journal of the week's events, and two columns of Private Correspondence' to be flown out by *Ballon Monté* (manned Balloon), for a subscription price of 20c. 'This week's edition of the *Balloon Post* gives clear and complete instructions on how to send and receive back Answer Postcards by which news and messages may be exchanged with all the departments of France.' With the kind permission of Grosvenor Auctioneers and Valuers

273 Letter dated 29 October 1870, rue St-Lazare, Paris, successfully sent by balloon to a firm of bankers in San Francisco, USA. With the kind permission of Siegel Auction Galleries, Inc.

274 *The Departure of Léon Gambetta in the L'Armand-Barbès from the place Saint-Pierre, 7 October 1870*, oil on canvas, by Jules Didier and Jacques Guiaud. © Musée de la Ville de Paris, Musee Carnavalet, Paris/Giraudon/Bridgeman Art Library

275 Léon Gambetta ballooning out of Paris, anonymous engraving, 1870. © Apic/Getty Images

276 Adapted photograph of the departure of Léon Gambetta, 1870, from *The Romance of Ballooning* by Edita Lausanne/Bibliothèque Cantonale et Universitaire, Lausanne

277 A balloon from Paris descends near Dreux pursued by Prussian cavalry, autumn 1870. © The Granger Collection/TopFoto

279 Balloon construction workshop at the Gare d'Orléans, drawing by A. Jahandier, from *Histoire d'un ballon*, by Gaston Tissandier, 1870

286 The projection and copying of the microfilmed siege letters by the Duboscq Megascope, from Jules Claretie's *Histoire de la révolution de 1870-7*, published by *Journal l'Éclipse*, 1872. Courtesy of Ashley Lawrence

287 *Le Pigeon*, by Pierre Puvis de Chavannes, oil on panel, 1871. © Private Collection/Archives Charmet/Bridgeman Art Library

289 Caricature of a defiant Victor Hugo as a hot-air balloon, with his various books cascading from the basket, by Georges Labadie Pilotell, 1870-71. © Victoria and Albert Museum, London

291 Memorial poster of balloon ascents during the siege of Paris, 1870-71, including a call-list of balloons keyed to a map of their landing places. An image of the Norwegian balloon appears bottom right, Prince's Atlantic balloon left. Lithograph by Grandjean et Gascard, c.1870s. © Library of Congress, Prints and Photographs Division, LC-USZC4-10775

296 Cover page of the first issue of *La Nature*, with an engraving by Albert Tissandier, edited by Gaston Tissandier, 1873

297 Gilt-bronze medal commemorating the Paris siege by Charles Jean-Marie Degeorge, released by the French Ministry for War, 1871-72. It shows (right,

342 Knut Fraenkel, Nils Strindberg and the dead polar bear, photographed by
Salomon Andrée, 1897. © Gränna Museum – Polarcenter/Swedish Society
for Anthropology and Geography

349 Towel recovered from the Andrée polar expedition camp at Kvitoya, now in
the Polarmuseet, Tromso, Norway. © Ealdgyth

Acknowledgements

———

For kind permission to consult and refer to manuscripts, rare editions, original illustrations, aerial objects and archives, my most grateful acknowledgements are due to the London Library; the British Library, London; the Science Museum, London; the National Aerospace Library, Royal Aeronautical Society, Farnborough; the Bibliothèque Nationale, Paris; La Musée de l'Air et de l'Espace, le Bourget, Paris; the Library of Congress, Washington, DC; the Smithsonian Library, Washington, DC; the National Air and Space Museum, Washington, DC; the Steven F. Undvar-Hazy Center, Dulles International Airport, Washington, DC; the Anderson-Abruzzo Albuquerque International Balloon Museum, New Mexico, USA; the State Library of New South Wales, Sydney, Australia; and the Andrée Expedition Polar Centre, Gränna, Sweden.

My warmest personal thanks go to Dr Tom Crouch at the NASM, Washington, DC, for his enthusiasm and technical advice, for his definitive work on American ballooning *The Eagle Aloft*, and for shoehorning me into that hot-air balloon at Albuquerque; to Dr Leonard Bruno at the Library of Congress, for all his patience and kindness in the archives; to Lila Vekerdy, at the Smithsonian Library, Washington, DC, for her support and her cocktails; and to Dr Marilee Nason, the inspired Director of the International Balloon Museum, at Albuquerque, New Mexico.

I would also like to thank Doug Millard, at the Science Museum, London; Dr Nancy Gwinn, Director of the Smithsonian Library, Washington, DC; Pierre Lombarde, Directeur, Centre de Documentation at le Bourget; Barbara Kiser, features editor at *Nature*; and to send an airborne greeting to the Thursday Night Group at Café Central, Pennsylvania Avenue, Washington, DC.

I have special scholarly debts to L.T.C. Rolt for his brilliant survey *The Balloonists*; to Professor Clare Brant, King's College, London, for her wonderful 'Ballomania' lecture at the Royal Society in 2007; to Professor Stephan Bann, editor of the stimulating collection *Seeing from Above*, who generously let me see a number of papers before their publication; to Keith Moore, polymathic head of the Library and Information Services, the Royal Society, London; to Dr Tom Spencer at Magdalene College, Cambridge, for hosting the 'Aerial View' session at the Festival Conference in November 2011; to Dr Philip Ball, masterly science writer and lecturer for his swift fly-by of chemical and other matters; and to the Master and Fellows of Churchill College, Cambridge, for electing me to an Honorary Fellowship, and reminding me that engineering and imagination must, more than ever, go arm in arm towards the big ideas for our global future.

I would like to thank several balloon companies for taking me safely aloft, notably Norwich Balloons, Norfolk; Balloons Aloft, Canberra, Australia; Le Dragon Volant, 30360 St-Hippolyte de Caton, France; and above all the Albuquerque International Balloon Fiesta, New Mexico, USA. Among the many skilful and meticulous balloon pilots that it has been my privilege to meet, on *terra firma* or above it, I would particularly like to express my appreciation to Julian Nott; and to Barbara A. Fricke and Peter J. Cuneo, together placed third in the historic 2004 Gordon Bennett Race, and twice joint winners of the America's Challenge Gas Balloon Race, in 2001 and 2010.

A number of wise and learned friends have encouraged me to stay afloat in the strange but fascinating *stratos* between the arts and the sciences: my old colleague Professor Jon Cook, who regularly ascends with me – literally or metaphorically – above the flatlands of East Anglia; Richard Mabey, who inspires me with meteorological lore; my brother Adrian, of Holmes Hobb Marcantonio, and my sister Tessa, of the Elephantpress, for their shrewd advice on design and presentation; Tim Dee of BBC Bristol; Alan Judd of Intelligence Reformed; Professor Kathryn Hughes, director of the UEA Life Writing MA; and my old and valued mentor Professor George Steiner in Cambridge. I also send greetings to my uncle, Squadron Leader D.C. Gordon, now at maximum altitude.

Here on the ground I have again been immensely lucky in my outstanding publishing team at HarperCollins. My thanks and appreciation go to Robert Lacey (words), Joe Zigmond (pictures), Jo Walker (design), Helen Ellis (upper-air trajectories), Douglas Matthews (the king of indexers), and above all to my visionary editor Arabella Pike, who does not suffer from any kind of vertigo. Best thanks to my agent David Godwin in London, and to Dan Frank at Pantheon, New York. Finally, greetings to the now far-flung wild Delancey boys (including the Hong Kong division), and to all at the New Balloon Centre at Queens Park, London. To my beloved Rose Tremain: a heartfelt earthly thank you.

R.H.

Bibliography

ARCHIVES AND MUSEUMS

The London Library
The British Library, London
The Royal Society, London
The Penn-Gaskell Collection, Science Museum, London
The Cuthbert-Hodgson Collection, National Aerospace Library, Royal
    Aeronautical Society, Farnborough, Hampshire
The British Balloon Museum and Library. Internet: http://www.bbml.org.uk/
The Ashby de la Zouche Museum, Leicestershire
Musée de l'Air et de l'Espace, Le Bourget, Paris
Château de Balleroy, Musée de Ballons, 14490 Balleroy, Normandy, France
The State Library of New South Wales, Australia
The Tissandier Collection, Library of Congress, Washington DC
The National Air and Space Museum, Washington DC
The Steven F. Udvar-Hazy Center, Dulles International Airport, Washington
    DC
The Smithsonian Library, Washington DC
The Anderson-Abruzzo Albuquerque International Balloon Museum, New
    Mexico, USA
The Andrée Expedition Polar Centre, Gränna, Sweden

BIBLIOGRAPHY

Aeronautica – see Monck Mason
Salomon Andrée, The Andrée Diaries, Being the Diaries and Records of S.A. Andrée,
    Nils Strindberg and Knut Fraenkel ... Discovered on White Island in 1930.
    Translated by Edward Adams-Ray, London, 1931
The Marquis d'Arlandes, 'My Ascent with Pilâtre de Rozier', Paris, 21 November
    1783; partly reprinted in Astra Castra, and Fontaine
Astra Castra – see Charles Hatton Turnor

Robert Baldick, *The Siege of Paris*, London, 1964

Thomas Baldwin, *Airopaedia, or Narrative of a Balloon Excursion from Chester in 1785*, London, 1786

R.M. Ballantyne, *Up in the Clouds*, London, 1870

Joseph Banks, *The Scientific Correspondence of Joseph Banks*, Vol. 2, 1782–1784, ed. Neil Chambers, Pickering & Chatto, London, 2007

Stephan Bann, 'Nadar's Aerial View', *Seeing From Above: The Aerial View in Visual Culture*, I.B. Tauris online publisher, 2012

Cyrano de Bergerac, *The Comical History of the Empires of the Worlds in the Moon and the Sun*, 1687; from the original French, *Histoire comique ... de la Lune*, 1657; available as *Journey to the Moon*, translated by Andrew Brown, Hesperus Classics, London, 2007

Jean-Pierre Blanchard, *Journal of my Forty-Fifth Ascension and the First in America*, Philadelphia, 1794; partly reprinted in *Astra Castra*

David L. Bristow, *Sky Sailors: True Stories of the Balloon Era*, Farrar Straus Giroux, 2010

F.L. Bruel, *Histoire aéronautique par les monuments*, Paris, 1909

Jean Bruno, *Les Aventures de Paul enlevé par un ballon*, illustrations by J. Desoudré, Paris, Bernardin-Bechet, 1858

Tiberius Cavallo, FRS, *A Treatise on the History and Practice of Aerostation*, London, 1785

Jason Chapman et al., 'Vertical-Looking Radar: A New Tool for Monitoring High-Altitude Insect Migration', *BioScience*, Vol. 53, No. 5, May 2003

Dr Alexandre Charles, *Representation du Globe Aérostatique ... avec le récit de son voyage aérien*, Paris, December 1783; partly reprinted in *Astra Castra*, and Fontaine

I.F. Clarke, *Voices Prophesying War: Future Wars 1763–3749*, OUP, 1993

John D. Cox, *The Storm Watchers: The Turbulent History of Weather Prediction*, 2003

Henry Coxwell, *My Life and Balloon Experiences*, London, 1887

Stephen Crane, *The Red Badge of Courage*, 1895

Tom D. Crouch, *The Eagle Aloft: Two Centuries of Ballooning in America*, Smithsonian Institution, 1983

Michael J. Crowe, *The Extraterrestrial Life Debate 1750–1900*, Dover Books, 1999

General George C. Custer, 'War Memoirs', *The Galaxy: A Magazine of Entertaining Reading*, Vol. XXII, November 1876, pp.685–7; partly reprinted in Crouch, and Evans

Victor Debuchy, *Les Ballons du siège de Paris*, Editions France-Empire, Paris, 1973

Charles Dickens, 'Vauxhall Gardens by Day', *Sketches by Boz*, 1836

Charles Dickens, 'Lying Awake', *Household Words*, 30 October 1852

Linda Donn, *The Little Balloonist*, (a novel about Sophie Blanchard), Penguin USA, 2006

Mark Dorrian, 'On Google Earth', *Seeing From Above: The Aerial View in Visual Culture*, I.B. Tauris online publisher, 2012

Michael Doughty, 'James Glaisher's 1862 Account of Balloon Sickness: Altitude, Decompression Injury, and Hypomexia', *Neurology*, No. 60, 25 March 2003

Arthur B. Evans, *Jules Verne Rediscovered: Didacticism and the Scientific Novel*, Greenwood Press, 1988

Charles M. Evans, *War of the Aeronauts: A History of Ballooning in the Civil War*, Stackpole Books, USA, 2002

John Fisher, *Airlift 1870: The Balloon and Pigeon Post in the Siege of Paris*, Max Parrish, 1965

Camille Flammarion, *L'Astronomie populaire*, Paris, 1880

Camille Flammarion, *L'Atmosphère*, Paris, 1888

Camille Flammarion, see James Glaisher, *Travels in the Air*

Kate Flint, *The Victorians and the Visual Imagination*, CUP, 2000

Raymonde Fontaine, *La Manche en ballon*, Paris, 1982

Wilfrid de Fonvielle, *Aventures aériennes et expériences mémorables*, Paris, 1876

Wilfrid de Fonvielle, see James Glaisher, *Travels in the Air*

Tom Fort, *Under the Weather*, Arrow Books, 2007

Elaine Freedgood, *Victorian Writing About Risk*, CUP, 2000

Théophile Gautier, *Tableaux du siège*, Paris, 1871

Théophile Gautier, *Les Plus belles lettres*, Paris, 1962

Charles Gillispie, *The Montgolfier Brothers*, Princeton UP, 1983

James Glaisher, with Camille Flammarion, Gaston Tissandier and Wilfrid de Fonvielle, *Travels in the Air*, with 125 illustrations, London, 1871

James Glaisher, 'Address to the Young Men's Christian Association', *Good News*, 1875; reprinted in *Astra Castra*

Thor Hanson, *Feathers: The Evolution of a Natural Miracle*, Basic Books, 2012

David Hempleman-Adams, *At the Mercy of the Winds*, Bantam, 2001

J.E. Hodgson, *The History of Aeronautics in Great Britain*, Oxford 1924

Richard Holmes, *Sidetracks: Explorations of a Romantic Biographer*, HarperCollins, 2000

Richard Holmes, *The Age of Wonder*, HarperPress, 2008

Richard Holmes, 'Joseph Banks Goes Ballooning', in *Seeing Further: The Story of Science and the Royal Society*, ed. Bill Bryson, The Royal Society and HarperPress, 2010

Alistair Horne, Chapter 8, 'A Touch of Verne', in *The Fall of Paris*, London, 1965/1981

Richard Hengist Horne, 'Ballooning', *Household Words*, Vol. IV, 25 October 1851

Victor Hugo, *L'Année terrible*, Paris, 1871

Victor Hugo, *Choses vues*, Paris, 1887

J.L. Hunt, 'James Glaisher', *Journal of the Royal Astronomical Society*, Vol. 37, 1996

John Jeffries, A Narrative of Two Aerial Voyages with Monsieur Blanchard as
    Presented to the Royal Society, London, 1786; partly reprinted in Astra
    Castra
Jane Loudon, The Mummy! A Tale of the Twenty-First Century, London, 1827
Thaddeus C. Lowe, My Balloons in Peace and War, 1890. Manuscript Library of
    Congress, 1931; facsimile published USA, 2009
Vincent Lunardi, My First Aerial Voyage in England, London, September 1784;
    and Five Aerial Voyages in Scotland, 1785; partly reprinted in Astra Castra
Michael J. Lynn, The Sublime Invention: Ballooning in Europe 1783–1820, New York,
    2010
George MacBeth, Anna's Book, London, 1983
Ian McEwan, Enduring Love, London, 1997
Richard Mabey, Turned Out Nice Again: On Living With the Weather, Profile Books,
    2013
Paul Maincent, Genèse de la poste aérienne du siège de Paris, Paris, 1951
Paul Maincent, Textes et Documents sur … les ballons du siège, Paris, 1952
Fulgence Marion, Wonderful Balloon Ascents, Paris, 1874
Monck Mason, Aeronautica, or Sketches Illustrative of the Theory and Practice of
    Aerostation, London, 1838
Guy de Maupassant, 'Mlle Fifi', 'Boule de Suif', 'Deux Amis', in Selected Short
    Stories, Penguin Classics, 1971
Album Maupassant, Pléiade, Gallimard, 1987
Henry Mayhew, 'A Balloon Flight over London', Illustrated London News, 18
    September 1852; edited version reprinted in Astra Castra
Major John Money, 'A Balloon flight from Norwich to the North Sea', 23 July
    1785; in Astra Castra
Félix Nadar, Mémoires du Géant, Paris, 1864
Félix Nadar, Sous l'incendie, Paris, 1871
Félix Nadar, Quand j'étais photographe, Paris, 1894
Fridtjof Nansen, Furthest North, 1897
Douglas Palmer, The Complete Earth: A Satellite Portrait of Our Planet, Quercus,
    2006
La Part du rêve: De la Montgolfière au Satellite, Grand Palais exhibition catalogue,
    Paris, 1983
Edgar Allan Poe, 'The Unparalleled Adventure of One Hans Pfaall', New York,
    1835; reprinted in Astra Castra
Edgar Allan Poe, The Trans-Atlantic Balloon Hoax, New York, 1844
John Poole, 'Crotchets in the Air', London, 1838; reprinted in Astra Castra
Jean Prinet and Antoinette Dilasser, Nadar, Collection Kiosque, Librairie
    Armand Colin, Paris, 1966
George Putnam, Andrée: The Record of a Tragic Adventure, New York, 1930

Rudolf Erich Raspe, *The Adventures of Baron Munchausen*, 1786; expanded
   English editions 1809, 1896

Brian Holden Reid, *The Civil War and the Wars of the Nineteenth Century*, series
   editor John Keegan, Smithsonian Books, USA, 2006

Joanna Richardson, *Paris Under Siege: A Journal of the Events of 1870–1871*, London,
   1982

Ann Rinaldi, *Girl in Blue*, USA, 1988

Graham Robb, *Victor Hugo*, Picador, 1997

L.T.C. Rolt, *The Aeronauts: A Dramatic History of the Great Age of Ballooning*,
   London, 1966 (republished as *The Balloonists*, 2006)

Faujas de Saint-Fond, *Description des expériences de la Machine Aérostatique*, Paris,
   1784

Mary Shelley, *The Last Man*, London, 1825

Percy Bysshe Shelley, 'A Defence of Ballooning', in 'Shelley at Oxford', by T.J.
   Hogg, *New Monthly Magazine*, 1832; republished in his unfinished *Life of
   Shelley*, 1858

Percy Bysshe Shelley, 'On a Balloon Laden with Knowledge', poem, 1812

Percy Bysshe Shelley, 'The Witch of Atlas', poem, 1820

Dolly Shepherd (with Peter Hearn), *When the 'Chute Went Up: Adventures of a
   Pioneer Lady Parachutist*, London, 1970

Patrick Stephens, *The Romance of Ballooning: The Story of the Early Aeronauts*,
   Patrick Stephens Ltd with Edita Lausanne, 1971

D.R. Stoddard, *Coral Reefs: Research Methods*, UNESCO, 1987

Per Olof Sundman, *Ingenjör Andrées luftfärd*, Stockholm, 1967

Marie Thébaud-Sorger, 'Thomas Baldwin's Airopaidia, or the Aerial View in
   Colour', *Seeing From Above: The Aerial View in Visual Culture*, I.B. Tauris
   online publisher, 2012

Gaston Tissandier, 'Histoire d'un ballon', *Le Magasin pittoresque*, Tome XXXVIII,
   1870, Tissandier Collection, Library of Congress, Washington DC

Gaston Tissandier, *Voyages aériens*, Paris, 1870

Gaston Tissandier, 'Les Ballons du siege de Paris', *Le Magasin pittoresque*, Paris,
   1872, Tissandier Collection, Library of Congress, Washington DC

Gaston Tissandier, *Histoire de mes ascensions*, Paris, 1878

Gaston Tissandier, *Les Ballons dirigibles*, Paris, 1885

Gaston Tissandier, *Histoire des ballons et des aéronautes célèbres*, 2 vols, Paris,
   1890

Gaston Tissandier, *see* James Glaisher, *Travels in the Air*

Julien Turgau, *Les Ballons: Histoire de la locomotion aérienne*, avec introduction
   par Gérard de Nerval, Paris, 1851

Christopher Hatton Turnor, *Astra Castra: Experiments and Adventures in the
   Atmosphere*, London, 1865

Jules Verne, *Cinq semaines en ballon*, Paris, 1863; translated as *Five Weeks in a Balloon: A Voyage of Discoveries in Africa by Three Englishmen*, London, 1864; Wordsworth Classics, 2002

Jules Verne, *The Mysterious Island*, 1875

Jules Verne, *The Clipper of the Clouds*, 1887

H.G. Wells, *The War in the Air*, 1908, Penguin Classics edn, 2005

Alec Wilkinson, *The Ice Balloon*, Fourth Estate, London, 2012

John Wise, *A System of Aeronautics*, USA, 1850

John Wise, *Through the Air: A Narrative of Forty Years as an Aeronaut*, USA, 1873

William Wordsworth, *Peter Bell*, poem 1819

Stan Yorke, *Weather Forecasting Made Simple*, Countryside Books, 2010

References

CHAPTER 1: THE FALLING DREAM

1 Ovid, *Metamorphoses*, translated by David Raeburn, Penguin Classics, 2004, pp.303–6
2 Carole Rawcliffe and Richard Wilson (eds), *Norwich Since 1500*, Hambledon and London, 2004; Hilaire Belloc, 'A Norfolk Man', *On Something*, 1925; *New Oxford Dictionary of National Biography*, John Money; J.E. Hodgson, *The History of Aeronautics in Great Britain*, Oxford, 1924, pp.179–85
3 *Norwich Since 1500*, pp.80–2
4 Ibid., pp.80–3
5 L.T.C. Rolt, *The Aeronauts: A Dramatic History of the Great Age of Ballooning*, London, 1966 (republished as *The Balloonists*, 2006), p.95
6 J.E. Hodgson, *The History of Aeronautics in Great Britain*, Oxford, 1924, p.183
7 Don Cameron, Preface to L.T.C. Rolt
8 John Milton, *Paradise Lost*, Book 2, lines 1049–55
9 See Julian Nott, official website
10 Sources: 'Freedom Balloon' by John Dornberg, *Popular Mechanics*, February 1980; *Ballonflucht*, Günter Wetzel official website; *Night Crossing*, film treatment
11 Airey Neave, *They Have Their Exits: The First Briton to Make the Home Run from Colditz*, London, 1970; and Pat Reid, *The Latter Days at Colditz*, London 1953
12 See Raymonde Fontaine, *La Manche en ballon*, Paris, 1982, pp.67–72
13 Thomas Baldwin, *Airopaedia, or Narrative of a Balloon Excursion from Chester in 1785*, London, 1786, p.204
14 J.E. Hodgson, *The History of Aeronautics in Great Britain*, Oxford, 1924, pp.132, 131
15 Blanchard, *Journal of my Forty-Fifth Ascension*, Paris, 1793
16 Anderson-Abruzzo Albuquerque International Balloon Museum, Albuquerque, New Mexico

17 David King-Hele, *Erasmus Darwin*, London 1999, p.187

18 Joseph Banks, *The Scientific Correspondence of Joseph Banks*, Vol. 2, 1782–1784, Neil Chambers, Pickering & Chatto, London, 2007, Letter 380

19 Richard Holmes, *The Age of Wonder*, London, 2008, pp.135–6

20 Joseph Banks, Letter 377

21 Richard Hamblyn, *Terra: Tales of the Earth*, London, 2010, p.121

22 See 'Balloon Flying Handbook', *Federal Aviation Authority*, official website; and 'Balloon Lift Statistics', *Chemistry Hawaii*, official website

23 Tiberius Cavallo, FRS, *A Treatise on the History and Practice of Aerostation*, London, 1785, pp.164–5

24 Ibid., pp.192–3

25 Ibid., pp.144–7

26 Ibid., p.189

27 Ibid., p.323

28 Cyrano de Bergerac, *Journey to the Moon*, translated by Andrew Brown, Hesperus Classics, London, 2007, pp.36, 34

29 Cyrano de Bergerac, *Comical History of the Moon*, partially reprinted in *Astra Castra*, pp.390–5 (see Bibliography)

30 See Robert Poole, *Earthrise: How Man First Saw the Earth*, Yale, 2008, pp.1–2

CHAPTER 2: FIERY PROSPECTS

1 Sophia Banks's scrapbook, British Library LR.301.3

2 Ibid., item no. 48c

3 Ibid., items nos 41t, 43t, 43b

4 Reported by Lieutenant G.E. Grover, *Military Ballooning*, 1862, p.9

5 See Wilfrid de Fonvielle, 'La Premiere Compagnie des Aérostiers', *Aventures aériennes*, pp.117–39

6 See I.F. Clarke, *Voices Prophesying War: Future Wars 1763-3749*, OUP, 1993

7 Rolt, pp.104–9

8 Hodgson, pp.218–20

9 Christopher Hatton Turnor, *Astra Castra: Experiments and Adventures in the Atmosphere*, London, 1865, p.115

10 Jacques Garnerin, *Three Aerial Voyages*, 1803; Patrick Stephens, *The Romance of Ballooning: The Story of the Early Aeronauts*, Patrick Stephens Ltd with Edita Lausanne, 1971, p.78

11 John Wise, *Through the Air*, 1873, pp.127, 129

12 Rolt, p.108

13 Raspe, ''The Frolic', Chapter XII, *The Surprising Adventures of Baron Munchausen*, London, 1895

14 *La Part du rêve: De la Montgolfière au Satellite*, Grand Palais exhibition catalogue, Paris, 1983; *Dictionnaire universelle*, 1854, p.216

15 J. Martin, *The Almanac of Women and Minorities in World Politics*, HarperCollins, 2000, p.466

16 A facsimile of this letter, together with several of Lisa Garnerin's posters, appears in F.L. Bruel, *Histoire aéronautique*, 1909

17 *Gentleman's Magazine*, Vol. 89, Part 2, July 1819, pp.76–7

18 John Poole, *Crotchets in the Air*, 1838, pp.79–80; reprinted in *Astra Castra*, pp.399–414

19 Monck Mason, *Aeronautica, or Sketches Illustrative of the Theory and Practice of Aerostation*, London, 1838, pp.261–2; Hodgson, pp.223–4; Rolt, pp.115–16

20 *Aeronautica*, p.263

21 Ibid., p.262; Hodgson, pp.223–4; Rolt, pp.115–16

22 Hodgson, p.224

23 *La Part du rêve*; and Hodgson, p.207, Fig. 65

24 Jane Loudon, *The Mummy!*, London, 1827, pp.50, 123, 217

25 Ibid., p.83

CHAPTER 3: AIRY KINGDOMS

1 Elaine Freedgood, *Victorian Writing About Risk*, CUP, 2000, pp.74–81

2 Mary Shelley, *The Last Man*, Vol. 1, 1825

3 See Mary Shelley, *Journal*, 4 August 1816

4 Mayhew in *Astra Castra*, pp.223ff. The largely satirical accounts of flying with Green, by Poole and Smith, can also be found in *Astra Castra*

5 Rolt, p.120

6 *Aeronautica*, pp.151–2

7 Rolt, pp.117–21

8 New Zealand government: Transport Accident Investigation Commission (TAIC) Interim Report No. 12-001, Carterton, 7 January 2012

9 *Aeronautica*, p.49; Rolt, p.124

10 *Aeronautica*, p.33

11 Ibid., p.40

12 Ibid., p.46

13 Ibid., p.49

14 Ibid., pp.52–3

15 Gaston Flammarion, in *Travels in the Air*, London, 1871, p.207

16 *Aeronautica*, p.55

17 Ibid., p.57

18 Ibid.

19 Ibid., p.59

20 Ibid., pp.59–60; there is a slightly different version in Stephens, p.89

21 Edgar Allan Poe, 'The Unparalleled Adventure of One Hans Pfaall', 1835; in *Astra Castra*

22 For the view that Poe rather than Mary Shelley was the father/mother of science fiction, see Adam Roberts, 'An Infinity Plus Introduction to Hans Pfaall', 2002. Internet website *Infinityplus*

23 *Aeronautica*, pp.62–3

24 Ibid., p.65

25 Ibid., pp.68, 70. William Parry made four major expeditions into the Arctic, in 1819, 1821, 1824 and 1827

26 Ibid., pp.66–7

27 Ibid.

28 Ibid., pp.66–8 and footnote

29 Ibid., p.77 and the not entirely reassuring footnote by Mason

30 Rolt, p.124

31 Thomas Hood, 'Ode to Messrs Green ...', poem, 1836

32 *Aeronautica*, pp.175–86

33 Ibid., p.171

34 Ibid., p.183

35 Elaine Freedgood, *Victorian Writing About Risk*, CUP, 2000, pp.74–81

36 John Poole, ' Crotchets in the Air', 1838; reprinted in *Astra Castra*, pp.408–10

37 Tennyson, 'Locksley Hall', 1842, lines 119ff

38 *Aeronautica*, p.26

39 Ibid., p.21

40 Charles Green interview, *Astra Castra*, pp.179–80

41 Daniel Burgoyne, 'Coleridge and Poe's Scientific Faith', *Romanticism on the Net*, February 2001

CHAPTER 4: ANGEL'S EYE

1 Mayhew, *London Labour and the London Poor*, 1851, pp.295–7

2 *Illustrated London News*, 18 September 1852

3 Charles Dickens, 'Vauxhall Gardens by Day', *Sketches by Boz*, 1836

4 Charles Dickens, *Household Words*, Contents Index, British Library X981/10221

5 Richard Hengist Horne, 'Ballooning', *Household Words*, Vol. IV, 25 October 1851

6 Charles Dickens, 'Lying Awake', *Household Words*, 30 October 1852

7 Anderson-Abruzzo Albuquerque International Balloon Museum, Albuquerque, New Mexico

8 'Newcastle on Fire', *Illustrated London News*, 14 October 1854

9 David Coke and Alan Borg, *Vauxhall Gardens: A History*, Yale UP, 2011

CHAPTER 5: WILD WEST WIND

1 Tom D. Crouch, *The Eagle Aloft: Two Centuries of Ballooning in America*, Smithsonian Institution, 1983, pp.222–4

2 Ibid., p.224

3 Ibid., pp.234–5

4 Ibid., pp.227–9

5 Wise, *Through the Air*, 1873, pp.27–31

6 Ibid., p.248; Crouch, pp.183–4

7 Wise, *A System of Aeronautics*, 1850, pp.260–1; Crouch, pp.186–7

8 Wise, 1850, p.261

9 Charles M. Evans, *War of the Aeronauts: A History of Ballooning in the Civil War*, Stackpole Books, USA, 2002, p.31

10 Crouch, p.189

11 Ibid., pp.197ff

12 Wise, 1873, Chapter XLV, p.530

13 Wise, 1850, p.261

14 Rolt, p.141; Crouch, p.248

15 Brian Holden Reid, *The Civil War and the Wars of the Nineteenth Century*, series editor John Keegan, Smithsonian Books, USA, 2006

16 Crouch, p.249

17 Rolt, pp.141–2

18 Wise, 1873, p.494

19 Crouch, p.689

20 Wise, 1873, p.510

21 Crouch, p.254

22 Wise, 1873, pp.493–4

23 Ibid., p.499

24 Ibid., p.508

25 Charles Dickens diary – letter to John Forster, see Forster, *Life of Charles Dickens*, Vol. 3, 1872–74, pp.240–60, 1842

26 Stephens, p.96

27 Wise, 1873, p.504

28 Ibid., pp.504–7

29 Ibid., p.507

30 Ibid., p.508

31 Stephens, p.96; Rolt, p.141

32 Wise, 1873, p.508; Crouch, p.252

33 Wise, 1873, p.509

34 Wise, 1873, pp.508–10; Crouch, pp.252–3;

35 Wise, 1873, p.513

36 Ibid., p.510
37 Ibid., p.514
38 Ibid., p.518
39 Crouch, p.254
40 Wise, 1873, pp.517–18
41 Crouch, pp.255–61
42 Evans, *War of the Aeronauts*, p.1
43 Lowe, *My Balloons in Peace and War*, p.3, quoted Evans, *War of the Aeronauts*, p.39
44 Lowe, *My Balloons*, quoted Crouch, p.264
45 Crouch, p.264
46 Ibid., p.275
47 Ibid., p.276
48 Lowe, *My Balloons*, pp.32–4
49 Crouch, p.277

CHAPTER 6: SPIES IN THE SKY

 1 Evans, *War of the Aeronauts*, p.63
 2 Reid, p.77. West Virginia refused to secede and was admitted to the Union in 1863
 3 Ibid., p.86
 4 Ibid., p.81
 5 Evans, *War of the Aeronauts*, p.261
 6 Crouch, p.368
 7 Evans, *War of the Aeronauts*, p.9
 8 Crouch, p.277
 9 Ms of Lowe's letter held in Library of Congress, and quoted Crouch, p.346
10 Lowe, *My Balloons*, p.69; Evans, *War of the Aeronauts*, p.73
11 Evans, *War of the Aeronauts*, p.85
12 Crouch, pp.343–4
13 Evans, *War of the Aeronauts*, p.130
14 Ibid.
15 Ibid., pp.98–9, 130
16 Ibid., p.143
17 Lowe, *My Balloons*, p.194
18 Reid, pp.77–80
19 Evans, *War of the Aeronauts*, p.133
20 For more on Lowe's wartime experiences, see Gail Jarrow, *Lincoln's Flying Spies*, USA, 2010; and Stephen Poleskie, *The Balloonist: The Story of T.S.C. Lowe*, USA, 2007, a partly fictionalised biography
21 Lowe, *My Balloons*, p.113

22 *Detroit Press*, 'The Yankee Balloon', 1886, quoted ibid.

23 George Townsend (newspaper reporter), 'Fitzjohn Porter Views the Confederate Army from a Balloon', quoted in Henry Steele Commager, *The Blue and the Gray*, Vol. 1, 1950

24 Lowe, *My Balloons*, pp.29–30

25 Evans, *War of the Aeronauts*, p.184

26 Lowe, *My Balloons*, p.32

27 Ibid., passim

28 Lowe, 'The Balloons with the Army of the Potomac: A Personal Reminiscence', p.3; quoted in Evans, *War of the Aeronauts*, p.329. See also the Lowe website at civilwarhome.com

29 Lowe, *My Balloons*, p.86

30 Henry Coxwell, *My Life and Balloon Experiences*, London, 1887, p.178

31 Evans, *War of the Aeronauts*, p.167. However, it is difficult to believe that some aerial photographs will not eventually be found in the ever-expanding Civil War archives

32 Lowe, 'The Balloons with the Army of the Potomac', p2

33 Ibid., p.1

34 Reid, p.92

35 Crouch, p.387

36 Evans, *War of the Aeronauts*, pp.168–9

37 *St James's Magazine*, 1863, pp.96–105, quoted Crouch, p.387

38 Lieutenant G.E. Grover, *Military Ballooning*, 1862, p.21; also quoted in Coxwell, p.178

39 Frederick Beaumont and George Grover, *On Balloon Reconnaissances*, 1863, British Library X639/1795

40 General George C. Custer, 'War Memoirs', *The Galaxy: A Magazine of Entertaining Reading*, Vol. XXII, November 1876, pp.685–7; see also Evans, *War of the Aeronauts*, pp.184–8; Crouch, pp.383–6 and note p.699

41 Evans, *War of the Aeronauts*, pp.205–6. The documentary discovery of the real 'silk dress balloon' must largely be credited to the brilliant archival research of Charles M. Evans

42 Cheeves to Longstreet, 1896, from ibid., p.206

43 Evans, *War of the Aeronauts*, p.236

44 Crouch, p.394

45 Lowe, *My Balloons*, p.143

46 Evans, *War of the Aeronauts*, p.237

47 Ibid., p.222

48 Ibid., p.223

49 Ibid., p.235

50 Ibid., p.228; Lowe, *My Balloons*, p.135

51 Evans, *War of the Aeronauts*, p.240

52 Opinion of Major-General A.W. Creeley, cited in Lowe, *My Balloons*, p.140

53 Evans, *War of the Aeronauts*, p.243

54 Ibid., pp.242–3

55 Lowe, *My Balloons*, p.145

56 Ann Rinaldi, *Girl in Blue*, USA, 1988, pp.199–200

57 General James Longstreet, *Century Magazine*, 1896, quoted Crouch, p.393

58 Walt Whitman, *Song of Myself*, Part 33, composed c.1867, partly about the Civil War: 'By the cot in the hospital reaching lemonade to a feverish patient ...'

59 Stephen Crane, 'The Price of the Harness', *Scribner's Magazine*, September 1898

60 Stephen Crane, *The Red Badge of Courage*, 1895

61 Evans, *War of the Aeronauts*, p.299

62 Lowe, *My Balloons*, pp.204–5

63 Ibid., p.208

64 Evans, *War of the Aeronauts*, p.308

65 Ibid., p.309

66 Ibid.

CHAPTER 7: GIGANTIC VOYAGES

1 Richard Holmes, 'Monsieur Nadar', *Sidetracks: Explorations of a Romantic Biographer*, HarperCollins, London, 2000, pp.57–8

2 Jean Prinet and Antoinette Dilasser, *Nadar*, Collection Kiosque, Librairie Armand Colin, Paris, 1966, p.124

3 Holmes, 'Monsieur Nadar'

4 Félix Nadar, *Quand j'étais photographe*, 1894, p.121

5 Ibid.; Prinet and Dilasser, pp.134–5

6 Nadar, *Quand j'étais photographe*, pp.113–14; see Stephan Bann, 'Nadar's Aerial View', *Seeing from Above; The Aerial View in Visual Culture*, I.B. Tauris online publisher, 2012

7 Prinet and Dilasser, p.135

8 Nadar, *Quand j'étais photographe*, 1894, p.121

9 Prinet and Dilasser, p.140

10 Félix Nadar, *Mémoires du Géant*, 1864, pp.24–7

11 Prinet and Dilasser, pp.145, 142; cover of *L'Aéronaute* in *La Part du rêve: De la Montgolfière au Satellite*, Grand Palais exhibition catalogue, Paris, 1983, p.141

12 Nadar, from *L'Aéronaute*, 1863, quoted R.M. Ballantyne, *Up in the Clouds*, 1870

13 Prinet and Dilasser, p.150

14 Rolt, pp.145–50; *L'Aéronaute*, 1863, quoted in R.M. Ballantyne, *Up in the Clouds*, London, 1870

15 *L'Aéronaute*, 1863, quoted in Ballantyne

16 Prinet and Dilasser, p.150; Stephens, p.101; *La Part du rêve* p.141; Nadar, *Mémoires du Géant*

17 Prinet and Dilasser, p.150

18 Rolt, p.147

19 Nadar, *Mémoires du Géant*; see commentary by Stephen Bann

20 Nadar, *Mémoires du Géant*; Stephens, p.103

21 Nadar, *Mémoires du Géant*; Stephens, p.101

22 Prinet and Dilasser, p.154

23 Nadar, *Mémoires du Géant*, pp.352–84

24 *Notre Dame de Paris*, 1831; Prinet and Dilasser, pp.154, 163

25 Ibid., pp.155–61

26 Ibid., p.157

27 Ibid., p.158

28 Hugo quoted ibid.; see I.F. Clarke, *Voices Prophesying War: Future Wars 1763–3749*, OUP, 1993, p.3

29 Quoted Prinet and Dilasser, pp.157–8, translation Richard Holmes

30 Quoted ibid., p.158

31 Quoted ibid., p.160

32 Quoted ibid., pp.158–9

33 Quoted ibid., p.161

34 Quoted ibid., p.162

35 Fonvielle, in *Travels in the Air*, London, 1871, p.233

36 Ibid., pp.251–3

37 Prinet and Dilasser, p.161

38 Arthur B. Evans, *Jules Verne Rediscovered: Didacticism and the Scientific Novel*, Greenwood Press, 1988, pp.18–20

39 Stephens, p.160

40 Arthur Evans, *Jules Verne Rediscovered*, p.20

41 See Arthur B. Evans, 'The "New" Jules Verne', *Science-Fiction Studies*, XXII: 1, No. 65, March 1995, pp.35–46

42 Arthur Evans, *Jules Verne Rediscovered*, p.21

43 Ibid., p.20

44 Jules Verne, *Five Weeks in a Balloon*, 1864; Wordsworth Classics, 2002, pp.166–7

45 Ibid., pp.172, 177

46 Ibid., pp.184, 185

47 Ibid., p.166

48 Ibid., p.188

49 Ibid., p.176

50 Percy Bysshe Shelley, 'A Defence of Ballooning', in 'Shelley at Oxford', by T.J. Hogg, *New Monthly Magazine*, 1832; republished in Hogg's unfinished *Life of Shelley*, 1858

51 Verne, *Five Weeks in a Balloon*, Wordsworth Classics, 2002, p.226

52 Ibid., p.207

53 Ibid., p.354

54 Ibid., p.213

55 Ibid., p.354

CHAPTER 8: VERTICAL EXPLORATIONS

1 Cornelius O'Dowd (sic) in *Blackwood's Magazine*, October 1864; reprinted in *Astra Castra*, p.434

2 Rolt, pp.188–9

3 Glaisher, *Travels in the Air*, pp.27–8; Hodgson, p.21; Rolt, pp.188–9

4 Rolt, p.191

5 Fonvielle, in *Travels in the Air*, London, 1871, p.329. Many of Charles Green's papers are now in the Cuthbert-Hodgson Collection, National Aerospace Library, Royal Aeronautical Society, Farnborough, Hampshire

6 Ibid., p.330

7 Ibid.

8 Ibid., pp.330–1

9 Hodgson, p.268; Rolt, p.191

10 J.L. Hunt, 'James Glaisher', *Journal of the Royal Astronomical Society*, Vol. 37, 1996; Rolt, p.192

11 Glaisher, *Travels in the Air*, pp.34–5; Rolt, Appendix, pp.248–50

12 Glaisher, *Travels in the Air*, p.42; *Astra Castra*, pp.228–9

13 Glaisher, *Travels in the Air*, p.22

14 Ibid.

15 Rolt, p.192

16 Tom Fort, *Under the Weather*, Arrow Books, 2007, pp.220–2

17 Rolt, pp.190, 192; J.L. Hunt, p.328

18 Glaisher, *Travels in the Air*, p.31

19 Ibid., pp.31, 33

20 Ibid., pp.38–40

21 Rolt, p.250

22 J.L. Hunt, p.327

23 Ibid.

24 Glaisher, *Travels in the Air*, p.43

25 Ibid., pp.44–7

26 Stan Yorke, *Weather Forecasting Made Simple*, 2010, p.46

27 See ibid., chapter 'Old Tales', pp.56–7

28 Edmund Halley's weather chart is held by the National Meteorological Archive, Great Moor House, Exeter

29 Dove's 'Law of Storms', 1858, described in John D. Cox, *The Storm Watchers: The Turbulent History of Weather Prediction*, 2003, p78

30 John Ruskin, quoted in Fort, p.234

31 Fort, pp.218–19

32 Ibid., p.225

33 Ibid.

34 Glaisher, *Travels in the Air*, pp.50–5; *Astra Castra*, pp.385–9

35 *The Times*, 10 September 1862, p.10

36 J.L. Hunt, p.329

37 Glaisher, *Travels in the Air*, p.54

38 *The Times*, 10 September 1862, p.10

39 Ibid.

40 Glaisher, *Travels in the Air*, p.57

41 *The Times*, 10 September 1862, p.10

42 See Michael Doughty, 'James Glaisher's 1862 Account of Balloon Sickness: Altitude, Decompression Injury, and Hypomexia', *Neurology*, No. 60, 25 March 2003

43 Glaisher, *Travels in the Air*, pp.60–1

44 *The Times*, 11 September 1862, p.8

45 Gaston Tissandier, *Histoire de mes ascensions: Récit de vingt-quatre voyages aériens*, Paris, 1868–77

46 Glaisher, *Travels in the Air*, pp.71–2

47 Ibid., pp.62, 50

48 J.L. Hunt, p.327

49 Glaisher, *Travels in the Air*, pp.84–94

50 J.L. Hunt, p.329

51 John D. Cox, *The Storm Watchers*, pp.20–3

52 Michael Doughty, 2003

53 Glaisher, *Travels*, pp.61–2

54 Hatton Turnor to *The Times*, 12 July 1863, in *Astra Castra*, p.245

55 Glaisher, *Travels in the Air*, p.62

56 Ibid., pp.79, 100

57 Ibid., pp.99–100

58 Ibid., pp.81–2

59 James Glaisher, 'Address to the Young Men's Christian Association', *Good News*, 1875; reprinted in *Astra Castra*, pp.387–8

60 Ibid., pp.386–9

CHAPTER 9: MARINERS OF THE UPPER ATMOSPHERE

1 Fonvielle, in *Travels in the Air*, London, 1871, p.265
2 Gaston Flammarion, in *Travels in the Air*, London, 1871, p.112
3 Ibid., p.120
4 Michael J. Crowe, *The Extra-Terrestrial Life Debate*, 1998, pp.378–9
5 R.H. Sherard, 'Flammarion the Astronomer', in *McClure's Magazine*, 1894
6 Crowe, p.386
7 Ibid., p.383
8 Camille Flammarion, *L'Astronomie populaire*, Paris, 1880
9 Camille Flammarion, *L'Atmosphère*, Paris, 1888, p.163
10 Sherard, 1894
11 Flammarion, in *Travels in the Air*, London, 1871, p.105
12 Ibid.
13 Ibid., p.122
14 Ibid.
15 Ibid., pp.123, 142, 106
16 Stan Yorke, *Weather Forecasting Made Simple*, Countryside Books, 2010, pp.6–8
17 Flammarion, in *Travels in the Air*, London, 1871, p.111
18 Ibid., p.123
19 Ibid., p.179
20 Ibid., pp.123, 136
21 Ibid., p.174
22 Ibid., p.136
23 Ibid., p.140
24 Jason Chapman et al., 'Vertical-Looking Radar: A New Tool for Monitoring High-Altitude Insect Migration', *BioScience*, Vol. 53, No. 5, May 2003
25 Flammarion, in *Travels in the Air*, London, 1871, pp.154, 160
26 Ibid., pp.128, 140, 136
27 Ibid., pp.169, 166
28 Ibid., pp.179, 183–4
29 Ibid., pp.120, 147
30 Ibid., pp.147–8
31 Tissandier, in *Travels in the Air*, London, 1871, p.291
32 Ibid.
33 Ibid., p.292
34 Ibid.
35 Ibid., p.295
36 Ibid.
37 Ibid., p.296
38 Ibid.

39 Ibid., pp.296-7
40 Ibid., p.301
41 Ibid., pp.305-6
42 Ibid., p.311
43 Ibid., p.322
44 Ibid., pp.350-3
45 Gaston Tissandier, 'Histoire d'un ballon', *Le Magasin pittoresque*, Tome XXXVIII, 1870, Tissandier Collection, Library of Congress, Washington
46 Ibid. 'Histoire d'un ballon', Chapitre XII, 'Les Courants d'air'
47 Gaston Tissandier, *Voyages aériens*, Paris 1870
48 Tissandier, in *Travels in the Air*, London, 1871, p.398
49 The copy of *Travels in the Air*, with Glaisher's manuscript comments, is held at the Royal Astronomical Society; J.L. Hunt, p.332

CHAPTER 10: PARIS AIRBORNE

1 Paul Maincent, *Genèse de la poste aérienne du siège de Paris*, Paris, 1951, p.58
2 *The Times*, original paper copies held in the London Library archive
3 Robert Baldick, *The Siege of Paris*, Batsford, 1964, p.28
4 *The Times*, 17 September 1871
5 Maincent, p.60
6 Henry Labouchère, in Baldick, p.28
7 Baldick, pp.29-31
8 Gaston Tissandier, 'Les Ballons du siège de Paris', *Magasin pittoresque*, Tome XL, 1872, p.1, Library of Congress Archive, Tissandier Box 11
9 Baldick, p.128
10 Théophile Gautier, *Les Plus belles lettres de Théophile Gautier*, présentées par Pierre Descaves, Calmann-Levy, Paris, 1962, p.139
11 Baldick, p.145
12 *The Times*, 3 October 1870, p.9
13 Victor Debuchy, *Les Ballons du siège de Paris*, Paris, 1973, Annexe: 'Envols des ballons', pp.408-22
14 Gaston Tissandier, 'Les Ballons du siège de Paris', *Magasin pittoresque*, Paris, 1872, p.i
15 Prinet and Dilasser, pp.173, 187; Maincent, p.56
16 Prinet and Dilasser, p.175; Maincent, p.66
17 Prinet and Dilasser, p.174; Maincent, p.60
18 Maincent, p.66
19 Ibid., p.91
20 Debuchy, p.408
21 Prinet and Dilasser, p.173
22 Maincent, p.102

23 Fonvielle, *Aventures Aériennes et expériences mémorables*, 1876, p.364

24 John Fisher, *Airlift 1870: The Balloon and Pigeon Post in the Siege of Paris*, Max Parrish, 1965, p.21; Fonvielle, *Aventures aériennes*, p.364

25 *The Times*, 28 September 1871, p.5, partly quoted in Prinet and Dilasser, pp.179–80

26 Fisher, p.23

27 Tissandier, 'Les Ballons du siège de Paris', p.2

28 Prinet and Dilasser, p.184

29 Ibid., pp.182–3

30 Fisher, p.34

31 Hugo, *Choses vues*, in Joanna Richardson, *Paris Under Siege: A Journal of the Events of 1870–1871*, London, 1982

32 Fisher, p.36

33 Gambetta's despatch quoted Stephens, p.107

34 Account taken from *Moniteur Universel* (Tours edition), 7 October 1870; Stephens, pp.106–7; and Tissandier, 'Les Ballons du siège de Paris'

35 Rolt, p.174

36 For a complete list of all siege balloons, with their launch dates and a landing map, see Stephens, pp.106–10

37 Prinet and Dilasser, p.184

38 Baldick, p.126

39 David L. Bristow, *Sky Sailors*, Farrar Straus Giroux, 2010, pp.89–91

40 Debuchy, pp.236–8; Rolt, p.176

41 Théophile Gautier, *Tableaux du siège*, Paris, 1871, p.42

42 Tissandier, 'Les Ballons du siège de Paris', p.11

43 Baldick, pp.114–15

44 Prinet and Dilasser, p.186

45 Baldick, pp.116–18

46 Ibid., pp.117–18; Fisher, pp.70–2; Rolt, p.175; Prinet and Dilasser, p.186. Not surprisingly, each gives slightly different figures (see n.48, below). The whole process is described by Nadar in *Quand j'étais photographe*

47 Prinet and Dilasser, p.186; Baldick, p.116

48 Statistics from Baldick, p.118, and Rolt, p.177. Victor Debuchy estimates 407 carrier pigeons released, carrying 95,642 individual messages, with seventy-three pigeons making it home. But he makes no specific claim for the total number of messages actually delivered by them. As with all wartime statistics, these figures have to be treated with caution. Debuchy, pp.402–4

49 Baldick, p.120

50 Richardson, p.92

51 '*Une Bombe aux feuillantes*', in Victor Hugo, *L'Année terrible*, Paris, 1871, p.126

52 Hugo, pp.116–18

53 Fisher, p.128

54 Hugo, pp.116 –17; also Fisher, p.129

55 This irony of war is discussed by Graham Robb, *Victor Hugo*, Picador, 1997, p.145

56 Fisher, p.139

57 Debuchy, Annexe: 'Les Ballons'; Stephens, flight list, pp.106–10; Rolt, p.176

58 Coxwell, pp.179–82

59 Rolt, p.176

60 Debuchy, pp.224–36

61 Coxwell, p.181; Rolt, p.176; Debuchy, pp.398–9

62 Prinet and Dilasser, p.184

63 Fulgence Marion, *Wonderful Balloon Ascents*, Paris, London and New York, 1874, p.220

64 J.L. Hunt, p.333

65 Rolt, p.177

66 Fisher, fig. X

67 Rolt, pp.198–9; *La Part du rêve*, 1983

68 Rolt, p.152

CHAPTER 11: EXTREME BALLOONS

1 See Thor Hanson, *Feathers: The Evolution of a Natural Miracle*, Basic Books, 2012

2 Rolt, p.214

3 *Album Maupassant*, Pléiade, Gallimard, 1987, pp.248–9, 186

4 H.G. Wells, *The War in the Air*, 1908, Chapter 3, 'The Balloon', Penguin Classics, 2005, p.53

5 Ibid., Chapter 4, 'The German Air-Fleet', p.79

6 All posters from the Anderson-Abruzzo Albuquerque International Balloon Museum, Albuquerque, New Mexico

7 Dolly Shepherd (with Peter Hearn), *When the 'Chute Went Up: Adventures of a Pioneer Lady Parachutist*, London, 1970. See also the Ashby de la Zouche Museum, Leicestershire. Dolly lived to raise a family, serve in the First World War, and fly (but not jump) with the Red Devils parachute team in the 1980s

8 Ibid., p.129

9 Ibid., pp.48–9

10 Salomon Andrée, *The Andrée Diaries, Being the Diaries and Records of S.A. Andrée, Nils Strindberg and Knut Fraenkel ... Discovered on White Island in 1930*. Translated by Edward Adams-Ray, London, 1931, pp.6–10

11 Alec Wilkinson, *The Ice Balloon*, Fourth Estate, 2012, p.39

12 Ibid., p.15
13 David Hempleman-Adams, *At the Mercy of the Winds*, Bantam, 2001
14 *Andrée Diaries*, pp.9–11
15 Rolt, p.153
16 George Putnam, *Andrée: The Record of a Tragic Adventure*, New York, 1930, p.87
17 *Andrée Diaries*, p.28
18 Putnam, pp.88–118
19 *Andrée Diaries*, p.30
20 Ibid., pp.35–6
21 Ibid., pp.36, 34
22 Ibid., p.36
23 Ibid., pp.29–31; Wilkinson, p.88; Rolt, pp.153–5
24 Rolt, pp.153–4
25 *Andrée Diaries*, p.31
26 Ibid., p.35
27 Ibid., p.38
28 Putnam, pp.75–6
29 Nansen quoted ibid.; and Wilkinson, p.130
30 Wilkinson, p.162
31 *Andrée Diaries*, p.68
32 Ibid., p.ix
33 Strindberg almanac, ibid., p.419
34 Ibid., p.422
35 *Andrée Diaries*, p.111
36 Ibid., pp.76–7
37 'Height of the Balloon during its Flight', a graph profile, *Andrée Diaries*, p.81
38 Strindberg almanac, ibid., p.428
39 Ibid.; *Andrée Diaries*, p.84
40 *Andrée Diaries*, p.348
41 Strindberg, 'Letter to Anna', ibid., p.431
42 Strindberg almanac, *Andrée Diaries*, pp.428–9; *Andrée Diaries*, p.112
43 *Andrée Diaries*, p.353
44 Ibid., p.352
45 Ibid., p.353
46 Ibid., pp.352–3
47 Ibid., p.88
48 Rolt, p.155
49 Strindberg almanac, *Andrée Diaries*, p.433
50 *Andrée Diaries*, p.92
51 Ibid., p.95; Strindberg almanac, ibid., p.434

52 Rolt, p.157

53 Strindberg almanac, *Andrée Diaries*, p.443

54 Strindberg 'Letter to Anna', ibid., p.451

55 *Andrée Diaries*, p.184; and see 'Andrée Polar Expedition' map

56 Ibid., p.189

57 Strindberg 'Letter to Anna', ibid., p.451

58 *Andrée Diaries*, pp.189, 444

59 Ibid., pp.ix, 169

60 Ibid., p.199

61 Ibid., p.412

62 Strindberg almanac, ibid., p.434

63 *Andrée Diaries*, pp.414–15; Wilkinson, p.221

64 Strindberg almanac, *Andrée Diaries*, p.435

EPILOGUE

1 As so often in ballooning, her true story has partly been told in fiction, in the novel *Anna's Book*, by George MacBeth, 1983

2 This remarkable phenomenon can be explored at various internet sites, such as HotAirBalloonEvents.org

3 *Coral Reefs: Research Methods*, D. R. Stoddard, UNESCO, 1987

Index

387.73209 HOLMES
Holmes, Richard,
Falling upwards :
R2001074315 PALMETTO

ODC

Atlanta-Fulton Public Library